Contents

For

Lorraine, Evan,
and Meredith

Preface

This book, like its predecessor, is intended primarily for the use of undergraduates and graduates in urban geography courses, although it will also be of interest to students of urban planning, urban economics, and urban sociology. The major theme of the book and the feature that continues to distinguish it from other texts in the field is the *integration of substantive and methodological material*. The substantive material is arranged in such a manner as to allow the statistical techniques to be introduced in an orderly fashion, beginning with the simplest and following through to the more complex. No previous training in statistics is required, equations are kept to a minimum, and the various techniques are explained in conjunction with the empirical testing of particular models and theories. It is the author's experience that the majority of students can easily grasp the major elements of these concepts if they are not burdened with mathematical proofs and are provided with appropriate schematic diagrams. More important, there is such a fine line between theory and methodology in contemporary urban geography that it is almost impossible to appreciate the one without the other. The traditional device of relegating methodological and technical discussions to an appendix is no longer appropriate, as theory development is integrally related to the calibration and testing of models via statistical techniques.

Two other themes are also represented in the book. First, the approach is *multidisciplinary*, in the sense that appropriate theory is culled from the major social science disciplines of economics, geography, psychology, and sociology. In all cases, however, these theories are discussed from a geographical perspective, as it is not just the processes themselves, but also their spatial manifestations, that are of primary interest. For example, particular attention is paid to the spatial patterns associated with the urban land market and to the spatial outcome of the decision-making process involved in residential mobility. Second, the *model-building* paradigm is emphasized throughout the book, and the ways in which mathematical or symbolic models can be used to represent processes and associated spatial patterns are discussed in detail. For example, models to explain the distribution of land values, the distribution of housing values, and the choice process in consumer spatial behavior are all examined at some length. It is in conjunction with the discussion of models such as these that the statistical procedures are introduced and that the mutual reinforcement of theory and methodology is exemplified.

The first two chapters introduce the student to the overall orientation of the book. Chapter 1 discusses the different approaches to urban geography, while Chapter 2 focuses on the scientific method and the process of model building. Chapters 3 through 7 deal with the internal structure of cities, especially with respect to the underlying socioeconomic processes and their spatial manifestations. Chapter 3 explores land use and land value theory, using the traditional approach of neoclassical economics, and bivariate correlation and regression techniques are introduced in order to test the major hypotheses generated by that theory. More complex, multivariate models of land values are developed in Chapter 4, and the calibration of these models necessitates an introduction to multiple correlation and regression. The characteristics of the urban housing market are also examined at this juncture, as a prelude to discussing multivariate housing value models. These models are couched within the framework of causal models and path analysis and so provide a natural methodological extension to the previously described multiple correlation and regression models. In particular, the notion of indirect effects between the explanatory variables is explored.

Chapter 5 deals with two components of urban spatial structure that are inextricably linked to the underlying pattern of land values. First, the urban retail structure is described and central place theory is used as a conceptual framework for analyzing the spatial distribution of shopping centers within cities. Second, the spatial pattern of population density is considered and the use of trend surface analysis, a form of multiple regression analysis, is explored. Having discussed the distribution of people in general, Chapter 6 addresses the question of whether different kinds of people tend to live in different parts of the city. The process of spatial sorting, according to social and economic differentiation, involves the generation of distinctive social areas within cities, and the characteristics of these social areas are identified through the use of factor analysis. Finally, Chapter 7 is concerned with the industrial structure of cities, and the economic base concept and input–output analysis are used to describe and calibrate the various interrelationships within an urban economy.

Chapter 8 introduces the ideas of environmental perception and cognitive maps. With these concepts as a backdrop, Chapters 9 through 11 discuss various approaches to understanding movement patterns within cities. More specifically, the discussion of psychological concepts such as imagery and preferences provides a fuller understanding of human behavior within an urban setting and leads naturally to the issue of spatial choice. Different approaches to choosing between alternatives within the context of shopping behavior are discussed in Chapter 9, while Chapters 10 and 11 address the processes involved in residential mobility. In Chapter 10, which considers aggregate behavior, the elaboration of multivariate statistics is extended by the introduction of simultaneous equation models and unobserved variables. Chapter 11, which involves the individual decision-making process, includes an introduction to various kinds of log-linear models.

Chapters 12 and 13 represent a change of scale, as the system of cities, rather than their internal structure, is now the focus of analysis. In Chapter 12 the attributes of size, growth, and housing value are all considered in some detail, while at the same time introducing the multivariate technique of discriminant analysis and providing further examples of simultaneous equation models. Interregional migration is the focus of Chapter 13, with particular attention being devoted to the determinants of migration and the potential for utilizing more dynamic approaches, such as time-series techniques and event history analysis. Chapter 14 concludes the book by discussing the role of state intervention in the form of urban and regional planning.

The previous version, published ten years ago, has been expanded and revised in a number of ways. In particular, it has been significantly enlarged by the addition of four new chapters. First, a new introductory chapter captures the current era of theoretical pluralism within urban geography. Besides providing a detailed discussion of the traditional perspectives, such as neoclassical economics, considerable space is also devoted to the political economy approach, with the phenomenon of residential differentiation being used to illustrate how the insights from these various approaches can be combined to develop overlapping and mutually reinforcing explanations. In addition, students are introduced to the contemporary debates involving structure and agency, postmodernism, and feminism. Second, the original chapter on residential mobility has been split into two chapters in order to fully cover both macro and micro perspectives. Finally, two chapters devoted to the system of cities have been added, thus broadening the range and scope of the book.

Extensive revisions have also been made to the existing chapters. For example, concepts and issues such as the following have been added at appropriate points in the discussion: polycentric models of land use and population density, rent control, gentrification, the spatial mismatch hypothesis, and the urban homeless. Similarly, in order to remain methodologically current, LISREL and log-linear models have been added to the presentation, while more space has been devoted to the explication of such multivariate statistical techniques as path analysis, simultaneous equations, canonical analysis, and discriminant analysis. Finally, more recent empirical examples of various models, such as those involving land use patterns, housing values, population density, and factorial ecology, are provided.

Parts of this book first appeared in academic journals and monographs, and I would like to thank the following journals and publishers for allowing me to reproduce some of my previously published material: *Area* (Institute of British Geographers), *Environment and Behavior* (Sage Publications), *Environment and Planning* (Pion), *Geografiska Annaler, Growth and Change, Professional Geographer* (Association of American Geographers), *Progress in Human Geography* (Edward Arnold), Routledge, Chapman and Hall, *Transactions of the Institute of British Geographers* (Institute of British Geographers), University of Wisconsin Press, and *Urban Geography* (V. H. Winston and Son).

I would especially like to thank Brian Berry, John Everitt, and John Paul Jones for their detailed suggestions on a previous draft of the manuscript and Ray Henderson of Prentice-Hall for his encouragement and patience throughout the project. Over the years I have benefited from the advice and friendship of John Everitt and Paul Knox. My colleagues in the Department of Geography and the Graduate School at the University of Wisconsin, Madison, have also been very supportive. Finally, like its predecessor, this book is dedicated to Lorraine, Evan, and Meredith.

Chapter 1

Urban Geography

1.1   Traditional Approaches

Urban geography is concerned with the spatial patterns and processes associated with urban areas. More specifically, it involves answering the following kinds of *questions*: What is the spatial distribution of land values within cities? What is the spatial distribution of population density within cities? To what extent are cities segregated according to race or ethnicity? Why do some cities grow faster than others? How is the spatial structure of urban areas perceived by individual residents? Why is there more migration between some pairs of cities than others? How might we describe the industrial structure of a city? What regularities can be identified with respect to commuting patterns within cities? All of these questions, and many more, are addressed by urban geography.

Traditionally, most of the research in urban geography has been carried out within the context of an empirically based, hypothesis-testing approach to social science. As a result, the *model-building paradigm*, especially as exemplified by neoclassical economics, has been dominant. This approach to research, which involves a positivist perspective and a belief in the scientific method, is described in Chapter 2. Before discussing the model-building approach in detail, however, it is important to consider the different theoretical frameworks that have been utilized in investigations of urban geography, as these conceptualizations have played a major role in the formulation and interpretation of empirical analyses (Bassett and Short, 1989; Marston et al., 1989).

Contemporary work on urban geography is characterized by *theoretical pluralism* and epistemological debate (Cadwallader, 1988). Neoclassical economics, political economy, behavioral and institutional perspectives, postmodernism, and feminist theory have all provided theoretical contexts for empirical analyses. Each of these perspectives can only provide a partial explanation of urban phenomena, and there is no epistemology that can be used as a yardstick against which all other approaches can be measured for scientific adequacy. Despite these differences, however, it has been strongly argued that there is a basic similarity in the procedures for assessing empirical work (Chouinard et al., 1984).

With this context in mind, the present chapter briefly describes and evaluates some of the major theoretical approaches to urban geography. To this end, we begin by discussing the more traditional approaches represented by neoclassical economics and the behavioral and institutional perspectives. In the next section Marxist theories of political economy will be discussed, and the following section will be devoted to demonstrating how these different perspectives can provide complementary insights within the context of residential differentiation and the urban housing market. The chapter is concluded by investigating the structure–agency debate and by providing a brief introduction to the recent postmodern and feminist critique of urban geography.

Neoclassical Economics

Neoclassical economics, developed in the latter half of the nineteenth century, signaled a shift from the emphasis on the production side of economic systems to a fuller consideration of consumer preferences (Myers and Papageorgiou, 1991). Indeed, a fundamental axiom of neoclassical economics is that individual preferences help to shape both the nature of the economy and the characteristics of the larger society within which that economy is embedded. Individual freedom, consumer sovereignty, and a reverence for market mechanisms are all implied within the framework of neoclassical economics, and this form of economics still constitutes the orthodox approach in most of Western Europe and North America.

In general, neoclassical economics conceives of the economy as comprising a large number of *individual households and firms*. Firms purchase land, labor, and capital, the factors of production, from the factor market. They then combine these factors of production in such a way as to maximize their profits. In particular, the firm must choose a specific productive process and a specific volume of output, or scale of production. The finished products are then sold on the product market, where they are purchased by households, which are attempting to maximize utility or personal satisfaction. Households obtain the money to buy goods on the product market by selling their land, labor, and capital on the factor market. The interaction between supply and demand regulates prices on both factor and product markets.

Throughout, the neoclassical analysis emphasizes *equilibrium* conditions and views the market economy as a harmonious, self–regulating system. Households, firms, product markets, and factor markets are all inextricably interrelated, such that a change in one part of the economy will have implications for all the other parts. For example, the price of labor on the labor market affects the costs of production, which in turn affect prices on the product market, and so on. In this sense, the phenomenon of excessive unemployment is often simply viewed as a temporary aberration that will eventually be corrected by market mechanisms.

A number of general *assumptions* are incorporated within the neoclassical perspective. First, individual firms and households are assumed to be too small to influence prices on either the factor market or the product market independently. Second, individual households allocate expenditures on the product market in order to maximize utility, subject to budgetary constraints and the prevailing set of prices. Third, individual firms behave in order to maximize profits. Fourth, both households and firms are assumed to possess perfect information about market conditions. Fifth, interchanging the factors of production is assumed to be unproblematic. Of course, subsequent modifications of neoclassical economics have addressed the unrealistic nature of these assumptions. For example, monopoly and oligopoly situations have been considered, as it has been recognized that buyers and sellers are often large enough to influence prices. Also, the category of public goods has been introduced to account for those goods and services, such as police protection and education, that are not generated by market conditions but for which society accepts a collective responsibility.

The Behavioral Approach

The emergence of the behavioral movement in urban geography is at least partly due to the search for alternative *models of human beings* (Golledge and Stimson, 1987). In particular, behaviorists have attempted to replace the behaviorally unsatisfactory concept of "economic man" and its attendant assumptions of profit maximization and perfect information with a more realistic counterpart. Following Simon (1957), some have suggested that a more realistic model of human beings would combine the principles of satisficing behavior and bounded rationality. In other words, it is assumed that people are sometimes satisfied with less than optimal profit levels and that decisions are often made in a context of incomplete knowledge and uncertainty.

The rejection of economic man as a workable descriptive model of human beings appears eminently reasonable, as acting with complete rationality in the real world is but one of a range of possible behaviors. This line of reasoning does not deny that economic man is a powerful tool in a normative context, but simply emphasizes that the behavioral approach is concerned with identifying regularities in actual, not optimal, behavior. As Harvey (1981) has pointed out, however, care must be taken to replace this conceptualization with an alternative that is both amenable to operationalization and theoretically useful. As yet, largely due to the difficulties involved in determining aspiration levels, the satisficer concept has proved very difficult to operationalize. In comparison, the principle of bounded rationality appears to be more promising, as it takes into account our simplified and distorted view of reality, thus attempting to explain behavior based on individual perceptions.

Given this interest in *individual differences*, it is hardly surprising that the behavioral approach is characterized by the use of individual, or microlevel data. Such data are prerequisites for those researchers interested in the variability of individual behavior, although this does not imply a preoccupation with the unique as opposed to the general. Rather, behaviorists accept that ecological fallacies can often occur when inferences about individual behavior are made from aggregated data. More important, however, the aim is to understand the individual decision-making process within a social psychological context.

Despite its theoretical promise, the behavioral approach has been beset by difficulties. Behaviorists have postulated the importance of environmental images, but it has proved exceedingly difficult to capture and measure the properties of those images. Such calibration efforts are confounded by *aggregation problems*. Although working at the level of individual data, behavioral researchers argue that there are significant similarities between individuals that can be used as a foundation for meaningful aggregation. The appropriate form and degree of aggregation, however, depend on the type of behavior being analyzed.

The assumption that specific images can be legitimately extracted from the totality of such images has been seriously questioned (Bunting and Guelke, 1979). Images are *holistic*, and the cognition of any particular environment or place is embedded within a set of cultural and ideological attitudes. It is exactly this kind of problem, however, that the empiricist tradition in cognitive psychology has attempted to address. A related problem involves the language of discourse. The categories traditionally used to characterize the real environment may have little relevance for the perceived environment.

Perhaps most important, criticisms of the behavioral approach have tended to focus on the implied assumption of *subject–object separation* (Cox, 1981), the idea that the world can somehow be divided into an objective world of things and a subjective world of the mind, thus allowing the observer to be separate from the observed. Such a division is less critical, however, if one emphasizes the interrelationships between subject and object and stresses that cognitive representations are formed by transactions between the two. This viewpoint has given rise to a more humanistically oriented behavioral approach, focusing on human feelings and values, like the sense of place. Such an approach, however, with its illustrative use of facts and examples, is not easily integrated with the hypothesis-testing orientation of most urban geography. Rather, the humanist philosophy provides an important form of criticism that helps to counter some of the extreme abstractive tendencies of the positivist tradition.

Because of its preoccupation with *measurement* and highly formalized methodology, the behavioral approach has sometimes been viewed as a mere appendage to the positivist tradition of neoclassical economics (Ley, 1981). In all fairness, however, behavioral geographers never intended to produce a different disciplinary subfield, but rather to incorporate behavioral variables and concepts within the explanatory schema (Golledge, 1981a). Indeed, in many respects behavioral research has moved well beyond its original positivist underpinnings (Couclelis and Golledge, 1983). With the recognition of mental reality as a primary objective of study, the tendency toward a reductionist philosophy of human behavior lost much of its impetus. The image of an objectively defined reality being passively observed by an impartial scientist has been replaced by an acceptance of the indivisibility of fact and value.

Like neoclassical economics, however, the behavioral approach is heavily oriented toward *consumer preferences* and the demand side of the economy. The behavioral viewpoint has thus been attacked for overlooking the societal constraints on human behavior. For example, it can be argued that an understanding of residential mobility cannot be divorced from a consideration of the urban housing market, as it is the operation of such a market that provides the larger context for mobility. In particular, the demand for and supply of housing are influenced by the interaction of various institutions. It is the motives and behavior of these institutions, such as mortgage lending companies and real estate agents, that comprise the subject matter of the institutional approach to urban geography. The institutional approach thus provides a useful antidote to the consumer sovereignty implied by the behavioral perspective.

The Institutional Approach

The institutional approach focuses on the behavior and motivations of various kinds of institutions that operate in the urban arena. Among other issues, this literature considers the relative role of private versus public sector institutions and examines the extent to which institutional ideologies are rooted within particular political systems. Various categories of institutions have been identified, including those involved in housing production, such as builders, developers, and planners, those involved in housing consumption, such as financial institutions and insurance agents, and those involved in exchange, such as real estate agents and appraisers (Bassett and Short, 1980:58). Within this context, the managerial perspective and the role of the state have been given particular emphasis.

A provocative book by Rex and Moore (1967) provided an important impetus to what later became known as the *managerial approach*. In formulating a detailed sociological analysis of an inner-city area in England, Rex and Moore drew upon the work of the sociologist Max Weber. In particular, they focused attention on the existence and nature of allocative structures within the urban housing market. The managerial perspective was later extended and codified by Pahl (1969), who suggested that previous research had been preoccupied with understanding consumer choice, while giving insufficient attention to the associated constraints. As a result, he argued that research should emphasize the interplay of spatial and social constraints that determines the differential access to resources such as housing, education, and transportation. Moreover, he believed that these social constraints could be best understood by examining the activities and values of the managers of the social system, including landowners, builders, real estate agents, and mortgage companies.

This kind of analysis owes a great deal to Weberian sociology, which emphasizes the motivations of individual actors or institutions in various kinds of social systems. The issue of *power* was central to Weber's sociology, as conflicts inevitably arise when individual actors attempt to actualize their differing goals (Saunders, 1981:123). Power emerges from the interrelationships between actors and is manifested when any one actor is able to realize his or her will over the opposition of others. This power over others is either economic or political in origin; unlike Marx, Weber argued that these two arenas of domination remain analytically distinct. Also unlike Marxist thought, the Weberian perspective identifies actors rather than classes as the basic units of analysis, and the Weberian notion of an ever increasing bureaucracy lends confirmation to the growing importance of managers (Leonard, 1982; Wilson, 1989).

There have been two major *criticisms* of the institutional, or managerial, perspective. First, managerialism should be considered a framework for study rather than a coherent theory (Williams, 1978). An empirical object, managers, is the central focus of the analysis, rather than any theoretically derived social process. At worst, institutional analysis can degenerate into a mindless empiricism, involving exercises of gathering descriptive data on those individuals who are assumed to occupy important positions (Saunders, 1979:168). There is little conceptual basis for the identification and selection of urban managers, and it is exceedingly difficult to assess their interrelationships and relative importance. It is especially difficult to isolate the management role in the private sector from the larger capitalist economic system. For example, real estate agents are both managers and capitalists. The distinction is somewhat easier to make in the public sector, as long as one is willing to accept a substantial degree of economic and political autonomy.

This latter issue, concerning the extent of autonomy in social systems, provides the basis for the second major criticism of managerialism. It has been repeatedly argued that managerialism should be related to more general theories concerning the political economy of cities in a capitalist society. To what extent do managers behave as independent units, and to what extent are they constrained by the overall socioeconomic structure within which they operate? While managers are responsible for the allocation of scarce resources, they are not responsible for creating scarcity in the first place (Bassett and Short, 1980:52). It is only possible to interpret, rather than merely describe, the actions of managers by placing them within the larger political economy of which they are a part (Williams, 1982). As a result, Marxists have tended to dismiss managerialism as merely diverting attention from more fundamental structural issues.

In this context it is possible to distinguish two *stages* in the development of the managerial perspective. Initially, Pahl (1969) seemed to suggest that managers could be analyzed as autonomous units; he saw managers as the independent variables in the explanation of any given pattern of resource distribution. In later analyses, however, he viewed managers more as intervening variables, mediating between the pressures of private profit and public welfare and between the central government and local populations (Pahl, 1979). This later conceptualization stresses the danger of ascribing too much independent authority to the managers of the system, while at the same time recognizing that the peripheral agents of a centralized state inevitably enjoy a certain amount of discretion in policy implementation. The extent of this autonomy is still unresolved, however, and the circumstances under which managers might be expected to exercise significant discretion remain unclear.

The issue of local autonomy is most clearly articulated in the burgeoning literature concerning various *theories of the state* (Johnston, 1982a). Almost all aspects of production and consumption are now profoundly influenced, either directly or indirectly, by state policies at the local, regional, national, and international levels. It is scarcely surprising, therefore, that much effort has been devoted to constructing viable theories of the state within capitalist society. Definitions of what is meant by the "state" vary. Miliband (1969) suggested that the term stands for a number of institutions, including the government, the judiciary, and the police. Harvey (1976), on the other hand, conceptualized the state as a process for the exercise of power, rather than as a circumscribable entity.

As one might expect, then, there are many theories of the state (Jessop, 1990; Barrow, 1993). Perhaps most simply, the state has been viewed as a supplier of public goods and services and as a mechanism for regulating and facilitating the operation of the market economy. The state can also be seen as taking on the role of arbitrator in disputes between different interest groups. Marx did not develop any rigorous theory of the capitalist state, but neo-Marxists have suggested a number of theories ranging from the instrumentalist model, which implies the existence of a conspiracy between the ruling class and the state, through the structuralist perspective, which suggests that the functions of the state are determined by the structure of society itself, rather than by a few key people within that society. Clark and Dear (1984:33) have argued for a materialist theory of the state, in which the state is viewed as an actor in its own right, while at the same time being embedded in the social relations of capitalism.

These theories are by no means mutually exclusive, however, and there is considerable consensus concerning the general role of the state in capitalist society. The state provides a framework for exchange, while guaranteeing property rights and facilitating social reproduction by providing housing, education, health care, and social services. It regulates competition by various antitrust laws and regulates the employment of labor through legislation on minimum wages and maximum hours of employment (Clark, 1992). The state also undertakes the production of public goods, such as roads and bridges, that are necessary for capitalist production and exchange but that individual capitalists are unable to provide at a profit. Finally, the state acts as a legitimator of the capitalist mode of production (Johnston, 1984). It protects a particular set of production relations by propagating an ideology that emphasizes business enterprise and profit maximization.

1.2   Political Economy

Marxist theories, which represent one particular form of political economy, provide a framework for analyzing urban phenomena, although Marx himself was not especially interested in urbanism per se (Peet and Thrift, 1989). Marx argued that social science should penetrate the *realm of appearances* in order to uncover the underlying relations that give rise to those appearances. He suggested that classical economists obscured the distinctions between social appearances and social reality by concentrating on objective laws governing commodity exchange within market economies. In contrast, he attempted to explore the structures of different modes of production by emphasizing the interrelationships among a variety of theoretical entities that are not themselves directly observable. In pursuing the nature of these interrelationships, Marx stressed that no single aspect of reality can be analyzed independently of the social system of which it is a part.

His commitment to *dialectical analysis*, with its concern for the perpetual resolution of opposites in which each resolution generates its own contradiction, made him extremely suspicious of any mode of analysis that failed to relate the part to the whole. Thus Marx never treated any concept as fixed; rather, each concept was reinterpreted when juxtaposed with other concepts. In other words, the relationships between the concepts are what really count (Harvey, 1982:2). Thus, according to Marx, our insights shift as we successively probe different conceptual relationships; in contrast, the building-block approach of much contemporary social science isolates and scrutinizes individual components of the overall system before making them fixed foundations for further inquiry.

In particular, Marxists argue that social phenomena result from the prevailing *mode of production*, the way in which societies organize their productive activities. The mode of production is an analytical construct comprising the productive forces (labor, resources, and instruments of labor) and the associated social relations of production. The social relations of production are embodied in class conflict between those controlling the means of production and the laborers. Latent contradictions within any particular mode of production are periodically manifested as crises, and these crises eventually result in a transition from one mode of production to another. Historically, four successive modes of production, with their associated social relations, have been identified: primitive communism, slavery, feudalism, and capitalism. Within the capitalist mode of production, the means of production are controlled by a capitalist class, or bourgeoisie, and production is carried out by a class of wage earners, or proletariat. The economic structure of a society, as represented by its particular mode of production, provides the basis for a superstructure of social, political, and legal forms.

According to Marx, the need for *capital accumulation* governs capitalist behavior. This rule, enforced by competition between capitalists, is the hallmark of individual behavior. The extraction of surplus value from labor generates the process of capital accumulation. Much of this capital is recycled into new production, thus expanding the bases of production, as there are limits to conspicuous consumption on the part of capitalists. Only if the proletariat has the purchasing power to consume what is produced, however, will the capitalist be able to realize surplus value in money form. Capitalists must realize their profits in this way in order to keep employing labor and producing.

For capitalism to survive, both the bourgeoisie and the proletariat must reproduce themselves, thus passing on class positions, skills, and attitudes to succeeding generations. Besides the family, social institutions such as schools, clubs, and churches play a key role in preparing members of each new generation for their places in society (Edel, 1981). Informal social relationships at the neighborhood level can also serve this function. Residential differentiation in contemporary cities and the resulting distinctiveness of different types of neighborhoods provide a setting for the process of *social reproduction*.

Crises

Marxists argue, however, that the insatiable quest for capital accumulation creates conditions that are inconsistent with the further accumulation of capital and the reproduction of class relations. In other words, the capitalist system is inherently unstable and subject to periodic crises. Three such crises deserve particular attention. First, a basic contradiction exists between the pressure to increase surplus value by maintaining low wages and the need for people to buy products so that the surplus value can be realized. If workers cannot afford new appliances, cars, and homes, an *underconsumption*, or realization, crisis will occur. Capitalists must pay sufficient wages to ensure that the working class possesses the effective demand needed to reproduce itself, but individual capitalists are forced to compete continually with each other and thus keep wages at a minimum.

Second, a decline in the *rate of profit* can occur due to the substitution of machinery for labor, thus reducing the rate of surplus value derived from labor and increasing unemployment, which further exacerbates the aforementioned underconsumption crisis. Capitalists tend to substitute capital for labor in order to compete more effectively with their fellow capitalists, so the basic process of technological innovation tends to undermine the stability of the capitalist system. Individual capitalists, acting in their own self-interest under the social relations of capitalist production, generate a technological mix that threatens further accumulation and the reproduction of the capitalist class. In short, individual capitalists necessarily act in a way that destabilizes capitalism (Harvey, 1982:188).

Third, these inherent contradictions within the capitalist system lead to various forms of state intervention and associated *fiscal crises*. Public spending is often needed to cope with the by-products of accumulation, such as urban congestion, pollution, and social tensions.

Labor is diverted to sectors that, although they provide employment, are unproductive in terms of any increase in the production of surplus value. State intervention in providing commodities such as streets, subway systems, and bridges and various kinds of social services helps to maintain the capitalist system. Class struggle is regulated by appeasing labor with employment and economic concessions, while at the same time combating underconsumption crises. However, while the growth of state intervention in the sphere of collective consumption helps to reproduce labor and appease class struggle, the more the state is required to provide various social services, the greater the gap between its revenues and expenses; hence the fiscal crisis of the state, most clearly exemplified by the near bankruptcy of New York City.

Criticisms

A variety of important criticisms have been leveled at Marxist theory. First, although many varieties of contemporary Marxist discourse exist (Agnew and Duncan, 1981), one of the most pervasive criticisms of the Marxist tradition concerns its commitment to some form of *economic determinancy*. Duncan and Ley (1982), for example, charge that to reduce the range of social experience to the surficial manifestations of some underlying economic structure is to present an excessively impoverished view of the social, cultural, and political realms of life. Some neo-Marxists have attempted to distance themselves from what Gregory (1981) has called "brute economic determinism" by emphasizing the variety of social formations that can be associated with a dominant mode of production. For example, as history has demonstrated, there is the potential for a good deal of cultural, institutional, and political variation under capitalism. Also, Marxists recognize that at any particular moment in history we are likely to find, alongside a dominant mode of production, the remains of previous modes and the beginnings of future modes. Marxists continue to contend, however, that, although a variety of political structures can coexist within a given form of economic structure, it is the economic structure, as represented by the mode of production, that has causal primacy (Wright, 1983). In contrast, critics of the Marxist tradition suggest that the developmental tendencies of economic and political structures are autonomous, in the sense that no general principles govern their interconnection (Giddens, 1981). In any specific historical situation, either one could be the major force behind social change, but there is no general priority of one over the other.

Second, Marxist analysis can be criticized for its functionalist approach to constructing social theory. *Functionalism*, a type of holistic explanation, views societies as composed of individual parts that both form and maintain the whole. In other words, various properties of society are explained by their functions within the social totality. Systems analysis is often invoked within this context, as it captures the interrelationships between the various elements of the total system. Harvey (1973:289), for example, espouses a functionalist viewpoint when he suggests that capitalism shapes the elements and relationships within itself in order to reproduce itself as an ongoing system. The Achilles' heel of the functionalist perspective, however, is that it falsely imputes "needs" to social systems. For example, Marxists often explain state policy as a response to the need for capital accumulation or unemployment as a result of the need of capitalism for a reserve army of labor. While functional descriptions may serve a useful purpose in social theory, the functionality of a given institution can only provide a partial explanation of that institution.

Third, Marxism often involves a *reification of a priori categories*. Reification occurs when mental constructs or abstractions are viewed as substantive phenomena with causal properties. Such constructs are often initially used as mere heuristic devices but later take on more concrete characteristics; hence reification has sometimes been called the fallacy of misplaced concreteness. Marxist social theory has tended to reify such abstract concepts as capitalism, mode of production, the state, and class. For example, the mode of production is seen as a driving force behind the development of social formations, or capitalism is described as maturing. Such statements suggest that these abstract, intellectual concepts, or categorizations, somehow have an independent existence of their own.

Fourth, and related to the problem of reification, Marxism can also be criticized for its dependence on *evolutionary* thinking. Societies are viewed as evolving through different modes of production, with their associated social relations. Giddens (1981) argues, however, that no empirical evidence exists that the forces of production develop throughout history; thus the mode of production cannot form the basis for a general trajectory of historical development. Moreover, societies do not have transhistorical imperatives to adapt to their material resources. Societies are not organisms, and it is dangerous to view them as evolving in the manner of biological organisms. In the extreme, theories of social evolution are sometimes grounded in teleological arguments, involving the inexorable development toward some kind of adaptively optimal end state. As an alternative to this implied succession of societies in a given sequence of stages, it might be more useful to view social change as involving a set of qualitatively distinct transitions that overlap and have no overall pattern of development.

Fifth, a standard criticism of Marxism is that it represents a *class reductionist* theory. Most Marxists retain a commitment to constructing an overall theory of historical development that revolves around the genesis and contradictions of class relations. Types of social formations are primarily rooted in the concept of class structure, which is itself based on the concept of mode of production. It is potentially misleading, however, to classify societies primarily by class structure, as societies are characterized by multiple forms of domination and exploitation, which cannot be reduced to the single principle of class. Similarly, Giddens (1981:242) argues that the relations among states, ethnic groups, and races all form alternative axes of exploitation. Following from this, Giddens (1981:108) makes an important distinction between class society, in which class is the central structural principle, and class-divided society, in which there are merely classes.

Sixth, within the context of *class structure*, Marxists have been criticized for their simplistic distinction between the bourgeoisie and the proletariat. One can argue that during the twentieth century there has been a growing interpenetration of the labor and capitalist classes. Workers have begun to join the owning class by purchasing shares in companies and thus deriving a part of their income from profits. It would be unfair to suggest, however, that Marxists have not been at least partially sensitive to the complexity of class structure. Harvey (1982:26–27, 74) notes that Marx began to disaggregate the capitalist class into separate strata: financial capitalists who live entirely off the interest on their money capital, landlords who live off the rent of land, merchant capitalists who circulate commodities, and industrial capitalists who organize the production of surplus value. Moreover, Harvey (1982:450) reminds us that class configurations should not be assumed a priori; rather, they are being constantly reproduced and thus take on particular manifestations at particular points in time and space.

Finally, structural Marxism in particular has tended to generate a *passive model* of the human being that underestimates the processes by which people change their economic and social environments. Even Harvey (1982:114) concedes that Marx himself tended to subjugate the authenticity of experience to the revelatory power of theory, thus underestimating the subjective dimensions of class struggle. As a result, a number of contemporary Marxists have recently begun to emphasize the role of human agency. Castells (1983), for example, suggests that Marxism has been unable to comprehend urban social movements because of its inability to accommodate the scope for human action in history. By way of criticizing his previous work, Castells admits that the Althusserian form of structural Marxism encourages researchers to interpret social findings in accordance with a preexisting theoretical system.

1.3   Residential Differentiation

Residential differentiation provides a useful framework for illustrating how the insights from these various approaches can be used to develop overlapping and mutually reinforcing explanations. Within the context of residential structure there are a variety of phenomena that have been investigated by urban geographers, such as the occurrence of fairly homogeneous social areas, suburbanization, and gentrification. The political economy approach involves explaining these spatial patterns in terms of some underlying socioeconomic structure. Harvey (1973:273) argued that the spatial patterns in urban residential structure should be regarded as the geographical expression of some structural condition within the capitalist economy, and so residential differentiation can be interpreted via the themes of accumulation and class struggle. He articulated this framework by identifying three *circuits of capital* (Harvey, 1978). The primary circuit of capital is involved in production, and the tendency toward overaccumulation is manifested in a number of ways: the overproduction of commodities relative to consumption rates, a falling rate of profit, and surplus capital. These problems encourage capital to switch to the secondary and tertiary circuits.

The secondary circuit involves fixed capital investment and consumption funds, while the tertiary sector comprises investment in research and technology and expenditures on a wide range of social needs, such as education and health, that facilitate the reproduction of labor power. Within the secondary circuit a certain proportion of capital flows into the urban built environment, thus providing the necessary physical infrastructure for production and consumption through the creation of factories, houses, and roads. The flow of capital into the secondary circuit and the associated investment in long-term assets are dictated by the rhythms of capital accumulation in the primary circuit, whereby periodic crises of overaccumulation lead to surpluses of both capital and labor. Investment in the built environment, however, is also related to the physical and economic life expectancy of the building stock. Thus the switch of capital from the primary to the secondary sector is manifested by cycles of building activity, which tend to exhibit a 15- to 25-year periodicity (Berry, 1991).

Suburbanization, according to Walker (1981), is an example of capital switching from the primary to the secondary circuit. The underconsumption problems of the 1930s were at least partly alleviated by massive suburban development, which generated a wide range of investment possibilities in single-family housing and related consumer items such as washing machines, refrigerators, and cars. Governmental intervention actively stimulated this process through Federal Housing Administration mortgage subsidies and the construction of highways. As Castells (1977:388) has observed, the single-family home in the suburbs became the perfect vehicle for maximizing capitalist consumption. This explanation of suburbanization thus reflects the importance of the suburbs in absorbing economic surplus and emphasizes the role of the state in encouraging car ownership and owner-occupied housing.

As might be expected, however, political economists claim that the suburbanization process contains within it a series of contradictions and can only be a short-term solution to the crises associated with capital accumulation. The political power of the suburbs, built as a response to the underconsumption problems of the 1930s, is now being used to inhibit growth and thus resist the further accumulation of capital in the built environment. In other words, after initially being a vehicle for capital accumulation, the suburbs have become a barrier to further accumulation. Redevelopment in the suburbs will not occur until previously invested capital has lived out its economic usefulness. As Harvey (1978) expresses it, capitalist development is continuously faced with the problem of preserving the exchange values of past investments in the built environment, while at the same time having to destroy the value of those investments to make room for further development and accumulation.

Viewed in this light, inner-city *gentrification* is merely a continuation of the forces and relations that led to suburbanization. In general, the logic behind uneven development suggests that the development of any area tends to create barriers to further development, but that the ensuing underdevelopment then creates opportunities for a new round of development. Thus gentrification involves a particular phase in what Smith (1982) has called the "locational seesaw," through which capital jumps from one place to another and then back again. In other words, capital investment undergoes locational switching as well as sectoral switching. As with suburbanization, the state and various kinds of financial institutions have played a major role in facilitating the redevelopment of inner-city neighborhoods (Smith and Williams, 1986).

Besides explicating the role of residential development as a short-term solution to accumulation crises, Marxist political economists have also stressed the importance of residential differentiation in terms of *reproducing class relations* within capitalist society. Individual neighborhoods or communities provide distinctive milieus within which households derive their values, expectations, and consumption habits and so become the primary source of socialization experience. This reproduction of labor and consumption classes is further facilitated by the differing access to educational opportunities that is provided by different neighborhoods. Marxists also argue that the localization of households into distinctive communities has served to fragment "class consciousness" and replace it with "community consciousness," thereby frustrating the class struggle to transform capitalism into socialism (Harvey, 1975).

Whatever the social implications, however, the previously described sectoral and locational switching of capital can only be achieved through the behavior of various kinds of private and public institutions. It is in this context that the institutional perspective becomes important and provides an additional set of insights. In the context of the urban housing market, the various managers of scarce resources include key actors from the spheres of finance capital, industrial capital, commercial capital, landed capital, and the public sector (Knox, 1987:227).

First, *finance capital* involves all the various savings and loan associations, banks, and mortgage companies that are involved in lending money for housing developments, purchases, and improvements. Such financial institutions play an important role in determining who lives where, as they regulate the flow of money into the housing market (Doling and Williams, 1983; Murdie, 1986). They often avoid neighborhoods perceived to have a high risk of declining property values, and the denial of funds to inner-city neighborhoods tends to encourage decay and stimulate decline. In its extreme form this practice is known as redlining (Jones and Maclennan, 1987) and involves financial institutions refusing to make mortgage funds available in certain parts of a city. In other circumstances, however, the policies of these financial institutions have revitalized inner-city areas by supporting the process of gentrification.

Second, *industrial capital* refers to builders and developers. Within the constraints of planning regulations, the decisions of various builders and developers greatly influence the type and quantity of housing supplied. The residential development process comprises three major stages: site selection, site preparation, and the housing construction itself (Baerwald, 1981). While small companies tend to focus on the construction of single-family dwelling units on infill sites, the larger companies emphasize extensive suburban developments. Most new construction reflects the demands of upper-income groups, as they are most able to afford new housing. Low-income groups tend to obtain housing via the filtering of housing process, whereby older housing is successively passed down to lower-income families.

Third, *commercial capital* involves professionals such as real estate agents who are engaged in the market distribution of housing. Because they make a profit by charging for their services as coordinators between buyers and sellers, real estate agents tend to encourage turnover in the housing market. Also, as they are a major source of information concerning vacancies, real estate agents can introduce a deliberate bias by channeling households into or away from specific neighborhoods. In some situations, real estate agents, through their gatekeeping role, try to maintain the social status of a neighborhood. In other situations, however, they can facilitate change, as when they stimulate the gentrification process by advising financial institutions to make money available for the purchase and rehabilitation of dilapidated property.

Fourth, *landed capital* refers to landowners and landlords. Access to rented housing is considerably influenced by the gatekeeping activities of landlords, although the behavior of individual landlords is often based on informal personal judgments, rather than the profit maximization policies that characterize large corporate landlords. Unlike homeowners, however, landlords are primarily interested in exchange value rather than use value. Nevertheless, only in those situations where all landlords maintain or improve their properties will an individual landlord maximize the returns on his or her investment. For this reason, landlords often try to slightly underinvest relative to other landlords in the neighborhood, thus accelerating the deterioration of the housing stock. In areas experiencing gentrification, however, landlords are more likely to cooperate with each other to protect their mutual interests.

Fifth, various actors within the *public sector* influence the operation of the housing market. Governmental policies relating to both private- and public-sector housing are generally formulated at the national level but implemented at the local level. As a result, significant regional variation in the implementation of policy exists. In general, however, national governments attempt to maximize economic growth by ensuring an orderly relationship between construction rates and new household formation. They also try to achieve economic stability by using the housing sector as a kind of Keynesian regulator.

Despite the constraints imposed by these supply-side considerations, there is still a significant role to be played by individual choice, or *consumer demand*. In general, capital is rather conservative and will not flow to certain kinds of housing markets until a sufficient level of consumer demand has been identified. Large-scale suburban development was not unrelated to the emergence of a distinct middle class, imbued with a consumerist ideology. Although this preference for suburban living was socially created to a certain extent through advertising and the availability of mortgage money, it also reflects some latent desire for the characteristics associated with suburban living.

Similarly, the contemporary process of gentrification is also partly a response to shifts in consumer preferences and demographic change (Hamnett, 1984; Ley, 1986; Bourne, 1993). Single-family suburban homes are less in demand as a result of the trend toward later marriages, fewer children, and two-career households. Instead, there is now a greater desire for neighborhoods that reflect the life-styles of young, upwardly mobile professionals, who live closer to the main centers of employment and entertainment (Williams, 1984). The increasing demand for accessibility by households having more than one wage earner is partly a response to the higher costs of running a car and to rising public-transportation fares.

While Marxist theories of gentrification have tended to focus on the production of gentrified dwellings, they have paid considerably less attention to the role of the *gentrifiers*, or occupiers of such dwellings (Rose, 1984; Hamnett, 1991). These gentrifiers, with their particular life-style and preferences, are crucial to the overall process of gentrification; they are not merely produced by the availability of gentrified housing. Consumption is not automatically determined by production, and to assert as much would imply a producer's sovereignty theory as one sided as the consumer sovereignty embedded within neoclassical economics. Rather, there is a symbiotic relationship between production and consumption that is mediated by the various kinds of institutions discussed above.

1.4   Contemporary Debates

The interaction of individual decision making, institutional behavior, and societal constraints that is suggested by the preceding analysis of residential differentiation is closely paralleled by the current debate concerning social structure and human agency, or what Gregory (1981) refers to as action and structure. In the extreme, society can be viewed as either conditioning all human activity, as in structural determinism, or as the product of unconstrained human action, as in voluntarism. It would be inappropriate, however, to view any social theory as completely embracing either of these two perspectives in their most exaggerated forms. Rather, most social theories are a mixture of both dimensions and are best represented as points along a continuum.

Structuration Theory

Giddens (1984) has proposed the most elaborate theory to date of social structure and human agency, in which social structures are viewed as both the medium and the outcome of the practices that constitute social systems. This theory of structuration is built around a concept of the duality of structure and agency (rather than a dualism), which emphasizes the recursive relationship between individuals and society. Social structure and human agency are combined in such way that structure is not merely a barrier to action but is actively reproduced by that action. Thus social practices are constituted by social structures and also produce those structures. In other words, structure is created by human action, while at the same time being the medium of that action, with "the accumulation of previous decisions creating the framework within which future decisions are made" (Johnston, 1984:483). Thus determinist and voluntarist approaches are dialectically synthesized, and in this way structuration theory provides a possible mode of analysis for transcending the debate about whether the social formation or individual human agency should be the ultimate basis of explanation.

Structuration theory also entails a number of other related postulates, or *axioms* (Johnston, 1983:104–5). First, agency and structure are temporally and spatially specific; societies and the individuals that form societies are uniquely located within a particular configuration of time and space. Structuration is therefore concerned with the connections between structure and agency during specific periods at specific places. Second, human agency is constrained by social structures; humans produce society, but only as historically located actors, not under conditions of their own choosing. Third, structures are not simply conceptualized as constraints, as they also enable human agency. Thus structure both constrains and enables, while simultaneously being reproduced and transformed by individuals.

This proposed marriage of structure and agency entails certain *problems*, however (Thrift, 1983). First, the notion of determination is not clearly specified, and unless one endorses a conception of the absolute randomness of society, social theory must involve some form of determination, whether it be based on the mode of production, individual behavior, or some more complex combination acknowledging the capacities of a bounded human agency. Second, there is no currently acceptable theory of social action that can be incorporated within the structurationist framework. Such a theory is exceedingly difficult to generate, as the reasons an actor gives for an action are not necessarily the real ones. Third, a theory of structuration must explicitly articulate how social structure is inextricably interwoven with spatial and temporal structure (Giddens, 1985). Fourth, as presently conceived, the synthesizing characteristic of structuration theory means that it is somewhat uncomfortably situated on several theoretical interfaces, such as Marxist and Weberian sociology, and is thus open to charges of theoretical eclecticism (Moos and Dear, 1986).

Moreover, it is one thing to appreciate the elegance and richness of structuration theory, but quite another to use it in substantive contexts. As yet, only a very limited amount of *empirical work* has been carried out within the guidelines provided by structuration theory (Gregson, 1986). In particular, empirical work is hampered by the need to preserve the recursive nature of the structure-agency duality; it is difficult to know how to cut into the data without emphasizing one side at the expense of the other. A structurationist explanation must attempt to balance structure and agency, without according a priori primacy to either side

(Smith, 1983; Duncan, 1988). In reality, however, attention has been focused on examples of voluntarism and determinism, which are then simply clamped together in a form of conceptual vise (Archer, 1982).

Thus far, perhaps the best application of structuration theory in an urban context has been provided by Dear and Moos (1986), who explored the ghettoization of former psychiatric patients in Hamilton, Ontario. Throughout their analysis, they took great care to explicitly focus on the interaction of structure and agency, without resorting to simply welding together two separate investigations. Neither agency nor structure was accorded primary ontological status, and Gidden's (1984:326) advice that the concepts of structuration theory be regarded primarily as sensitizing devices was duly followed. Although a variety of problems remain, such as the designation of "ideal types" in the context of agent and institutional categories, Dear and Moos managed to convey the impression that the empirical application of structuration theory can indeed generate a set of substantive insights that might not otherwise be obtained.

The nagging doubt persists, however, that structuration theory provides no clear and unambiguous rules of interpretation. Criteria for determining the adequacy of a given explanation or for generalizing results are simply not provided. As a result, it is the nature of the evidence and the relationship between theory and observation that continue to be problematic and of central concern in contemporary urban geography.

Theory and Observation

Any urban analysis must be sensitive to the complex relationship between theory and data. As Harvey (1985:xiv) stresses, theory must be confronted with experience, but there is no independent entity called experience that is unmediated by imagination. No data are independent of a priori conceptualizations; while observation is neither theory neutral nor theory determined, it is certainly theory laden (Sayer, 1984:78). As a result, *empirical observation* cannot be the sole adjudicator among competing theories, and empirical verification should not be regarded as a peculiarly privileged form of verification (Sheppard, 1988). Thus social theory sits uneasily within Karl Popper's system of falsification and refutation, lacking the controlled experimental situations in which to truly test the correspondence between theoretical propositions and empirical measurements (Chouinard et al., 1984). All research programs within urban geography face similar problems in evaluating theory via empirical research, as the relationships between postulated causal mechanisms and expected empirical outcomes are always muted or disguised by the presence of uncontrolled external factors.

Despite these problems, however, we should be committed to constructing theories of urban phenomena that are *empirically grounded* (Fincher, 1987). Failure to do so leaves us susceptible to tautological reasoning. Structural Marxists have been particularly criticized in this context. Saunders (1981) argues that they postulate universal tendencies, such as a declining rate of profit, but merely invoke counteracting tendencies, or contingencies, if such a tendency does not occur. Thus structural Marxists claim to identify underlying mechanisms in society, but are suspicious of any attempt to empirically validate these claims on the grounds that contingent factors may be in operation. At the extreme, this viewpoint implies that, because we cannot know the world directly, empirical research has little significance, and in those situations where theoretical concepts and categories are imposed on historical reality, that reality becomes merely illustrative, rather than contributing to the creation of theory.

This friction between abstract theorizing and empirical research has been characterized as a tension between the universal and the concrete. Most contemporary social scientists are sensitive to this problem and accept the need to explore the intersection between theoretical abstractions and actual historical configurations (Macmillan, 1989). Harvey (1982:xiv), for example, notes that Marxist theory should not be regarded as correct and sacrosanct, but should be continually modified through a thorough testing against the historical record. He urges Marxists to bridge the gap between the abstract theorists and those wishing to reconstruct the complex historical geographies of actual social formations (Harvey, 1982:450).

On the other hand, the inherent problems of intertwining theory and observation should not encourage us to retreat to the atheoretical posture that has characterized so much urban geography. Massey's (1985:19) observation that "the unique is back on the agenda" does not represent a call for concentrating on the uniqueness of events, but reflects a concern for *theoretically informed* investigations of specific localities. In this respect, the knife-edge we are attempting to negotiate, between what Johnston (1985:335) has called the generalization trap and the singularity trap, is a fine one. While there are obvious limitations to excessively abstract theorizing, there are equally obvious limitations to narrow, empirical case studies. Theorizing devoid of empirical content should not be replaced by mindless empiricism (Smith, 1989).

In recent years *realism* has become an increasingly popular methodological stance for mediating between theory and observation (Sayer, 1985; Johnston, 1991). Such an approach distinguishes between necessary and contingent relations and argues that abstract theory pertains to necessary relations, while observed geographic patterns are the product of localized contingent relations that mute and obfuscate the underlying mechanisms. That is, general causes can, through the presence of localized contingencies, produce different outcomes in different places. Thus realism treats laws as tendencies, rather than as empirical regularities, and in so doing emphasizes the need to view theory as conceptualization (Sayer, 1984:49). In this context, social scientists are admonished to unpack their "chaotic conceptions." For example, Urry (1987) suggests that the occupational characterization of the service sector is problematic, as it contains a wide variety of diverse and distinctive activities that deserve to be disentangled.

The realist approach, however, tends to shelter theory from the possibility of *empirical disconfirmation*, as it insists on distinguishing between "real" mechanisms and mere phenomenal appearances. In this way, causality can only be determined theoretically, and empirical research is reduced to the role of illustrating theory, as opposed to evaluating or reconstructing that theory. As expressed by Saunders and Williams (1986:395), "the 'necessities' are simply asserted while the catchall category of 'contingency' is used to mop up all the problems." Similarly, within Gidden's (1984) conceptualization of structuration theory, the dialogue between theory and observation also tends to be one-way. While theory is used to sensitize empirical research, there is apparently no place for the interrogation of theory through empirical application (Gregson, 1987). We need to be able to move backward and forward between the theoretical and the empirical, such that each activity sheds light on the other.

Postmodernism and Feminism

Most recently, urban geographers have been influenced by and have contributed to the postmodernist movement that originated in architecture and literary theory (Cooke, 1990; Marden, 1992; Ley, 1993). *Postmodernism* can be treated as both an object of study and also as a methodological attitude. As object, the increasing complexity and architectural variability of contemporary cities can be described as postmodern. People and places are tied together in a dynamic kaleidoscope of economic, social, and political spaces, generating the evocative fragments of a volatile and fragile city such as Los Angeles (Soja, 1989). Modern and postmodern landscapes are juxtaposed to provide a collage of forms and functions. For example, on the north side of False Creek in Vancouver, Canada, one is confronted by a modernist landscape composed of high-rise buildings whose minimalist geometry provides a backdrop for the sports stadium and conference center (Ley, 1987). The south side of False Creek, however, displays a far more diverse and chaotic postmodernist landscape associated with the process of gentrification.

This changing and restless urban landscape is the physical manifestation of new forms of investment, production, and consumption (Knox, 1991; 1993). More specifically, these new forms of economic activity represent a shift away from the Fordist approach, with its reliance on economies of scale and standardized mass production, to a more flexible strategy of production and accumulation. In the labor market, for example, there has been an increasing reliance on subcontracting and temporary or part-time employment arrangements, thus providing the necessary flexibility to react to rapid changes in patterns of consumption and to take advantage of technological innovations in production.

Of equal importance, however, is the postmodernist approach to explanation. Postmodernism rejects the generalizing tendencies of conventional social science and the associated fascination with grand theory, systematic order, and universal truths. Rather, postmodernists attempt to replace the meta-languages, meta-theories, and meta-narratives of modernism with a concern for *difference, disorder, and fragmentation* (Harvey, 1989a). The broad interpretative schemas deployed by Marx, Freud, and others are condemned as totalizing, as glossing over important disjunctions and details. In other words, there is a deeply held suspicion of any theoretical schema that claims to be comprehensive or any intellectual point of departure that assumes a preexisting social order that is somehow independent of our human efforts to conceptualize it. Indeed, the postmodernist approach denies that one can adjudicate between competing theories (Dear, 1988). Thus, in principle at least, all theoretical frameworks are deemed to be equally significant or insignificant.

The celebration of difference that is inherent within the postmodernist philosophy has encouraged urban geographers to devote more time to studying the experience of women in cities, although there are often tensions underlying the connections between urban geography, postmodernism, and *feminism* (Bondi, 1990; Bondi and Domosh, 1992; McDowell, 1992). Some forms of feminism are sympathetic to postmodernism, while others are not. For example, while many feminists embrace postmodernism's concern for difference, others emphasize the modernist concern for justice and equality. Meanwhile, empirical generalizations characterize women as becoming increasingly involved in the formal sector of the labor force, having shorter journeys to work than men, and making greater use of public transportation (Fincher,

1990). Within the occupational structure of contemporary capitalism, women tend to monopolize the fast-growing part-time component of the work force, although there is tremendous variation in the types of jobs that are held by women.

There is little doubt that gender and *gender relations* are reflected in the spatial structure of cities (England, 1991; 1993). The various social areas that can be identified in most large cities are less homogeneous with respect to socioeconomic status than the traditional models of residential structure would lead us to believe (Pratt and Hanson, 1988). For example, the increase in the number of two-earner households leads to greater neighborhood heterogeneity if the occupations of both partners are taken into account. Blue-collar, working class neighborhoods often contain white-collar female workers, and white-collar, executive neighborhoods often contain female spouses who work as secretaries. Particular individuals, especially women, thus experience different class positions in different parts of their lives. The family-class position, normally defined by the husband's occupation, may not coincide with their own class position as defined by their paid employment (Pratt and Hanson, 1991).

The role of women in terms of generating heterogeneity in neighborhood socioeconomic status undermines the assumed relationship between residential areas and the social reproduction of class (see Section 1.3), although it should be noted that neighborhood homogeneity is not necessarily the most important component of *social reproduction* (Huxley and Winchester, 1991). Similarly, the changing role of women has played an important part in the process of *gentrification*. While gentrification has obviously been stimulated by the increased participation of married women in the labor force, it is also true that female-headed households are among the most vulnerable to displacement as a result of gentrification (Bondi, 1991). The recognition of women as both agents and victims within this context highlights the need for further investigations of the interconnections between gender relations and other social relations, such as those associated with class and race.

Chapter 2

The Model-building Approach

2.1 The Scientific Method

One way of understanding the evolution of cities is to think of the urban future as being a response to historical and contemporary socioeconomic forces, plus the intervention of government agencies and factors outside the urban system itself. Within this context, the *urban system* can be conceived of as containing two components: the system of cities and the internal structure of cities (Herbert and Thomas, 1990). Investigations of the *system of cities* focus on cities as points in space, generally at either a national or regional level, while investigations of the *internal structure* of cities focus on the spatial arrangement of places and activities within those cities.

Governmental intervention in the operation of this system occurs via the activities of various local urban and regional planning departments and federal agencies such as the Department of Housing and Urban Development. These departments and agencies use devices such as land use zoning and environmental pollution standards to influence the distribution and characteristics of activities within cities. Generally beyond their control, however, are certain forces whose impact is generated outside the system. Examples of these outside forces are various natural phenomena, such as earthquakes and volcanoes, and the socioeconomic policies and problems of other countries.

Within this conceptual framework, the present book is primarily, although not exclusively, concerned with the internal structure of cities. In particular, it is concerned with statements that apply to cities in general rather than to specific, individual cities. The information used to identify these generalizations is usually collected in *matrix* form. A city is first divided into subareas, using spatial units such as blocks, census tracts, or traffic zones (Figure 2.1a). The most popular spatial unit tends to be the census tract, since it is used for reporting data in

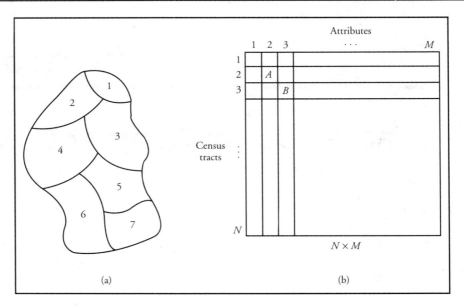

Figure 2.1   Representation of information in matrix form.

the U.S. Census of Population and Housing. Generally, a city of 250,000 people will have somewhere between 40 and 60 such census tracts.

Interest is then focused on various attributes, or characteristics, of these census tracts, such as the predominant type of land use, the average land value, and the population density. The information relating to these attributes is represented in a matrix by having the rows correspond to census tracts and the columns correspond to attributes (Figure 2.1b). For example, the cell labeled A might contain the average land value for census tract 2, while the cell labeled B might contain the population density for census tract 3. It is customary to designate the total number of cases, or census tracts, as N, and the total number of attributes as M, thus creating an $N \times M$ matrix.

The organizational structure of the present text revolves around the characteristics of the columns within this matrix, as we are interested in the different attributes, especially with respect to how their values vary from one census tract or one part of the city to another. For example, Chapter 3 is concerned with the characteristic patterns of land use and land value within cities. Chapter 4 investigates, among other things, the pattern of housing value; Chapter 5 is concerned with the distribution of population density; and so on. In contrast, an overriding interest in the rows within the matrix would entail the examination of a single census tract at a time, and such an approach would become excessively encyclopedic.

As a practical matter, most empirical research in urban geography has been dominated by a positivist perspective. The *positivist approach* was originally developed in the natural sciences and forms the basis of the *scientific method* (Johnston, 1989). It is characterized by a search for generalizations and laws as a means of explaining and predicting the phenomena of interest. Abstract modes of thought, especially the use of statistics and mathematics, are invoked to assist with the identification and representation of these generalizations. In this sense, positivism assumes that there is a material world and that there is an identifiable order to that material world.

Within the philosophical framework of the positivist approach to urban geography, or indeed to social science in general, the scientific method involves trying to identify *interrelationships* among significant variables. These interrelationships are often represented by statements of the following kind:

$$Y = f(X) \tag{2.1}$$

where Y is the dependent variable and X is the explanatory or independent variable. This equation states that the values of variable Y are in some way dependent on the values of variable X. The precise nature of this dependency is represented by the term f, which indicates the functional relationship linking X and Y.

Three major *questions* are asked about these relational statements. First, is there a relationship between the two variables? Second, if there is a relationship, what is the strength of that relationship? Third, what is the precise form of the relationship? The answer to the third question involves ascertaining whether the relationship is linear, curvilinear, or more complex. Imagine, for example, that the two variables Y and X represent income and education, respectively. That is, we are postulating that a person's income level is determined, to a certain extent at least, by his or her level of education. If the relationship between these two variables is linear, it means that, for every extra year of schooling, income increases by a certain, constant amount (Figure 2.2a). If, however, level of education increases income quite sharply at first but then diminishing returns begin to set in, we have a curvilinear relationship (Figure 2.2b). Finally, if increasing levels of education first cause an increase in income, then a decrease, followed by another increase, and so on, the relationship between the two variables is more complex (Figure 2.2c).

Spatial Patterns

Within this context, the positivist approach to urban geography tends to emphasize the spatial patterns of urban phenomena, both in terms of spatial distribution and spatial interac-

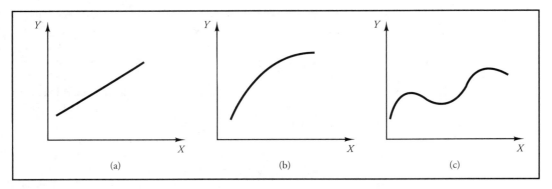

Figure 2.2 (a) Linear, (b) curvilinear, and (c) complex relationships.

tion. Patterns of *spatial distribution* can be categorized into four major groups: point patterns, networks, surfaces, and regions (Unwin, 1981). A *point pattern* is formed when the phenomenon of interest is represented as a series of points (Figure 2.3a). The distribution of banks, hotels, or supermarkets within a city constitutes a set of point patterns, and these patterns can be arranged along a continuum going from completely clustered at one end to completely dispersed at the other. Clustered patterns are generally produced by contagious processes. A measles epidemic, for example, would produce a spatially clustered pattern of victims. Dispersed patterns, on the other hand, are generated by competitive processes. For example, the distribution of supermarkets within a city generally conforms to a dispersed pattern, as the supermarkets are competing for customers.

A *network*, or line pattern, is used to represent such linear features as roads and railways (Figure 2.3b). We can quantitatively investigate the flow of traffic through the network, the relative accessibility of the different nodes (*a*, *b*, *c*), and the overall connectivity of the network. Network connectivity, for example, is amenable to analysis via the branch of mathematics known as graph theory.

Surfaces are formed by isolines, which join together points of equal value. The most common form of isoline is a contour, which joins together points of equal elevation. Within the urban context, population density can be represented as a surface. Such a surface would be cone shaped, with the highest densities toward the center of the city and the lowest densities toward the periphery (Figure 2.3c). Alternatively, if the values of the isolines are reversed, the surface might represent the distribution of income within a city, which is generally lowest at the center and increases toward the suburbs.

The final kind of spatial distribution is composed of *regions* within a city (Figure 2.3d). Geographers distinguish between two types of regions: uniform regions and functional, or nodal, regions. Uniform regions are areas that are relatively homogeneous with respect to certain specified characteristics. Social areas in a city constitute uniform regions, since they are relatively homogeneous with respect to the kind of people living there as measured by such variables as income, education, and family size. Functional regions, on the other hand, are characterized by a high degree of interaction. The market areas around supermarkets, for example, which circumscribe the areas within which people travel to purchase their weekly groceries, are classified as functional regions.

In addition to these patterns of spatial distribution, geographers are also interested in describing and explaining patterns of *spatial interaction*, or movement patterns, within cities. These movement patterns might be of a temporary nature, such as with the journey to work or the journey to shop, or they might be of a more permanent nature, as with residential mobility. Within the context of these different kinds of spatial patterns, the initial chapters in this book deal with the spatial distributions of particular variables, such as land use, land value, and housing value, while the later chapters are more concerned with patterns of spatial interaction, as represented by consumer behavior and residential mobility.

The form of *spatial positivism* represented by this approach to urban geography has not been immune from criticism (Schatzki, 1991), especially with respect to three of its major assumptions: first, the assumption that social science can generate scientific laws that are well-confirmed empirical regularities; second, the assumption that scientific discourse can somehow be value free or neutral; and third, the assumption that peculiarly spatial laws can

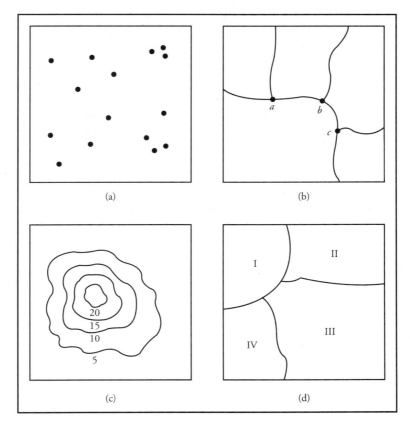

Figure 2.3   Patterns of spatial distribution: (a) point; (b) network; (c) surface; (d) region.

be generated in which space is considered independently from time and matter (Sack, 1980). Although these criticisms are certainly valid, they are to a certain extent caricatures of the positivist position, as most practitioners have tended to argue for a less dogmatic form of positivism (Hay, 1979).

Stages in the Scientific Method

In simplest terms, the scientific method is merely a strategy, or series of steps, for uncovering relationships between variables (Marshall, 1985). There are five main stages in this process: problem identification, hypothesis formulation, data collection, data analysis, and a statement of conclusions. The nature of these steps can be illustrated by using the example of residential mobility within cities.

The process of *problem identification* involves asking a particular question about the phenomenon under investigation. For example, we might ask why there is more migration between some pairs of neighborhoods within a city than between others. It often helps to conceptualize the problem by drawing a picture of it (Figure 2.4a). In this representation, five neighborhoods have been identified, and we are interested in investigating the amount of migration between one neighborhood and each of the other four. The actual volume of migration is represented by the thickness of the lines, so the research problem reduces to explaining why some of the lines are thicker than others.

The next step is to formulate a *hypothesis*. Hypotheses are tentative answers to the original question, so we might speculate that the amount of migration between neighborhoods depends on the distance separating them. Specifically, the amount of migration should decrease as distance increases, thus producing a negative, or inverse relationship between the two variables.

After formulating a hypothesis, *data* are collected in order to test that hypothesis. Because the hypothesis contains two variables, migration and distance, the associated data matrix will have two columns (Figure 2.4b). In addition, there are four rows, one for each potential destination. At this juncture, decisions have to be made with respect to how the

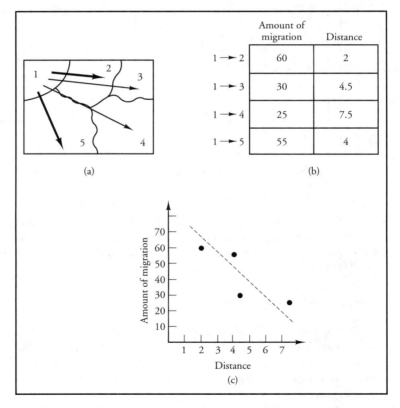

Figure 2.4 (a) Schematic representation of intraurban residential mobility, with (b) the associated data matrix and (c) graph.

variables should be measured. What time period, for example, is most appropriate for measuring the amount of migration? A week, a year, or perhaps even some longer period? Also, how should distance be measured? As straight-line distance, road distance, or in some other way?

Once such measurement issues have been satisfactorily resolved and the appropriate data collected, the process of data *analysis* begins. The precise details of this process will vary from study to study but in this particular instance we can at least draw a graph or picture of the relationship (Figure 2.4c). A graph involves plotting out the data in a two-dimensional space, where the dependent variable, migration, is placed on the vertical axis, and the independent variable, distance, is placed on the horizontal axis. Four data points are represented on the graph, as there are four cases, or pairs of neighborhoods, in the data matrix. A dashed line has been "eye-balled" through the four points in order to indicate that there is a generally negative relationship.

The final step is to arrive at some *conclusion* concerning the validity of the original hypothesis. In this instance, although there is some evidence of the hypothesized negative relationship between migration and distance, that relationship is not exceptionally strong as the points do not all lie on, or close to, the dashed line. If all the points were exactly on the line, the amount of migration between any two neighborhoods could be predicted exactly by merely knowing the distance between them. As it is, however, other variables must obviously be involved. Perhaps the population sizes of the neighborhoods and their relative income levels should be included in the analysis, thus requiring the hypothesis to be reformulated in terms of a multivariate rather than bivariate equation. A multivariate equation has more than two variables, so the new hypothesis is expressed as follows:

$$Y = f(X_1, X_2, X_3) \qquad (2.2)$$

where Y is the amount of migration, X_1 is distance, X_2 is population size, and X_3 is median income. In this equation there are three explanatory variables, so the data matrix expands to four columns. Also, more cases, or neighborhoods, will be needed if any kind of formal statistical analysis is to be undertaken.

This search for the interrelationships between variables often involves the *hypothetico-deductive spiral*. Initial hypotheses are usually found wanting when tested with appropriate data, so researchers repeatedly return to the stage of hypothesis formulation and begin anew the whole process of data collection and data analysis. It is only in this way that hypotheses and their associated generalizations become increasingly refined and useful in an explanatory and predictive sense.

Scientific Terms

A variety of technical terms are commonly used in conjunction with the scientific method (Hay, 1985). A *hypothesis*, which as we have seen plays an important role when identifying relationships among variables, can be formally defined as a statement whose truth or falsity is capable of being asserted. This condition of testability is crucial as it allows us to evaluate the empirical validity of competing hypotheses.

If a hypothesis is tested and found to be correct, a *scientific law* is established. These laws are of two major types: deterministic and probabilistic. A *deterministic law* can be stated as follows:

$$\text{If A, then } P(B) = 1.0 \tag{2.3}$$

This statement signifies that if event A occurs, the probability of event B occurring is 1.0, or 100 percent, as probabilities run between 0 and 1.0. Alternatively, a *probabilistic law* would take the following form:

$$\text{If A, then } P(B) < 1.0 \tag{2.4}$$

In other words, if event A occurs the probability of event B occurring is something less than 1.0. This situation entails a probabilistic relationship because the presence or absence of event B is not completely determined by the presence or absence of event A.

An example will help to make this distinction clearer. Suppose that we imagine a person living in the suburbs who can choose among three possible routes when commuting to work in the central city (Figure 2.5). These routes are labeled A, B, and C. Over a lengthy period of observation we might note that this person uses route A 33 percent of the time, route B 50 percent of the time, and route C 17 percent of the time. In other words, the probabilities associated with using routes A, B, and C are 0.33, 0.50, and 0.17, respectively. Thus, on any given day, if we predicted that our commuter would be using route B, for example, we would have a 50 percent chance of being correct.

These probabilities are also useful in an aggregate context. If there are 100 commuters who need to travel to the central city every day and we assume that they all have similar behavioral characteristics, we would expect 33 of them to use route A. Of course, we cannot say with any certainty which particular 33 persons will be involved as this will change from day to day. Note that if one of the routes is assigned the value 1.0 it means that a person uses that route to the exclusion of all others, so his or her behavior is completely predictable and thus describable in terms of a deterministic law. If all 100 commuters exhibit the same behavioral pattern, we also have a deterministic relationship at the aggregate level.

Most relationships between variables that have been identified by social scientists are probabilistic. In other words, social scientists deal with *generalizations*; event A is generally related to event B, but not in every instance. In contrast, physical scientists have identified some truly deterministic relationships, as exemplified by the statement that pure water freezes at a temperature of 32 degrees Fahrenheit. When a series of laws, or generalizations, are inter-

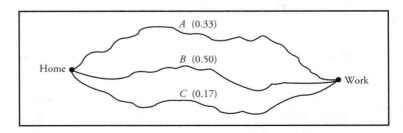

Figure 2.5 Journey to work in probabilistic terms.

connected in some way, they create a *theory*, and like the laws on which they are based, theories can be either deterministic or probabilistic. Empirically validated theories represent the culmination of the scientific method as they integrate previously isolated generalizations into broader systems of knowledge. The hypothetico-deductive spiral is never completed, however, as theories can also serve as vehicles for generating new hypotheses (Amedeo and Golledge, 1975:39).

2.2   Models

Many of the theories developed by social scientists are more accurately described as models, as they consist of a series of interconnected hypotheses, rather than a set of empirically validated laws (Macmillan, 1989). A model is often an idealized representation of reality, in order to demonstrate certain of its properties. Such idealized representations are abstractions of reality and thus omit certain unimportant details. The process of model building, therefore, represents a procedure for making these abstractions (Thomas and Huggett, 1980).

Types of Models

There are three general categories of models: iconic models, analog models, and symbolic models (Taylor, 1977:3). An *iconic model* is one in which reality is transformed primarily in terms of scale. Examples of such models include the scale models of buildings used by architects and the relief models of physical landscapes used by geographers. In both cases the models involve a reduction in scale. *Analog models* transform the properties of the real object or event, as well as changing the scale. The most common form of analog model is a map, where elevation is usually represented by brown contour lines, for example. Sometimes physical analogs are used to analyze social systems, as when stream networks are compared with transportation networks. Finally, *symbolic models* represent the real world in terms of symbols. A mathematical equation is an example of a symbolic model. For example, in the equation depicting the relationship between income and education shown in Figure 2.2, the symbol Y was used to represent income and the symbol X was used to represent education. These symbolic models are the most abstract of the three main types of models and thus the most easily manipulated.

Wherever possible, the models described in this book are expressed in equation form and then empirically calibrated and tested using appropriate data and statistical techniques (Figure 2.6). For example, we might construct the following *multivariate model* to explain the distribution of housing values in a city:

$$HV = f(HA, HQ, D) \tag{2.5}$$

where HV is housing value, HA is housing age, HQ is housing quality, and D is distance from the city center. To test this model, data might be collected at the census tract level so that measures of housing value, housing age, and housing quality could be taken from the U.S. Census of Population and Housing. The distance from the city center of each census tract can be measured using a map. For this particular type of model, the most appropriate statistical tech-

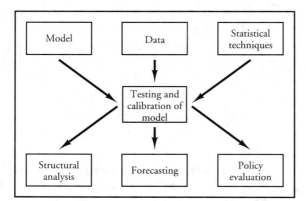

Figure 2.6 Model-building approach. (Based on Michael D. Intriligator, *Economic Models, Techniques, and Applications*, © 1978, p. 43. Reprinted by permission of Prentice Hall, Englewood Cliffs, N.J.)

niques for testing and calibration are multiple correlation and regression analysis. The multiple correlation coefficient indicates the extent to which the three explanatory or independent variables account for the variation in housing value, while the regression coefficients indicate the relative importance of the individual explanatory variables. A complete test of the model involves comparing its performance across a number of different cities.

The fully tested and calibrated model can then be used for three major purposes: structural analysis, forecasting, and policy evaluation (Intriligator, 1978:5). *Structural analysis* involves an attempt to understand the model for its own sake, simply for what it tells us about the quantitative relationships between the variables. *Forecasting* involves using a mathematical model to predict future values for particular variables, while *policy evaluation* involves using a model to simulate the potential outcomes of alternative policy formulations. In the latter context, for example, we might be interested in estimating what will happen to the housing values of a given neighborhood if the overall housing age is substantially altered by some kind of urban redevelopment project.

Structural Equation Models

Structural analysis is best undertaken in the context of structural equation models, which provide a powerful approach for examining the structure of relationships among a set of variables. As we shall see in the remainder of this book, such models include causal models, path analysis, and systems of simultaneous equations. These models are composed of exogenous and endogenous sets of variables. Each equation represents a causal link, rather than a mere empirical association, and each of these links generates a hypothesis that can be tested by estimating the magnitude of the relationship. In general, structural equation models involve the analysis of nonexperimental data, whereby statistical procedures are substituted for the conventional experimental controls that can be utilized under laboratory conditions. The notion of a system is also relevant, as the models usually consist of several equations that interact together.

The use of the word causal in the term *causal modeling* is not unproblematic, however. Bunge (1959) suggests that one central characteristic of the scientist's conception of causality is the notion of producing. That is, if X is a cause of Y, then a change in X should produce a change in Y, not simply that a change in X is always followed by a change in Y, or that a change in X is always associated with a change in Y. This idea of producing a change is similar to the concept of forcing as applied to an external stimulus that generates changes in a physical system. Unfortunately, however, a fundamental objection to the idea that causes involve a producing or forcing phenomenon is that such phenomena cannot be directly observed. We have only indirect evidence concerning covariation and temporal sequence.

Perhaps a more productive way to capture the nature of causal terminology is to focus on the ingredients of *causality*. As Kenny (1979:3) notes, at least three conditions must hold for a scientist to claim that X causes Y. First, a temporal sequence should be involved, as the idea of causality implies a process that takes place over time. For X to cause Y, X must precede Y in some kind of temporal sequence. An effect cannot precede its cause. Second, causality implies the presence of a functional relationship between cause and effect. Cause and effect are operationalized as variables, and these two variables should not behave independently of each other. That is, a known value of one variable for any particular observation should provide information about the value of the other variable for that same observation. Furthermore, we determine whether the relationship could be explained by chance, and a variety of statistical procedures can help us infer whether a sample relationship indicates a relationship within the population.

The third condition is more difficult to detect empirically: the relationship should be nonspurious. That is, there should not be a third variable Z that causes both X and Y such that the relationship between X and Y disappears when Z is controlled for. A difficulty here is that controlling for Z might mask the fact that Z is really an intervening variable, in that X might cause Z and Z might cause Y. In either event, controlling for Z will make the relationship between X and Y vanish, but while a spurious variable explains away a causal relationship, an intervening variable helps to elaborate the causal argument. Distinguishing between spurious and intervening variables is not always easy in an empirical context, and this third or excluded variable problem provides a major difficulty for causal analysts.

In practical terms, it is probably best not to attach any elaborate philosophical meaning to the use of the word causal in causal modeling, and the term *system modeling* might be an equally appropriate label. No specific definition of the word causal need be implied, as it merely corresponds to a hypothesized, unobserved process that represents the mechanisms embodied in a system of equations. As such, it is perhaps most useful to think of structural equation models simply as formal representations of the ideas that we have about a particular phenomenon. In this sense, we will not be in danger of losing sight of the fact that causality is merely a preconceived idea that we use to interpret our experience of phenomena.

In using observational data to *test* the predictions of any given model, we are far removed from the experimental ideal. It is thus useful to consider the differences between experimental and nonexperimental designs. In particular, Dwyer (1983:16) identifies three general classes of model-testing strategy: cross sectional, longitudinal, and experimental, where these three strategies are ordered in ascending internal validity. That is, the experiment can most definitively demonstrate a causal link, whereas nonrandomized cross-sectional designs are often the least definitive.

Cross-sectional designs are different from the other designs in two important ways. The observations are made at a single point in time, and no attempt is made to manipulate the observations along any of the measured dimensions. The inferential power of cross-sectional studies can be improved, however, by utilizing statistical rather than experimental controls, although the use of such statistical controls is most meaningful when all the variables relevant to a causal model have been identified and tested. In addition, of course, the empirical results of statistical controlling procedures are only reliable in those situations where the chosen indicators really measure the hypothetical constructs of interest. Nevertheless, cross-sectional designs provide a flexible methodology for exploring patterns of covariation among sets of variables, and the insights gained from such exploration can later be further elaborated using the often more costly designs associated with longitudinal and experimental studies.

Nonrandomized *longitudinal designs* involve either panel or time-series designs. In a panel design, the same sample of subjects is observed at more than one point in time, although the prohibitive cost of conducting large-scale panel studies usually means that there are not more than two or three waves. By contrast, a time-series analysis involves observing individuals, or spatial units, over an extended period of time. Such an analysis can help to uncover the direction and sequence of causal connections that might underlie a set of cross-sectional correlations.

The true *randomized experiment* is ideal in the sense that it provides the best opportunity for identifying spurious relationships. The presence in the real world of certain agents that produce changes in the states of systems is paralleled in the laboratory by the manipulations of the experimenter, who acts as just such an agent. Randomization provides a means of methodologically controlling, through research design, what cannot be controlled either physically or statistically. By appropriate manipulations, the experimenter attempts to partial out the separate effects of several explanatory variables that simultaneously influence some dependent variable. Individual subjects, or cases, are randomly assigned to treatments, thus ruling out self-selection and reducing the number of confounding variables.

In practice, however, one can never be certain that some variable, other than the one being intentionally manipulated, has not been inadvertently changed. That is, it is impossible to be sure that all confounding variables have been eliminated through the process of randomization. In this respect, the uncertainty about unmeasured third variables that arises in cross-sectional studies is often exchanged for the problem of irrelevant mediation that occurs in experimental studies. In addition, many causal models in the social sciences, including those concerned with urban geography, are not suited to experimental testing. For example, we are obviously forced to assess the effects of such variables as social class, family characteristics, and unemployment in a nonexperimental context.

Thus far we have focused on the problem of *internal validity*. That is, to what extent can it be demonstrated that an observed set of associations between two or more variables implies an underlying set of causal connections? Of equal importance, however, is the issue of *external validity*. To what degree can the conclusions generated by a particular study be generalized beyond the specific domain, or sampled population, associated with that study? There is often a trade-off between experimental and nonexperimental research strategies. Randomized experiments tend to minimize internal validity, but the costs involved with randomization and the manipulation of relevant variables often result in the use of rather small and restricted samples. By contrast, observational designs, although more problematic for internal validity, are generally able to utilize larger and more theoretically interesting samples.

In sum, although we should readily admit that causal laws can never be empirically demonstrated, even in an experimental context, it seems reasonable to maintain that the development of causal models is a useful heuristic device. In this sense, causal relations take on the role of working assumptions, rather than empirically verifiable statements about the system of interest. Causal models provide a highly useful theoretical tool, even though their implications are only indirectly testable. It would be extremely difficult for many social scientists to go about their daily work without clinging to the metaphysical assumption that something similar to causal relations is embedded within social systems. But most social scientists would also accept that such causal relations can at best be only indirectly apprehended, irrespective of whether the research design is experimental or nonexperimental in nature.

2.3   The Process of Model Building

Although the construction and testing of mathematical models has been a central theme in much contemporary urban research (Wilson, 1989a), the process of model building represents a creative endeavor that is in many ways more reflective of the artist than the scientist. The personal perspective that the model builder brings to his or her craft means that there are no hard and fast rules of model building. In most situations, however, the model-building process consists of a series of steps. Six such steps, or stages, are common to most models: (1) identifying the problem, (2) constructing a conceptual model, (3) translating the conceptual model into a symbolic model, (4) operationalizing the symbolic model, (5) testing, and (6) evaluation (Cadwallader, 1978a). These six steps can be illustrated by way of a simple model of consumer behavior.

Problem Identification

The initial step in any kind of model construction is to specify the purpose for which the model is being built. This involves identifying a scientific *problem* and then posing a *question* designed to clarify that problem. For example, we might ask: On what basis do consumers choose between alternative supermarkets? This question provides a good starting point for a general investigation, but the problem is not yet posed precisely enough for model-building purposes. At this juncture it is often a useful strategy to draw a picture of the problem. In the present context, we can identify the group of consumers in whose behavior we are interested and note that within a certain radius around them there are a finite number of supermarkets. These supermarkets represent the opportunity set, the boundary of which is defined by the greatest distance consumers are likely to travel in order to buy groceries (Figure 2.7).

If five supermarkets lie within the chosen radius, the problem becomes one of analyzing how the consumers choose among these five alternatives. More specifically, the aim is to construct a model that is capable of predicting the proportion of consumers who patronize each

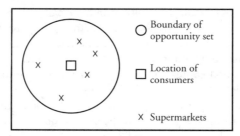

Figure 2.7 Opportunity set. (From M. T. Cadwallader, "The model-building process in introductory college geography: An illustrative example," *Journal of Geography*, 77, 1978, Fig. 1, p. 101.)

supermarket. Accomplishment of this goal suggests that the major variables involved in the decision-making process have been successfully identified. Also, the problem has now been formulated in sufficiently abstract terms for it to apply to a variety of types of behavior, not simply consumer behavior. For example, it could be used to represent a group of prospective migrants who are faced with choosing among alternative destinations or a group of vacationers who must choose among various recreational areas.

The Conceptual Model

After specifying the purpose for which the model is to be built, the next step is to select the main ingredients of the model. This involves selecting the most important variables in the process under investigation. These variables are chosen on the basis of either previous research or intuition. In the case of consumer behavior, conventional wisdom suggests that store attractiveness and distance to the supermarket are two of the major variables. After the variables have been chosen, it is often helpful to diagram the proposed model (Figure 2.8). The behavior to be accounted for, in this case the choice of a supermarket, is placed on the right side of the diagram, and the two major variables governing this decision, supermarket attractiveness and distance from the consumers, are placed on the left side. The arrows represent the hypothesized causal connections between the variables.

This particular model is very simple and there are a variety of ways in which it could be made more elaborate. For example, the possibility of change over time could be represented by a feedback effect between supermarket choice and supermarket attractiveness, as the perceived attractiveness of a supermarket may change after each visit. Similarly, an intervening variable representing the amount of information known about each supermarket might be inserted between the major variables and the choice of a supermarket.

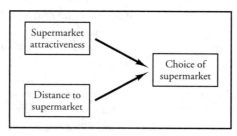

Figure 2.8 Conceptual model. (From M. T. Cadwallader, "The model-building process in introductory college geography: An illustrative example," *Journal of Geography*, 77, 1978, Fig. 2, p. 101.)

The Symbolic Model

After the proposed model has been understood in this rather general fashion, it should be written in a more precise mathematical form, using symbols rather than words. In this case, the symbolization might be as follows:

$$P_i = f(A_i, D_i) \tag{2.6}$$

where P_i is the proportion of consumers patronizing supermarket i, A_i is the attractiveness of supermarket i, and D_i is the distance to supermarket i. The symbolic model should then be rewritten to reflect the expected relationships among the variables:

$$P_i = \frac{A_i}{D_i} \tag{2.7}$$

where the notation is the same as in equation (2.6). This second formulation indicates that a positive or direct relationship is postulated between P and A, while a negative or inverse relationship is postulated between P and D.

At this point, following the precise specification of the model, the *assumptions* that have been incorporated into the model should be explicitly recognized. Two of the most obvious assumptions are, first, that the consumers are aware of all five supermarkets and, second, that no consumer will choose a supermarket that lies outside the opportunity set. Assumptions like these simplify the real situation, but are often a necessary part of model construction. The hope is that they can be relaxed when more insight has been gained into the process under investigation.

Operationalizing the Model

Once the model has been couched in symbolic terms it should be empirically tested. To do this, however, we must provide values for each of the constituent variables. Generally, there are a number of ways in which any model can be made operational, and the following procedure represents only one of the many ways in which the three variables in the present model might be calibrated. Supermarket *attractiveness* is measured in terms of four attributes: parking facilities, prices, quality of goods sold, and range of goods sold. Each consumer assesses each store on each of these evaluative dimensions by using a seven-point rating scale going from very unsatisfactory to very satisfactory. The resulting scale values for each store are then aggregated, using the median scale values, into an attractiveness matrix (Figure 2.9). Each element in the matrix represents the rating that the corresponding supermarket has been given with respect to one of the four attributes. For example, the value a_{12} is the rating given supermarket B with respect to parking facilities.

Each supermarket's overall attractiveness measure can be derived by simply adding the values in each column. However, such a procedure implies that the four attributes are equally important. If this assumption is thought to be too unrealistic, a *weighting vector*, indicating the relative importance of the attributes, can be obtained by asking the consumers to rank the attributes in order of their importance to them when selecting a supermarket. Each time an attribute is ranked first, it is given four points, each time it is ranked second, it is given three

	Supermarket				
	A	B	C	D	E
Parking facilities	a_{11}	a_{12}	a_{13}	a_{14}	a_{15}
Prices	a_{21}	a_{22}	a_{23}	a_{24}	a_{25}
Quality of goods	a_{31}	a_{32}	a_{33}	a_{34}	a_{35}
Range of goods	a_{41}	a_{42}	a_{43}	a_{44}	a_{45}

Figure 2.9 Attractiveness matrix. (From M. T. Cadwallader, "The model-building process in introductory college geography: An illustrative example," *Journal of Geography,* 77, 1978, Fig. 3, p. 101.)

points, and so on. The sum of the points associated with each attribute comprises the weighting vector. If the attractiveness matrix is premultiplied by this weighting vector, the resulting vector describes the relative attractiveness of each supermarket, taking into account the varying importance of the four attributes. This particular form of weighting vector assumes that the intervals between the ranks are equal. If this assumption is felt to be too simplistic, however, a more sophisticated weighting procedure could be devised.

The *distance* variable can be measured in a variety of ways, including cost, time, or mileage distance. It might also be measured in terms of cognitive distance, which represents how far people think the distance is. In this case, each consumer is asked to estimate the distance in miles to each supermarket, and then the average estimate for each supermarket is calculated. Finally, the actual *shopping behavior* of the consumers can be determined by simply asking them where they usually do their grocery shopping and then standardizing these values to reveal the proportion of consumers who normally patronize each of the five supermarkets. It is inadvisable to ask the subjects which supermarket they used on their most recent shopping expedition because that may only represent a minor fluctuation in their overall shopping strategy.

Testing

When each variable in the model has been measured, all that remains is to determine how closely the predicted proportion of consumers patronizing each supermarket approximates the actual proportion. In this case, the predicted proportions have been derived by dividing the attractiveness value by the distance value for each supermarket and then standardizing so that the values add to 1. Normally, such a comparison between the observed and the predicted involves a statistical test of some kind, and a wide variety of such tests are currently available. It is also interesting to test how well the model works in comparison with others that have been designed for the same purpose, and in the present instance the gravity model, which is discussed later in the book, would provide a suitable point of reference.

If the model does not perform as well as expected, it can usually be traced to one or more of three major *sources of error.* First, it might be due to sampling error, which would occur if the subjects chosen for the study were not representative of the general population. In this event the model should be retested by selecting further samples. A second source of error is measurement error, which occurs when the variables in the model have not been accurately calibrated. In the present example, the use of seven-point scales to measure supermarket attractiveness might be inappropriate, and the investigator would need to experiment with different ways of measuring supermarket attractiveness. Finally, the model might not perform satisfactorily because of specification error. This type of error occurs when the variables themselves, or their hypothesized relationships, are incorrectly specified. For example, in the present model some significant variable, such as the influence of advertising, might have been inadvertently omitted from the analysis.

Evaluation

Throughout this book the model-building approach to understanding urban processes and phenomena is emphasized (Batty, 1989). Although a rather catholic view of models is maintained, most of the models are presented in symbolic, or equation, form. These analytical models are only selective approximations of reality, in that incidental detail is omitted in order to generalize certain fundamental relationships. They are also representative of a structured, or pattern-seeking, viewpoint that emphasizes recurrent connections and interrelationships (Wilson, 1989b).

Such analytical models can serve a variety of *functions* (Haggett and Chorley, 1967). First, they are constructional in that they form stepping stones to the development of theory in a systematic progression of understanding. Second, by explicating the interrelationships between exogenous and endogenous variables, models allow one to predict future values of the endogenous variables. Third, models perform an organizational function with respect to data acquisition, in that they provide a framework for defining, collecting, and ordering information. Fourth, models can be used to generate hypotheses that both substantiate and extend the original theoretical structure. Finally, as it is hoped this book will illustrate, models can be used as pedagogical devices, in that they allow complex phenomena to be visualized and understood more easily (Macgill, 1989).

It is nevertheless true, however, that the model-building approach also entails certain *disadvantages* (Flowerdew, 1989). In particular, despite the recent advances involving dynamic models (Clarke and Wilson, 1985; Bertuglia et al., 1990), models can present an overly simplified view of reality that leads to unsuccessful predictions. Interestingly enough, however, Haggett (1990:155) suggests that increasing model complexity often leads to diminishing returns in terms of predictive capability or model fidelity, the ability to replicate faithfully a real-world system. It is also noteworthy, in this context, that most models are merely calibrated rather than tested. Calibration is usually achieved by searching for parameter values that optimize the goodness of fit between the model's predictions and observed behavior. This procedure is obviously rather different from validating the model, for which two sets of data are required, one for calibration and the other for testing (Polzin, 1990).

More important, in addition to these somewhat technical problems, analytical models fail to take into account certain influential determinants of urban structure. First, the effect of the overall political economy on urban patterns is very difficult to capture in any kind of formal model (Harvey, 1989b; 1991). In particular, the residential morphology of individual U.S. cities is partly a reflection of the class system within capitalist society and the associated mechanisms by which housing is assigned a value. Second, the model-building approach fails to capture the importance of different institutional structures and their associated constraints.

For example, Canadian and U.S. cities can be partly distinguished on the basis of their institutional structures, which have led to marked variations in patterns of investment and disinvestment (Goldberg and Mercer, 1986). Similarly, cross-cultural comparisons are only rather poorly developed within the model-building paradigm. Finally, analytical models seldom incorporate the historical context of urban development, thus failing to appreciate the role of changing patterns of land ownership, legislation, and government or institutional policy (Ward, 1989).

2.4   Problems of Urban Definition

The subject matter of this book is not easily defined, as it is sometimes difficult to distinguish between rural and urban settlements. This difficulty is reflected by the wide variation in population sizes used by different countries in order to categorize urban as opposed to rural settlements. In Sweden and Denmark, for example, settlements of only 200 people are counted as urban, whereas in Japan settlements have to contain at least 30,000 people before they are designated as urban (Hartshorn, 1992:3). These different definitions make it difficult to compare levels of urbanization across countries.

Perhaps the most rewarding approach for discussing the difference between urban and rural settlements is to think of them being arranged along a *continuum*, going from rural at one end to urban at the other. The settlement types along this continuum can be sequentially categorized as hamlets, villages, towns, cities, and metropolitan areas. At the rural end of this continuum, settlements such as hamlets are characterized by relatively low density living and agrarian-related occupations, whereas settlements at the other end of the continuum, such as metropolitan areas, are characterized by relatively high density living and nonagrarian-related occupations.

Once it has been decided which settlements are urban, there still remains the problem of defining the *spatial extent* of those settlements. For example, how far east does Los Angeles extend? In most cases the corporate boundary, or legal definition of a city, does not represent the true extent of that city. This problem has led to the identification of *underbounded cities*, where the legal city lies inside the real, or physical city, and *overbounded cities*, where the reverse is true.

Most cities in the United States are of the underbounded variety, as the corporate, or central city, is usually surrounded by a ring of suburbs. These suburbs use the facilities of the corporate city, such as museums and schools, but are not within the jurisdiction of that city. As one can imagine, this situation leads to severe problems in terms of governmental fragmentation and fiscal imbalance, as it is exceedingly difficult to coordinate the wide variety of local governments that represent the interests of individual communities (Cox and Jonas, 1993).

Local Government Fragmentation

The primary mechanism responsible for the evolution of underbounded cities is the *suburbanization process*, which dates back to the years immediately following World War II (Mills, 1992). The move to the suburbs, of both people and jobs, was precipitated by a variety of factors, including transportation improvements and a growing dissatisfaction with the quality of life in central cities. The housing boom of the late 1940s, fueled partly by cheap government loans, helped reinforce this decentralization process.

As the suburbs grew in population and the volume of industrial and commercial activity intensified, a whole range of government services were required. These services usually came to be supplied by local authorities rather than by the corporate city, and local government units proliferated at a rapid pace. For example, the Chicago Standard Metropolitan Statistical Area has a total of 1194 local government units (Yeates, 1990:273). This multiplicity of government units entails a wide variety of individual jurisdictions. Besides the elected governments of the various municipalities and counties, there are a whole range of special districts. These special districts are responsible for such services as fire protection, police protection, education, health facilities, and libraries, and they often overlap both each other and the underlying political districts.

A variety of problems are created by these underbounded cities and their associated plethora of local governments, perhaps the most critical of which is the problem of *fiscal imbalance*. In most cities the tax base of the central city has been steadily eroding in real terms as people and industry have moved out to the suburbs. Coincidental with these rapidly falling revenues is the equally rapid rise in central-city expenditures. The old and dilapidated building stock of these areas generates increased costs for fire protection, the higher crime rates bring requests for greater police protection, and the general poverty and high rates of unemployment induce increased demand for social services. In sum, there is a conspicuous disparity between the revenue resources and expenditure needs of the different political units within large cities. Central cities tend to exhibit the greatest disparity between revenues and expenditures, especially when compared to the more affluent suburban municipalities.

A second major problem arising from the fragmented pattern of local government is that often there are many districts administering *overlapping programs* of social services. The special districts set up to provide such services as water, gas, hospitals, libraries, and parks are generally unifunctional, and it is very difficult to coordinate their individual policies and interests. As a result, many of the larger issues in urban areas, such as land use planning and conservation, are neglected due to the inability to coordinate policy decisions made at the local level by very specialized administrative boards. It should be noted, however, that it has also been suggested that fragmentation is not necessarily inefficient (Cox and Nartowicz, 1980).

Possible Solutions

Solutions to these twin problems of fiscal imbalance and government fragmentation are not readily forthcoming, but responsible efforts have begun to be made (Barlow, 1991). Central cities receive increasing amounts in grant money from federal, state, and provincial governments as part of an overall policy of revenue sharing. Local governments are also attempting to develop additional *sources of revenue*. The present system of property taxation, for example, includes a provision for the value of improvements, which tends to discourage reinvestment in inner cities, thus contributing to the declining tax base. An alternative system of taxation might involve primarily taxing the value of the land. Such a system would force the owners of high-valued central-city land with deteriorated property to either sell or to improve the rent-producing potential of their property by making appropriate improvements.

Solutions to the coordination problems generated by the multiplicity of special districts have been sought by administrative reform involving various types of government consolidation. Some metropolitan areas have attempted to consolidate by merging city and county governments. In those cases where *city–county consolidation* has been approved, however, the unification does not usually extend beyond the core county. Such city–county consolidation is similar to the concept of a federation, where a two- or three-tiered local government system is developed in those situations where the metropolitan region spreads over a number of counties as well as municipalities.

It has also been suggested that daily *commuting patterns* might provide the basis for local government consolidation. This suggestion is attractive in the sense that those people who commute to the central city also use the services of the central city and therefore should be part of the tax base supporting those services. In practice, however, we are still faced with the overriding problem of where to actually draw the boundary. In the case of Madison, Wisconsin, for example, a small percentage of commuters travel to the corporate city from as far away as Milwaukee, while many of the residents of Dane County do not use Madison on a daily basis.

In general, Honey (1976) suggests a series of ways in which both the federal and state governments could strengthen the role of government at the metropolitan level. First, he contends that the federal government should (1) continue to augment metropolitan review of local government applications for federal funds; (2) provide revenue sharing for the metropolitan area; (3) assist in the creation and development of more powerful regional agencies in multistate metropolitan areas; and (4) institute a national land use planning policy that prevents the coalescence of individual metropolitan areas. Second, he suggests that, among other things, state governments should (1) provide financial assistance for metropolitan governments; (2) make local government participation in regional planning compulsory; (3) institute procedures to allow for the orderly amalgamation, incorporation, and annexation of urban areas; and (4) cooperate in the organization and planning of multistate metropolitan areas.

Chapter 3

Patterns of Land Use and Land Value

3.1   Agricultural Land Use Theory

The theoretical framework for explaining the distribution of land use and land value in urban areas is derived from the pioneering work of Johann Heinrich Von Thünen in agricultural land economics. Von Thunen was a German farmer in the early nineteenth century, and many of his ideas were generated by this practical experience and the associated cost accounting on his estate. He was concerned primarily with analyzing the patterns of land use and land value in agricultural areas, and the basic principle underlying his theory was the concept of economic rent, which later came to be known as location rent. In this chapter we first outline Von Thünen's agricultural theory and then apply the same kind of reasoning to urban areas (Papageorgiou, 1990). After discussing the effects of relaxing the simplifying assumptions and various other modifications, we conclude the chapter by considering the related problem of urban sprawl.

Assumptions

Like many economic theories, Von Thünen's agricultural land use theory is based on a set of simplifying assumptions. There are four major assumptions (Amedeo and Golledge, 1975:299). First, it is assumed that there is an agricultural area isolated from all other such areas, called the *isolated state*. Within this area there is a single city, or market, that is centrally located. This city is the only market for all surplus agricultural products produced within the region, and no products are imported from outside the region. The market price for a given product, or commodity, is the same for all farmers.

Second, it is assumed that this isolated state occupies a *uniform plain* that is completely homogeneous with respect to the physical environment. Topography, soil fertility, rainfall, and the like are assumed to be the same everywhere. As a result, the yield per unit area for any particular crop and the production costs associated with that crop are the same for all farmers throughout the region.

Third, it is assumed that transportation is equally available in all directions from the central city and that *transportation costs* are directly proportional to distance. That is, there is a positive, linear relationship between transportation costs and distance from the market, and the slope of the line is invariant across different directions of travel. Furthermore, transportation costs are a constant for any given agricultural product, although they vary from product to product and are borne by the farmer.

Finally, it is assumed that all farmers wish to *maximize profits* and are blessed with complete information and perfect decision-making abilities. Also, they are assumed to be capable of adjusting their agricultural operations in response to changing economic conditions. This concept of profit maximization and complete information is in stark contrast to the more behaviorally oriented satisficing concept. According to this latter conceptualization, men and women are assumed to be satisfied with less than optimal profits, and their decision-making ability is constrained by a lack of complete and accurate information.

Location Rent

Given these simplifying assumptions, the major mechanism behind the distribution of land use within the isolated state is represented by the concept of location, or economic, rent. Location rent is the total revenue received by a farmer for a particular product on a particular parcel of land, minus the production and transportation costs associated with that same product and parcel of land. Location rent reflects, therefore, the economic utility of a particular site for a particular product. Note that the term rent, as used in this context, does not correspond to its more popular usage as contract rent, which is the actual payment tenants make to others for the use of their land or property.

The location rent associated with any plot of land, for any particular crop or commodity, is calculated as follows:

$$LR = Y(p - c) - Ytd \qquad (3.1)$$

where LR is the location rent per unit of land; Y is the yield per unit of land; p is the market price per unit of commodity; c is the production cost per unit of commodity; t is the transportation rate per unit of commodity per unit of distance; and d is distance from the market.

This equation contains only two variables, location rent and distance from the market. All the other terms are constants that correspond to the initial simplifying assumptions. Because of the assumption of a homogeneous physical environment, the yield per unit of land and the production cost per unit of commodity will be the same everywhere. The assumption of a single market and a single price for each commodity ensures that the market price per unit of commodity is a constant, and, finally, the transportation cost is assumed to be a constant function of distance.

Imagine that we are interested in the location rent associated with a particular crop, say

Table 3.1 Differences in the Location Rent for Wheat at Various Distances from the Market

Distance (miles)	0	10	20	30	40
Location rent (dollars)	400	300	200	100	0

wheat, at a series of different distances from the market. Let the constants in the equation take the following values: Y is 100 bushels per acre; p is $10 per bushel; c is $6 per bushel; and t is 10 cents per mile per bushel. These values are then substituted into the location rent equation in order to determine the location rent at various distances from the market (Table 3.1). The relationship between location rent and distance (Figure 3.1) represents a *marginal rent curve*, as it reflects, for any given distance from the market, the rent that would accrue to an infinitesimal unit of land. In the present example, wheat will not be cultivated beyond a distance of 40 miles, as at that distance the transportation costs equal the net revenue resulting from the difference between market price and production costs.

Bid–Rent Curves

The location rent associated with a particular parcel of land represents a surplus to the farmer as long as the production costs include a payment for the farmer's labor and skills. This surplus can be paid to the landowner for the use of the land, and the farmer can afford to bid the full amount of the surplus and still remain in business in the long run. In this sense, the line expressing the relationship between location rent and distance from the market can also be called a *bid-rent curve* (Figure 3.1). The farmer is prepared to bid high amounts for the land right next to the market, as that land generates a large surplus, but he or she is prepared to bid far less for land farther away, as that land generates a far smaller surplus.

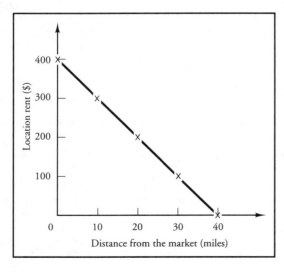

Figure 3.1 Relationship between location rent and distance from the market.

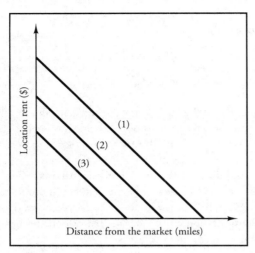

Figure 3.2 Family of bid-rent curves associated with high (1), medium (2), and low (3) market prices for wheat.

If the market price for wheat were to decrease for some reason, a new bid-rent curve would come into operation with lower values than the previous one, as the location rent would be less everywhere. In other words, a whole series, or family, of bid-rent curves can be conceived of, with each being related to a different market price for wheat (Figure 3.2). These curves are examples of *indifference curves*, as the farmer is indifferent with respect to the various possible locations on each particular curve, because his or her profits will be exactly the same at any point along the line.

We can now consider the situation where there are a number of competing crops or land uses, such as wheat, barley, and oats. Assume that the constants associated with these three types of land use, for substitution in the location rent equation, are as shown in Table 3.2. For the sake of simplicity, the yield for all three crops has been made exactly the same, although this condition is not necessary for the operation of the model. Given these values, the location rent for each crop can be determined for different distances from the market. For example, the location rent for barley at 10 miles from the market can be calculated by making the following substitutions in the location rent equation:

$$LR = 100\,(8-4) - 100\,(0.08)\,(10) \tag{3.2}$$

Table 3.2 Hypothetical Yields, Market Prices, Production Costs, and Transportation Rates for Three Different Crops

	Yield (bushels per acre)	Market Price (dollars per bushel)	Production Cost (dollars per bushel)	Transportation Rate (dollars per mile per bushel)
Wheat	100	10	5	0.20
Barley	100	8	4	0.08
Oats	100	6	3	0.02

Table 3.3 Location Rents for Wheat, Barley, and Oats at Different
Distances from the Market, Based on the Hypothetical Data in Table 3.2

	Distance (miles)		
	0	10	20
Wheat	$500	$300	$100
Barley	400	320	240
Oats	300	280	260

The location rent is $320, and this substitution procedure can be undertaken for all three
crops at distances of 0, 10, and 20 miles from the market (Table 3.3).

The bid-rent curves for the three land uses are generated by plotting the relationships
between location rent and distance from the market (Figure 3.3). Wheat has the steepest bid-
rent curve because, in this hypothetical example, the transportation costs associated with
wheat are the highest. Oats are the least sensitive of the three crops to transportation costs, so
they have the least steep bid-rent curve.

Assuming that each plot of land is used by the highest bidder, the land closest to the
market will be used for wheat, the intermediate land will be used for barley, and the land far-
thest from the market will be used for oats (Figure 3.4). More precisely, the land between the
market and distance *A*, which represents the intersection of the wheat and barley bid-rent

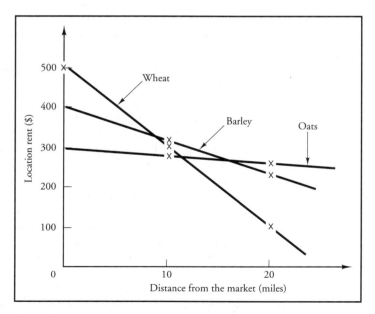

Figure 3.3 Bid-rent curves for wheat, barley, and oats based on
the hypothetical data in Table 3.2.

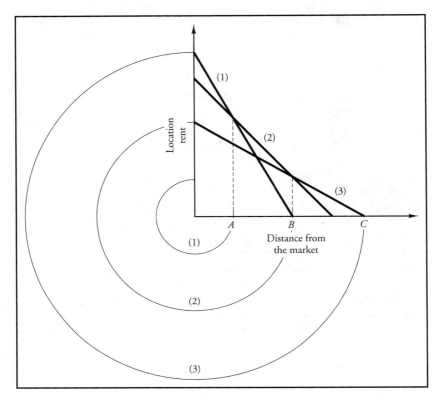

Figure 3.4 Concentric land use zones generated by the bid-rent curves for wheat (1), barley (2), and oats (3). Distances *A* and *B* represent the intersections of pairs of bid-rent curves.

curves, will be devoted to wheat, while the land between distances *A* and *B* will be used for barley, and the land beyond *B* as far as distance *C* will be used for oats. The resulting land use pattern can be expressed diagrammatically by swinging the graph around on its vertical axis and letting the distances *A*, *B*, and *C* trace out a series of three concentric land use zones.

It is instructive at this point to determine how the bid-rent curves will *change* if the constants in the location rent equation are altered in some way. For example, what if, due perhaps to some kind of government intervention, the market price for wheat rose from $10 per bushel to $11 per bushel? The location rent for wheat would increase to $600 right next to the market, $400 at a distance of 10 miles, and $200 at a distance of 20 miles. The new bid-rent curve would be to the right of its previous position, and the land use pattern would adjust accordingly (Figure 3.5). In particular, the amount of land devoted to wheat production would increase at the expense of that devoted to barley.

Also, in this context, we might ask what would happen if the transportation cost associated with wheat is increased from 20 to 25 cents per mile, per bushel. The location rent for wheat would then be $500 right next to the market, $250 at a distance of 10 miles, and $0 at

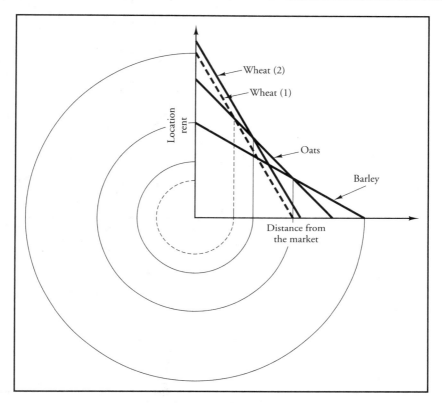

Figure 3.5 Land use pattern resulting from an increase in the market price of wheat. Wheat (1) represents the bid-rent curve for wheat before the price increase, and wheat (2) represents the bid-rent curve for wheat after the price increase.

a distance of 20 miles. In other words, the bid-rent curve would move to the left of its former position and have a steeper slope (Figure 3.6). This leftward shift of the bid-rent curve entails a corresponding decrease in the amount of land devoted to wheat and an increase in that devoted to barley. Note that, unlike changing the market price, the effect of changing transportation costs varies with distance from the market. Obviously, the farmers close to the market will be affected very little, while those farther away will be affected substantially. It is because of this varying sensitivity to changes in transportation costs that the new bid-rent curve is not parallel to the previous one.

Relaxing the Assumptions

Having analyzed the effect of changing the values of some of the constants, we can also determine what will happen if those constants are allowed to become variables. In other words, what happens if the simplifying assumptions on which the preceding analysis is based

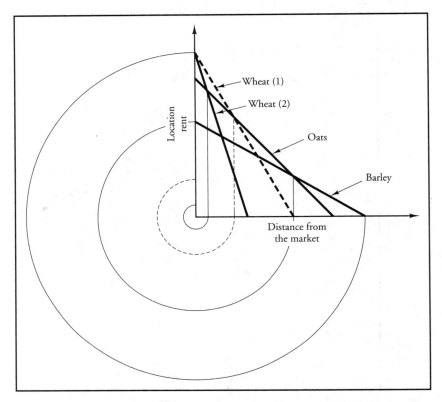

Figure 3.6　Land use pattern resulting from an increase in the transportation costs associated with wheat. Wheat (1) represents the bid-rent curve for wheat before the increase in transportation costs, and wheat (2) represents the bid-rent curve for wheat after the increase in transportation costs.

are relaxed? First, it was assumed that there is a *single market* at which all surplus agricultural products are sold. If a second market is introduced, the land use pattern will conform to a series of intersecting zones (Figure 3.7). Some of the land previously devoted to crop number 4 is now devoted to crop number 3, as the higher transportation costs associated with the latter crop are offset by the fact that it can be sold at the new market.

Second, it was assumed that the agricultural area under discussion occupies a *uniform plain* that is homogeneous with respect to the physical environment and thus contains no internal variation in terms of such attributes as soil fertility. If this assumption is relaxed, the bid-rent curve associated with any particular crop is unlikely to decline as a simple, linear function of distance from the market (Figure 3.8). Those areas characterized by fertile soils will produce a higher yield per unit area and thus generate higher location rents. On the other hand, those areas characterized by infertile soils will produce lower yields per unit area, or the yields will have to be increased by the use of fertilizers and thus involve an increase in production costs. In either situation, the location rent is lower than that in the fertile areas and is reflected by a trough in the bid-rent curve. Of course, soil fertility is not usually a simple

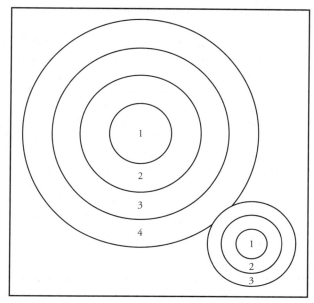

Figure 3.7 Effect on the land use pattern of competing markets.

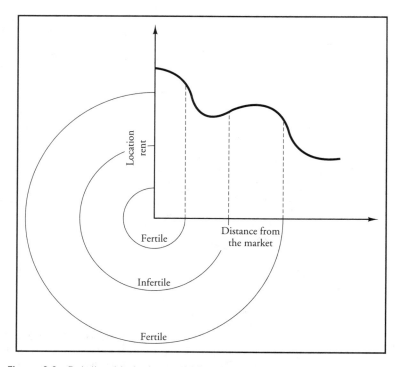

Figure 3.8 Relationship between bid-rent curves and soil fertility.

function of distance from the market, so variations in soil fertility tend to create a fragmented type of land use pattern.

Third, it was assumed that *transportation costs* are directly proportional to distance and that travel is equally easy in all directions from the market. In reality, however, transportation costs vary in response to such factors as the availability of roads or railways and the characteristics of the physical landscape. The bid-rent curves will be less steep in the direction of cheaper travel, meaning that the land use zones are no longer concentric circles, but are elongated in the direction of such features as major highways (Figure 3.9).

Finally, if the assumption of *profit maximization* is relaxed, the agricultural landscape will reflect a variety of irrational economic behavior. Farmers will not necessarily know which crop is the most profitable, and even if they do, they might not choose to grow that particular crop due to some kind of personal preference. Such idiosyncratic behavior tends to produce a highly segmented land use pattern.

In sum, the overall impact of relaxing the various simplifying assumptions is to create a land use pattern that is far more complex, and thus closer to reality, than the concentric land use zones generated by the concept of location rent. This difference between the model and reality, however, does not necessarily deny the validity of Von Thünen's argument. Beneath the apparent confusion of the agricultural landscape, the principle of location rent may well be in operation. However, the spatial pattern associated with this economic con-

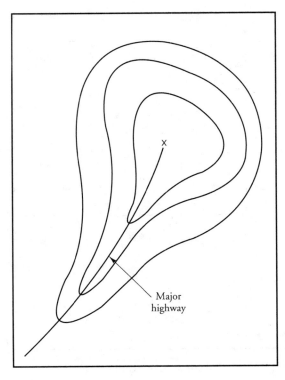

Figure 3.9 Effect of varying transportation costs in different directions from the market.

Major highway

cept varies from place to place due to the great variety of physical and socioeconomic environments.

Empirical Testing

Any empirical test of Von Thünen's model of agricultural land use distribution is rather problematic, as there is no region where the simplifying assumptions are completely met. One can only look for signs of concentric land use zones, rather than conclusive proof. In this context, the distribution of land use around the Sicilian village of *Canicatti* seems to be based on accessibility considerations (Chisholm, 1979:49).

Canicatti is a settlement of approximately 30,000 inhabitants, located 18 miles from the nearest settlement of similar size. The percentage of land devoted to different commodities varies with distance from Canicatti (Table 3.4). In particular, the production of vines is concentrated within the inner zone, between 0 and 4 kilometers, olives are more concentrated in the intermediate zone, between 2 and 5 kilometers, while unirrigated arable farming is especially prevalent in the area beyond 5 kilometers.

The rationale behind this land use sorting and the resultant land use zones seems to be related to the labor input associated with the different types of crops. This labor input can be measured by the average number of person-days per hectare expended on each crop. Values of 90, 45, and 35 for vines, olives, and unirrigated arable farming, respectively, strongly suggest that the crops with the highest transportation costs are located closest to the village. Note that transportation costs can be thought of either in terms of transporting the crop to market or in terms of getting labor to the crop, as is the case here.

Table 3.4 Percentage of Land Area in Various Uses and Annual Labor Requirements per Hectare for Canicatti, Sicily

Distance from Canicatti (km)	Percentage of Land Area:		
	In Vines	In Olives	Unirrigated Arable
0–1	15.8	—	19.7
1–2	18.0	8.4	15.9
2–3	2.3	14.4	23.6
3–4	13.3	0.6	18.1
4–5	5.1	2.4	43.4
5–6	6.3	1.6	64.1
6–7	3.3	—	68.7
7–8	4.0	—	62.4
Average number of person-days per hectare	90	45	35

Source: M. Chisholm, *Rural Settlement and Land Use: An Essay in Location*, 3rd ed., Hutchinson, London, 1979, Table 8, p. 50.

The relationship between labor input and distance is probably even greater than that indicated by Table 3.4, for two reasons. First, the data are based on the assumption that labor input is a constant for each crop, although the labor input, within any one crop type, probably decreases with increasing distance from the village. Second, the unirrigated arable land is left fallow more frequently at distances farther from the market, which would also be reflected by less labor input.

3.2   Urban Land Use and Land Value Theory

The ideas of Von Thünen have been formally applied in the urban context by William Alonso (1971). Like Von Thünen, Alonso based his analysis on the concept of economic, or location, rent and generated a series of land use zones from the intersections of different bid-rent curves. Also like Von Thünen, Alonso began his theoretical analysis by postulating a set of simplifying assumptions.

Assumptions

For the most part, the assumptions associated with urban land use theory parallel those of agricultural land use theory. First, it is assumed that the city has *one center*, or central business district. All employment opportunities are located within the central business district, and the buying and selling of goods also takes place at the city center. These last two points have the same effect as assuming a single market for all goods in the case of agricultural products.

Second, it is assumed that the city is located on a *flat, featureless plain*, which is similar to the assumption concerning a homogeneous physical environment in the agricultural case. No sites within the city have particular advantages or disadvantages with respect to such attributes as the underlying geology or an attractive view.

Third, it is assumed, as in the agricultural context, that *transportation costs* are linearly related to distance. More specifically, transportation costs increase with increasing distance from the city center, and the rate of that increase is the same in all directions. In this way the central business district is the most accessible location in the city, and accessibility decreases as one moves away from that location.

Finally, it is also assumed that each plot of land is sold to the *highest bidder*. This assumption implies that all land users have equal access to the land market, so there are no monopoly situations in terms of either buyers or sellers. Furthermore, there is assumed to be no intervention in the market economy on the part of government planning agencies and no restrictions due to legislation associated with land use zoning or environmental pollution standards.

Bid–Rent Curves

Given these simplifying assumptions, the urban land market operates in a similar fashion to its agricultural counterpart. Those uses that have the greatest to gain by locating on a particular plot of land will be able to bid highest for the use of that land. Thus each plot of land is sold to the *highest and best use*, highest in the sense of being the highest bidder and best in the sense of being the type of land use that is best able to take economic advantage of that particular plot. The term best, in this context, is used in a purely economic rather than a social sense.

The bid-rent curves and associated land use pattern can be analyzed in terms of three major types of urban land use: retailing, industrial, and residential (Figure 3.10). *Retailing* has the steepest bid-rent curve, as it is the most sensitive to accessibility considerations. The importance of high turnover rates makes a central location especially attractive, and the location rent declines with increasing distance from the central business district in response to decreasing accessibility and potential profits. As in the agricultural context, the bid-rent curve can be thought of as an indifference curve. Individual retailers are indifferent with respect to their location along the line, as their profits will be the same everywhere. Toward the center of the city, where potential profits are high, they can afford to bid a great deal for the use of the land, whereas toward the edge of the city, where potential profits are much less, they can afford to bid relatively little.

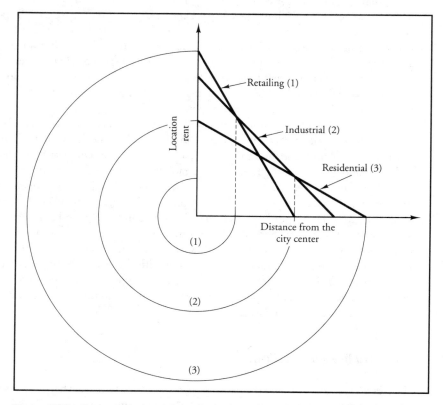

Figure 3.10 Concentric land use zones generated by the bid-rent curves for retailing, industrial, and residential land uses.

The bid-rent curve for *industry* also decreases with increasing distance from the city center, but at a slower rate than that for retailing. It is argued that industry is less susceptible to accessibility considerations than retailing, because many industrial products are sold outside the city, thus reducing the relative importance of location within the city. Industry still has some incentive to be close to the central business district, however, as in the present hypothetical city that is the most convenient location for their employees.

The bid-rent curve for *residential* use is the shallowest of the three, with the primary advantage accruing to central locations being associated with the shorter journey to work. This advantage, however, is not such that residential land use is able to outbid either retailing or industry. Those residential areas that are located toward the center of the city involve a greater capital investment than those farther out and will be characterized by higher-density living. This situation arises because, in order to obtain a satisfactory return on their investments, residential developers need to use intensively the more expensive land.

In some respects, however, the residential sector of the urban economy fits this type of analysis less well than do its retailing or industrial counterparts. Individual households are presumed to behave in a way that maximizes their overall satisfaction, as the assumption of profit maximization is not entirely appropriate in this context. It is argued that households make a trade-off between more living space and greater commuting costs. The poor, with relatively little disposable income, consume small amounts of space at the center of the city, where commuting costs are negligible. The rich, on the other hand, with much larger disposable incomes, can pay the same amount of money for a large plot of land near the edge of the city and still have enough left to cover commuting costs. This argument explains the apparent contradiction of relatively poor people living on relatively expensive land, as it points out that the poor use only small parcels of that expensive land.

As a corollary of the configuration of bid-rent curves, *land values* decrease with increasing distance from the city center. The thick line in Figure 3.11 denotes the highest price offered for any particular plot of land, and this line slopes away from the central business district in a curvilinear fashion, with land values decreasing rapidly at first and then more gradually.

In summary, five major points can be made concerning the interrelationships among bid-rent curves, land use, and land value (Alonso, 1971). First, a family of bid-rent curves can be derived for each user of land such that the user is indifferent to his or her location along any one of those curves. Second, the equilibrium rent for any location is determined by choosing the highest bid from among the various potential users. Third, because each plot of land goes to the highest bidder, those land uses with the steeper bid-rent curves capture the central locations. Fourth, through this competitive bidding among potential users, land uses determine land values. Fifth, land values also determine land uses, as the latter are distributed according to their ability to pay for the land.

3.3   Relaxing the Assumptions

As with agricultural land use theory, it is instructive to determine what happens to the pattern of land use and land value when the original simplifying assumptions are relaxed. If it is no longer assumed that the city has a *single center*, or central business district, then one can imag-

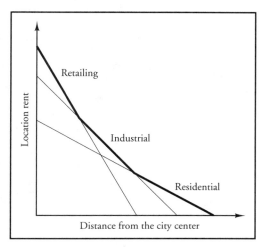

Figure 3.11 Relationship between land value and distance from the city center.

ine a series of subcenters, each generating its own set of concentric land use zones. The land use pattern in such a polynuclear city will be fragmented, as the zones surrounding individual subcenters will intersect each other. Los Angeles is perhaps the classic example of a polynuclear settlement, as it is really a series of cities within a city. Culver City, Century City, and Santa Monica, for example, each have their own distinctive land use patterns, which merge with each other to provide the overall pattern for Los Angeles. Recent research has attempted to place such polycentric cities within the same theoretical structure as that constructed for single-center or monocentric cities (Yinger, 1992).

If the assumption that the city is located on a *uniform plain* is relaxed, the land use and land value patterns will respond to variations in such things as the local topography and underlying geology. The nature of this response depends on what particular kind of land use is being considered. For example, if a plot of land is to be used for a multistoried office complex, it is important to have a suitable underlying geology. If, on the other hand, one is planning to construct a set of high-priced condominiums, the presence of an attractive view might be a more important consideration. In general, of course, such attributes as the underlying geology and the location of scenic views are not distributed in any regular fashion within cities. In fact, cities tend to be highly idiosyncratic with respect to the spatial distribution of these attributes, and that is why the latter are best treated as constants within the overall theory.

The assumption concerning *transportation costs* can be at least partially relaxed by imagining a city that is composed of a series of major highways radiating out from the center, with a second set of highways forming concentric circles, or ring roads (Figure 3.12). This transportation network is much more realistic than that implied by the original assumption that accessibility simply decreases with increasing distance from the central business district. In the present network, accessibility is greatest along the major radial and ring roads, with minor peaks of accessibility occurring at the intersections.

The associated pattern of land values has three main elements (Figure 3.12). First, in accordance with the original theoretical formulation, land values generally decrease as one

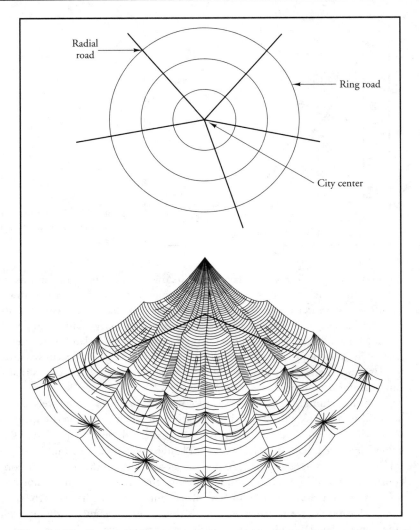

Figure 3.12 Transportation network based on radial and ring roads and the corresponding land value surface. (The land value surface is from B. J. L. Berry, *Commercial Structure and Commercial Blight: Retail Patterns and Processes in the City of Chicago*, Department of Geography Research Paper 85, University of Chicago, Chicago, 1963, Fig. 3, p. 14.)

moves away from the city center. Second, there are ridges of higher-valued land associated with both the ring roads and radials. Third, there are local peaks of higher land value that are coincidental with the intersections of the major traffic arteries.

In turn, the land use pattern responds to this new distribution of accessibility and land values. Retailing activity emerges at the intersections, often in the form of planned shopping centers, and along the major highways. These outlying shopping centers often become surrounded by apartment buildings and other kinds of multifamily dwelling units, as residential developers adjust to the increased cost of land by using it more intensively. In sum, a series of mutual readjustments take place within the system due to the complex interrelationships among accessibility, land use, and land value.

These adjustments are not instantaneous, however, so at any given point in time the land use pattern reflects a certain amount of *inertia*. In general, urban land is less flexible than rural land in terms of being adjusted to changing conditions. The short seed-to-harvest cycle of most agricultural crops is much more conducive to land use changes than is the lengthy effective life of the urban building stock. This comparative inflexibility of urban land, at least in the short run, means that a certain percentage of the city's land area is always underutilized, in the sense that it is not occupied by the highest and best use as defined previously.

Externalities present a second major difference between the urban and rural contexts. The value of a particular farm, for example, will not ordinarily depend on what is being produced in adjacent fields, whereas the value of an urban lot may depend significantly on the adjacent land uses. At one extreme, obnoxious facilities, such as meat-packing plants, tend to lower the value of the surrounding land. At the other extreme, the concentration of similar industries may generate positive externalities, as such concentrations reduce costs by facilitating subcontracting.

Because the assumption of a free-market economy does not hold for the urban land market, imperfections occur in the land use allocation process. A variety of different types of rent can be identified, one of the most important of which is *monopoly rent*. Monopoly rent occurs when the ownership of land becomes concentrated in the hands of a few individuals or corporations. In this situation, the amount of rent associated with a particular plot of land is not determined primarily by a competitive bidding process, but rather by the price at which the monopoly owner is willing to sell.

Finally, the process of land use allocation discussed by Alonso and others assumes the absence of *government intervention*. Such intervention, however, is fairly common and can sometimes have a profound impact on the distribution of land use and land value. For example, the presence of tax breaks just beyond corporate city limits has tended to encourage the suburbanization of industry and the development of outlying industrial parks. Of even greater importance, however, has been the compartmentalization of land uses associated with the proliferation of zoning ordinances.

Land Use Zoning

Comprehensive land use zoning first appeared in New York City in the early part of this century. In 1916 the New York State legislature delegated authority to the city of New York, which then enacted the first comprehensive zoning ordinance in North America (Goldberg and Horwood, 1980:3). This ordinance allowed the city to designate certain areas for specific types of land use, such as residential or industrial, whereas, prior to 1916, land use control was mainly limited to private actions related to the laws of nuisance. Since this initial legislation, the overwhelming majority of zoning cases that have been decided by federal and state appellate courts have involved suburbs rather than central cities. In this context, it is symptomatic that when the U.S. Supreme Court decided its first zoning case in 1926, the justices chose a case challenging a zoning ordinance in a suburb of Cleveland. This celebrated case, involving the village of Euclid, extended to zoning enactments a presumption of validity that they had not formerly received.

The demand for zoning arose because property owners were concerned about *externalities* in the land market, whereby the value of land, and property on it, was often significantly influenced by the characteristics and uses of adjacent land. High-rise office buildings or apartment structures, for example, often had a negative impact on neighboring properties because they blocked out the daylight. Zoning was a means by which negative externalities could be minimized, while positive externalities, such as the clustering of compatible land use types, could be maximized.

A variety of *zoning concepts* have been developed to promote specific land use planning goals (Hartshorn, 1992:353). Transfer zoning, for example, is designed to promote historic preservation. Impact zoning is used to manage urban growth. Percentage zoning encourages a certain prespecified mix of land use types, and agricultural zones are used to designate areas for agricultural use, thus preventing speculative development, especially at the edge of the city within the rural–urban fringe.

The practice of *exclusionary zoning* has been particularly closely scrutinized (Pogodzinski and Sass, 1990). Exclusionary zoning is associated with specific performance standards and is frequently used in suburban areas to ensure high levels of economic and land use uniformity. This kind of zoning tends to screen lower-income families from certain neighborhoods by regulating such things as lot size, floor space, and even the number of bedrooms. Equally troublesome is the potential use of exclusionary zoning for purposes of racial discrimination. Indeed, one judge maintained that zoning was nothing but a vast plan for segregation, while others have claimed that the single major effect of zoning legislation has been to exclude lower-income classes from certain neighborhoods, especially lower-income blacks.

As one might imagine, the *administrative structure* associated with land use regulation is often exceedingly complex. Zoning activity in a typical community may involve a planning board, a zoning commission, an appeals board, a zoning enforcement officer, a building inspector, and so on. Delays are inevitable when these agencies, or individuals, have a backlog of work, and the role of independent planning commissions has been subject to increasing attacks. These lay planners, who often administer the subdivision regulations and recommend zoning changes, have been criticized on two main grounds. First, they lack the expertise of professional planners. Second, unlike the city council or other legislative bodies, they lack political accountability.

In any event, all types of land use zoning tend to *fragment* the regular pattern of land use predicted by Alonso's model. In particular, each plot of land is no longer necessarily sold to the highest and best use. Due to some local zoning ordinance, for example, land that would ordinarily be used for industrial activity might instead be developed for housing (Figure 3.13). In other words, land that cannot be used by the highest bidder, or the land use with the highest bid-rent curve at that location, has to be used by the next highest bidder, or the land use with the second highest bid-rent curve at that location. The associated land value will drop accordingly, creating a discontinuity in the land value surface.

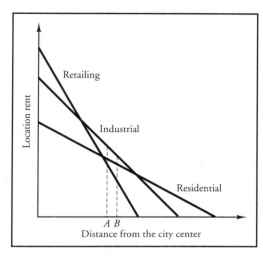

Figure 3.13 Effect of land use zoning on bid-rent curves. The land between distances *A* and *B* is zoned against industrial use and is therefore used for residential purposes.

Because each plot of land is no longer utilized by the highest and best use, zoning tends to reduce the overall tax base of the city, and land use zoning has also been criticized on a number of other grounds (Goldberg and Horwood, 1980:26–9). First, zoning creates a rigid division of uses that are unable to cope with the variety of tastes and demands that occur in modern society. Second, zoning has come to dominate the planning function as a whole. Third, zoning ordinances tend to preserve the status quo and thus impede change in the land use system. Fourth, zoning can encourage monopoly situations in the urban land market.

On the other hand, it can also be argued that zoning is a very beneficial form of land use regulation (Goldberg and Horwood, 1980:24–6). First, zoning protects property values by minimizing negative externalities. Second, zoning helps to maximize the public good by ensuring that sufficient amounts of land are available for different kinds of uses. Third, by identifying areas of historical importance, zoning can preserve the architectural character of a city. Fourth, zoning ensures that the majority of residents are in control of land use, such that one or two people cannot threaten the property values of their neighbors by indiscriminately adjusting their land use.

3.4   Empirical Testing

The urban land use and land value theory discussed above generates two testable propositions. First, it postulates that land values decrease in a curvilinear fashion with increasing distance from the city center. Second, it argues that land uses will be arranged in a series of concentric zones radiating out from the city center. It is possible to test both these propositions by the use of bivariate correlation and regression analysis and thus come to some conclusions concerning their empirical validity.

Introduction to Bivariate Correlation and Regression Analysis

Correlation analysis indicates the strength of the relationship between two variables, while regression analysis identifies the form of that relationship. Although one would ordinarily wish to know something about the strength of a relationship before becoming overly concerned about its form, for the purposes of exposition it is easier to begin with the rudiments of regression analysis.

The first step involves drawing a *scatter diagram*, whereby all the cases, or observations, are placed on a graph according to their values for the two variables X and Y (Figure 3.14). The *least-squares* line is placed through the center of these points, such that the sum of the squared deviations around that line is minimized. The deviations referred to are the vertical distances between each point and the line, two of which are shown in Figure 3.14. These deviations are then squared, thus eliminating all negative values. Finally, the squared deviations are summed, and that value is minimized.

Appropriate formulas are available to solve for the unknowns in the following *least-squares equation*, thus determining the position of the least-squares line (Blalock, 1979:392):

$$Y = a + bX \tag{3.3}$$

where Y is the dependent variable; X is the independent or explanatory variable; a is a constant representing the value of Y when X is zero; and b is a constant representing the slope of the line. That is, b measures the change in Y for a unit change in X. Note that the equation contains two variables, X and Y, and two constants, a and b. The term a is a constant because there is only one point at which the least-squares line intercepts the Y axis, and b is a constant because the line is straight and therefore has the same slope throughout the relationship.

In the same way that this equation expresses the form of the overall relationship, each individual case can be represented by its own particular equation. The value for any individual case i is described as follows:

$$Y_i = a + bX_i + e_i \tag{3.4}$$

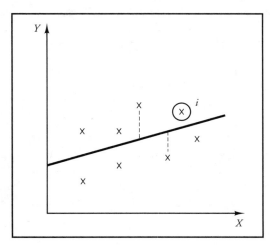

Figure 3.14 Scatter diagram and least-squares line.

where the terms are the same as in equation (3.3), except that the subscript i has been added and an error term, denoted by the symbol e, has also been included. The *error term* expresses the degree to which any case i is removed from the least-squares line and so is not exactly predicted by the least-squares equation. In Figure 3.14, for example, the error term associated with case i, or the *residual*, as it is often called, is positive because i lies above the line. In other words, the predicted value of i, based on the least-squares equation, must have something added to it in order to equal the actual value of i. Similarly, those cases lying below the line have negative error terms, or residuals.

A variety of factors are represented by the error term. First, it might include measurement error in variable Y, although least-squares analysis is based on the assumption that there is no measurement error in variable X. Second, the error term might reflect the fact that the relationship is not really linear, although, again, least-squares analysis is based on the assumption of a linear relationship. Third, and most important, the error term includes all the other variables, in addition to variable X, that influence the dependent variable Y in some way.

In sum, the least-squares equation, representing the least-squares line, is a predictive equation relating the two variables X and Y. Once the two constants in this equation have been estimated, values of Y can be predicted for particular values of X simply by inserting the value of X in the right side of the equation. The accuracy of these predictions will depend partly on the strength of the relationship between the two variables, as determined by correlation analysis.

The *correlation coefficient* (r), which measures the strength of the relationship between two variables, runs between -1.0 and +1.0. A negative value indicates a negative or inverse relationship, while a positive value indicates a positive or direct relationship. Note that, for any pair of variables, the signs associated with b, the slope of the line, and r are always the same. A negative slope is always reflected by a negative correlation, and vice versa.

The strength of the relationship and the associated correlation coefficient are measured by the degree to which the points are scattered around the least-squares line (Figure 3.15). If all the points lie exactly on the line, the relationship is perfect, and the value of r will be +1.0 if the slope is positive and -1.0 if the slope is negative. The greater the scatter of points around the line, or the larger the residuals, the closer the correlation coefficient will be to zero. Due to the linearity assumption in least-squares analysis, however, a low correlation coefficient does not necessarily signify the absence of a relationship. A perfect curvilinear relationship would not be reflected by an r of 1.0 unless some transformation of the variables had previously been undertaken (Figure 3.15).

A related statistic to the correlation coefficient is the *coefficient of determination* (r^2), which is simply the square of the correlation coefficient. The coefficient of determination runs between 0 and 1.0 and represents the amount of variation in variable Y that is accounted for or statistically explained by the variation in variable X. If the variation in Y is completely accounted for by the variation in X, the coefficient of determination will be 1.0.

The use of a bucket-and-sponge analogy helps to clarify the interpretation of the coefficient of determination (Figure 3.16). Imagine that the water in the bucket represents the amount of variation associated with Y, while the sponge represents the amount of variation associated with X. When dipped into the bucket, the sponge soaks up a certain amount of water, and this water represents the percentage of variation in Y accounted for by the variation

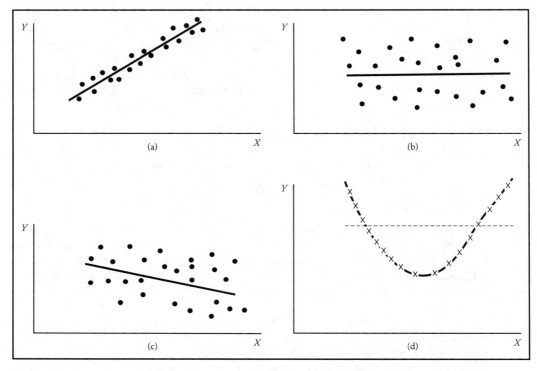

Figure 3.15 Examples of the relationship between the scatter diagram and the associated correlation coefficient: (a) strong positive relationship; (b) no relationship; (c) weak negative relationship; (d) perfect nonlinear relationship for which $r = 0$. (From H. M. Blalock, Jr., *Social Statistics*, Revised Second Edition, McGraw-Hill, New York, 1979, Fig. 17.7, p. 397, and Fig. 17.8, p. 398. Reproduced with permission.)

in X, that is, the coefficient of determination. The water remaining in the bucket represents the variation in Y that is left unaccounted for by X or, in other words, the residual variation in Y. This residual variation must be explained by other variables not presently being considered.

The Pattern of Land Value

Alonso's proposition that land values decrease with increasing distance from the city center has been empirically tested for Seattle, Washington, by Seyfried (1963). Seyfried considered a variety of functional relationships, but of greatest interest are the results associated with a simple linear function and a power function. The *linear function*, in this particular context, can be expressed as follows:

$$LV = a - bD \tag{3.5}$$

where LV is land value, D is distance, and a and b are constants representing the value of the intercept and the slope of the line, respectively. The unknowns in this equation and the

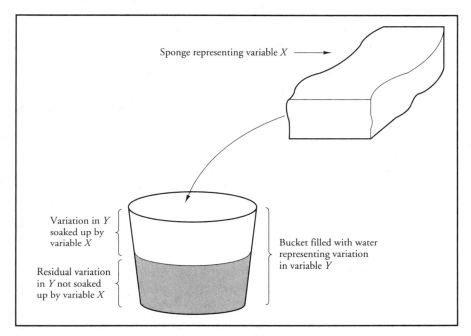

Figure 3.16 Bucket-and-sponge analogy for interpreting the coefficient of determination. (Based on R. Abler, J. Adams, and P. Gould, *Spatial Organization: The Geographer's View of the World*, © 1971, Fig. 5.23, p. 129. Reprinted by permission of Prentice-Hall, Inc., Englewood Cliffs, N.J.)

strength of the relationship between land value and distance can be determined via the previously described least-squares analysis.

The *power function*, on the other hand, is expressed as follows:

$$LV = aD^{-b} \tag{3.6}$$

where the notation is the same as in equation (3.5). This power function provides a closer representation of Alonso's prediction, as it describes land values as decreasing at a decreasing rate. In other words, it is a curvilinear rather than linear function. In order for equation (3.6) to be investigated using least-squares analysis, however, it must first be transformed into a linear equation by taking the common logarithms, to the base 10, of both variables. The new equation is as follows:

$$\log LV = \log a - b \, (\log D) \tag{3.7}$$

where the notation is the same as in equation (3.5), and log represents common logarithms to the base 10.

To evaluate the relative appropriateness of the linear and power functions, Seyfried collected data on the mean assessed value per square foot for street intersections radiating outward from the central business district of *Seattle*. The gradients associated with these assessed values were determined for the four major compass directions, and the results provide strong support for Alonso's hypothesis (Table 3.5). First, all the b values, representing the slopes of the lines, or land value gradients, are in the expected negative direction. Second, the correlation coefficients suggest that the curvilinear, or power function, is more appropriate than its linear counterpart. Indeed, the correlation coefficients associated with the power function are exceptionally high, indicating that, at least in three of the four directions, approximately 80 percent of the variation in land values is accounted for by distance from the city center.

The results obtained for the westerly direction are rather surprising, in that the correlation coefficient for the linear function is as high as that for the power function. On closer inspection, however, this apparent anomaly is explained by the fact that both these coefficients are based on only six observations, due to the proximity of the ocean. In general, a small sample size tends to inflate the correlation coefficient. At the extreme, when dealing with bivariate correlation, a sample size of 2 will always yield a perfect correlation of 1.0, as one can always draw a straight line through two points.

The Pattern of Land Use

Alonso's second proposition, that land uses will be sorted into concentric zones radiating out from the city center, has been investigated by the economist Edwin Mills (1972:40). Mills analyzed the density gradients associated with different types of land use patterns for a sample of 18 U.S. Metropolitan Areas. This sample was not chosen randomly, but rather with four factors in mind. First, the cities had to be approximately circular or semicircular in shape, thus excluding those cities with highly irregular political boundaries. Second, a wide range of population sizes was included. Third, the cities were chosen from different regions of the country, and, fourth, a variety of historical growth rates were included.

A negative *exponential function* was used to identify and compare these density gradients. This function is expressed as follows:

$$Y = ae^{-bD} \tag{3.8}$$

Table 3.5 Correlation and Regression Results for Land Value Gradients in Seattle, Washington

	North	South	East	West
Linear function				
r	−0.275	−0.356	−0.454	−0.959
b	−0.240	−0.048	−0.244	−8.610
Power function				
r	−0.795	−0.928	−0.895	−0.914
b	−0.807	−1.078	−1.576	−2.343

Source: W. R. Seyfried, "The centrality of urban land values," *Land Economics*, 39 (1963), Table I, p. 280.

where *Y* is the dependent variable, such as employment density in retailing; *D* is distance from the city center; *a* and *b* are the empirically derived constants; and *e* is the base of natural logarithms. The negative exponential function is extremely similar to the previously described power function and, like the power function, it must be logarithmically transformed before it can be analyzed via least-squares analysis. In this instance, the transformation is achieved by taking the natural logarithm of the dependent variable, thus generating the following equation:

$$\ln Y = \ln a - bD \tag{3.9}$$

where the notation is the same as in equation (3.8), and ln represents natural logarithms.

Mills averaged the density gradients associated with each type of land use, based on the value of *b* in equation (3.9), across the 18 cities (Table 3.6). The values were computed for four different time periods in order to assess the change over time. Two major points emerge from this analysis. First, for all four time periods, as predicted by Alonso, the density gradient associated with retailing is the steepest, while that associated with population is the gentlest. In other words, retailing dominates the central locations, while residential land use dominates the peripheral locations. Second, all three density gradients have become less steep over time and also less distinguishable from each other. The declining gradients are due to the postwar suburbanization process, while the increasing similarity between the gradients reflects the fact that both retailing and manufacturing have become more peripherally located in recent years, as witnessed by the growing number of outlying shopping centers and industrial parks.

Criticisms

In summary, Alonso's hypotheses concerning the spatial arrangement of land use and land value within cities seem to enjoy substantial empirical support (Coulson, 1991). The patterns could never be exactly as hypothesized, as no city completely conforms to the various simplifying assumptions upon which the theoretical analysis is based. Despite this empirical support, however, Alonso's overall approach to understanding urban spatial structure has been subjected to increasing criticism. In large part these criticisms can be applied to neoclassical economics in general, of which Alonso's model is a representative example (Ball, 1985).

At least six major criticisms have been leveled at the *neoclassical approach* to understanding land use and land value patterns within cities. First, buyers and sellers in the urban land market do not really possess perfect information. Also, the emergence of powerful corpora-

Table 3.6 Average Density Gradients Associated with Different Types of Land Use

Land Use	1948	1954	1958	1963
Population	−0.58	−0.47	−0.42	−0.38
Manufacturing	−0.68	−0.55	−0.48	−0.42
Retailing	−0.88	−0.75	−0.59	−0.44

Source: E. S. Mills, Studies in the Structure of the Urban Economy, Johns Hopkins University Press, Baltimore, 1972, Table 12, p. 42. Published for Resources for the Future by the Johns Hopkins University Press.

tions and the increasing intervention of urban governments cast serious doubt on the idea of a free market in land. Second, little is known about how households actually make the complex trade-offs between housing expenditures, travel expenditures, and the many other household budget items. Third, Alonso's approach tends to ignore the supply side of the equation, as under the assumptions of profit maximization and perfect competition, supply is deemed to follow automatically from the structure of demand (Evans, 1983). Such considerations are not beyond the scope of the general approach, however, as Muth (1985) has been able to incorporate the production of housing into his more sophisticated version of the trade-off model.

Fourth, neoclassical models tend to be ahistorical in nature and overlook the inertia from the past that has left many activities in suboptimal locations. In this respect, such models provide no insight into the process of historical change and thus do not directly address the question of how we came to our present urban condition. Fifth, the neoclassical theory of consumer choice pays insufficient attention to the constraining influence of the larger political-economic structure in which the land and housing markets are embedded. For some, the social relations of production must obviously precede, in theoretical terms, the system of consumption (Scott, 1980:76). Finally, the neoclassical paradigm reduces the factors of production, such as labor and capital, to the status of abstract and mutually substitutable technical inputs and thus assumes that they are politically neutral (Scott, 1980:81).

The Polycentric Model

As previously noted, many U.S. cities, such as Los Angeles, are now best described in terms of some kind of polycentric model (Berry and Kim, 1993). Such models involve a series of subcenters, rather than a single, dominant central business district. The central business district may still be the most accessible location for the entire urban area, but it is now joined by a number of suburban subcenters. These subcenters generate their own set of land use zones (Figure 3.17), thus generating the polycentric version of the monocentric city represented in Figure 3.10. In most cases these outlying centers are located near the intersections of major highways.

Data that Erickson (1986) has collected for some of the largest metropolitan statistical areas suggest that suburban municipalities with 5000 or more employees have become more numerous in recent years (Table 3.7). For example, using this definition of a subcenter, the number of *suburban nucleations* in Chicago increased from 31 in 1967 to 44 in 1977, while the number in Philadelphia increased from 14 to 22 over the same time period. As a result, these suburban nucleations account for an increasingly large share of the total employment in the urban area (Table 3.7). For example, the percentage of the total employment associated with suburban nucleations in Chicago rose from 21.6 in 1967 to 30.2 in 1977, while the figures for Philadelphia indicate an increase from 16.6 percent to 28.0 percent. As one might expect from our discussion of monocentric cities, these employment subcenters have a significant impact on the surrounding land values (McDonald and McMillen, 1990).

As yet, there is no single polycentric *theory* of urban spatial structure. Rather, there is a distinction between theories that prespecify the location of subcenters and theories that postulate the conditions under which subcenters might emerge, without explaining where those centers will be located. The most important element of a satisfactory polycentric model is the endogenous generation of subcenters. For example, Wieand (1987) extends the monocentric

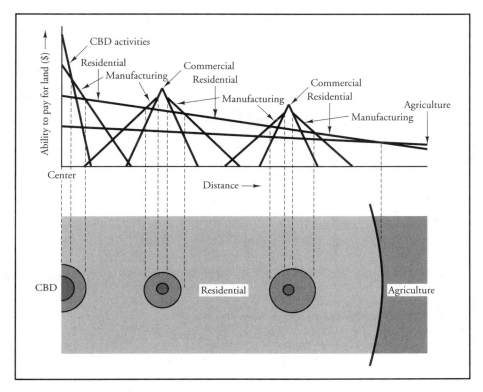

Figure 3.17 Hypothetical land-use zones in a multicentered city. (From M. Yeates, *The North American City*, Fourth Edition, Harper & Row, Publishers, Inc., copyright © 1990, p. 132. Reprinted by permission of HarperCollins Publishers, Inc.)

spatial framework by exploring the formation of a suburban employment node along a ray extending outward from the central business district.

It is unlikely, however, that the transformation of a monocentric city into a polycentric city is associated with the sequential addition of individual subcenters. A more likely scenario is that the central business district is successfully challenged when several subcenters develop simultaneously (Richardson, 1988). In simplest terms, the driving force behind the emergence of a polycentric city can be conceptualized by a trade-off between the agglomeration economies that can be found in the central business district versus the associated congestion costs. Ultimately, however, the generation of subcenters must be explained in terms of the myriad decisions of individual firms and developers.

3.5 Urban Sprawl

The phenomenon of urban sprawl is one of the major problems associated with efficient land use utilization in contemporary U.S. cities, and this problem can be usefully analyzed within the context of urban land use and land value theory. Urban sprawl refers to the continuous

Table 3.7 Concentration of Employment in Large Suburban Nucleations[a]

	1967		1977	
MSA	Number of Suburban Nucleations	Percent of MSA Employment	Number of Suburban Nucleations	Percent of MSA Employment
Boston	29	45.0	34	48.0
Chicago	31	21.6	44	30.2
Cleveland	11	21.7	15	26.8
Detroit	22	38.5	26	45.5
Los Angeles–Long Beach	39	38.9	44	40.9
New York	11	6.5	13	8.4
Newark	21	45.8	25	51.6
Philadelphia	14	16.6	22	28.0
Pittsburgh	5	5.2	8	6.0
San Francisco–Oakland	15	32.9	18	38.7

Source: R. A. Erickson, "Multinucleation in metropolitan economies," *Annals of the Association of American Geographers,* 76 (1986), Table 1, p. 332.
[a]Suburban employment nucleations are defined as municipalities having 5000 or more jobs in manufacturing, retail trade, wholesale trade, and selected service industries combined.

expansion around large cities, whereby there is always a zone of land that is in the process of being converted from rural to urban use. There are three major *forms* of sprawl (Harvey and Clark, 1971). The first, and perhaps least offensive, is the low-density continuous development that surrounds most large cities. Second, ribbon development sprawl is characterized by segments of development that extend axially from the city. Third, leapfrog or checkerboard sprawl is characterized by discontinuous patches of urban development, and it is usually this type of sprawl that is attacked as being economically inefficient and esthetically unattractive.

Causes

A variety of factors have been held responsible for the proliferation of urban sprawl (Harvey and Clark, 1971). Perhaps the most obvious of these is that sprawl is sometimes a response to the *physical terrain* surrounding a city. Mountains, rivers, and swamps, for example, often necessitate discontinuous urban development.

Sprawl is also encouraged by different *administrative policies* in the rural–urban fringe, as opposed to the core of the city. If building codes are stricter within the corporate city than outside it, residential developers tend to be attracted to suitable locations just beyond the city limits. Similarly, land use zoning regulations are often more stringent within the city than outside it.

The leapfrog or checkerboard type of sprawl appears to be particularly related to the number of *independent land developers*. The greater the number of independent firms operating within a given area, the greater the number of fragmented projects. More collusion among these developers or greater guidance from the local planning agencies would lead to a more ordered progression of development.

The *tax laws* in the United States also tend to encourage discontinuous urban development. The real property tax ensures that as soon as agricultural land is platted it is taxed at urban rates, thus encouraging subdivision developers to avoid excess platting in advance of actual development. In addition, income tax methods also encourage this kind of piecemeal development, as developers tend to spread their building programs over a number of years to avoid occupying a particularly high tax bracket for any single year.

Finally, *transportation routes* are catalysts of urban sprawl, and, historically, trolley and bus lines were largely responsible for the ribbon development within the rural–urban fringe. More recently, freeway construction has generated ribbon-type sprawl at the edge of cities, and these ribbons have been extended and reinforced when associated with rapid transit lines. Closely related to these freeways and rapid transit lines is the development of outlying shopping centers. Such shopping centers are often constructed before the surrounding area is fully developed in order to take advantage of cheap land. Once in existence, however, they add impetus to the residential development process.

Consequences

Unchecked urban sprawl has a variety of undesirable consequences (Sargent, 1976). One major drawback of urban sprawl, especially the checkerboard type, is that potentially productive *agricultural land* is often left idle. Parcels of land at the urban fringe tend to be used inefficiently or sometimes left completely unused if their main value lies in speculation.

Sprawl can sometimes have a deleterious effect on *inner-city neighborhoods*, as commercial and industrial enterprises are siphoned off into the urban fringe. Not only is the tax base of the central city decreased, but the distribution of employment opportunities is also profoundly altered. At the same time, access to public land surrounding the city is made more difficult, as residential developments reduce the number of entry points to parks, forests, and beaches.

The costs of providing various kinds of *public services* are also generally much higher in areas of low density or discontinuous development. For example, the costs associated with police and fire protection are greater per person, as are the costs of constructing sewer, gas, and electricity lines. It is also argued that low-density sprawl is a major impediment to the development of economically viable mass transit systems.

Possible Solutions

A number of different strategies have been suggested as possible solutions to the problem of urban sprawl (Sargent, 1976). First, *tax reforms* might be instituted to reduce the present comparative advantage of land speculators. In particular, the capital gains provision might be modified, and local communities should consider taxing vacant land more heavily and structures less heavily.

Second, *local planning agencies* could make greater efforts to control the expansion of cities by manipulating the location and availability of such services as water lines, sewage lines, and roads. This manipulation could be construed as an orderly growth rather than a no-growth strategy, although there are certain drawbacks. The unethical influencing of local politicians, for example, would probably increase, and the restrictions on new housing development would inflate the price of existing housing.

Third, *zoning ordinances* and *subdivision controls* might be modified and strengthened. Such modification and strengthening, however, would require the cooperation of all the local administrative agencies, so it should be preceded by some kind of governmental consolidation.

Fourth, a fairly radical strategy would be to increase substantially the amount of *local government purchased land* within the rural–urban fringe. This land could be acquired from the present owners, held, and then sold at a later date to private developers who would have to adhere to certain conditions concerning its use. Federal loans and subsidies might be created to facilitate the initial acquisition of the land.

Fifth, the delicate balance between public and private interests should be monitored by a sophisticated *information system*. Major land transactions and associated prices should be immediately recorded and made publicly available so that buyers and sellers are more fully aware of the prevailing market situation. By maximizing the potential success of individual development decisions, unnecessary or premature subdivision and the related loss of agricultural land would be minimized.

In summary, this chapter began by outlining the basic arguments involved in Von Thünen's agricultural land use theory. These arguments were then adapted to the urban context, and the effects of relaxing the simplifying assumptions and considering other factors such as externalities and inertia were considered. Simple bivariate correlation and regression analysis was then introduced in order to assess the empirical validity of the major theoretical propositions, and the chapter was concluded by discussing one of the major problems associated with the urban land market, the phenomenon of urban sprawl.

Chapter 4

Multivariate Land Value and Housing Models

4.1   Multivariate Land Value Models

The land value theory described in Chapter 3 is essentially bivariate in nature, as it addresses the relationship between land values and distance from the city center, while assuming that other variables are held constant. This chapter extends this discussion in a number of ways: first, by investigating the relationships between land values and a set of other variables within the context of multiple correlation and regression analysis; second, by discussing the idea of a housing market and associated multivariate housing value models; and third, by introducing the ideas of causal models and path analysis and using them in the context of housing quality and size. Before the multivariate models can be described in detail, however, the basic elements of multiple correlation and regression analysis must be considered.

Introduction to Multiple Correlation and Regression

Multiple correlation and regression analysis is used in those situations where one is investigating the interrelationships among more than two variables. In this particular multivariate context, there is one dependent variable and a series of explanatory, or independent, variables. The general relationships are expressed as follows:

$$Y = f(X_1, X_2, \ldots, X_m) \tag{4.1}$$

where Y is the dependent variable and X_1 through X_m are the independent variables. Two major questions with respect to equation (4.1) can be answered by multiple correlation and regression analysis. First, how strong is the overall relationship between the independent variables and the dependent variable? Or, in other words, how well, in combination, do they account for the variation in Y? Second, what is the relative importance of the independent variables in terms of accounting for the variation in Y?

As in the bivariate situation, a *scatter diagram* can be conceptualized, and this is most easily thought of in the three-dimensional case where there is one dependent variable and two independent variables (Figure 4.1). Each axis represents an individual variable and all three axes are at right angles to each other. The individual case, or observation i, has a particular value for each of the three variables, giving it a unique location in this three-dimensional space. Obviously, when more than three variables are involved it is impossible to draw the associated scatter diagram, although the location of observations in a multidimensional space can still be described algebraically.

A plane is fitted through the observations in three-dimensional space such that the sum of the squared deviations around this plane is minimized (Figure 4.2). This *least-squares plane* is analogous to the least-squares line in the bivariate case and is represented as follows:

$$Y = a + b_1 X_1 + b_2 X_2 \tag{4.2}$$

where Y is the dependent variable; X_1 and X_2 are independent variables; and a, b_1, and b_2 are constants. The constant a represents the value at which the plane intercepts the Y axis, and b_1 and b_2 represent the slopes of the plane in the directions of X_1 and X_2, respectively. Note that in the $(m + 1)$-dimensional space, corresponding to m independent variables, we would be fitting an m-dimensional hyperplane, although such a figure can hardly be imagined.

The *multiple correlation coefficient* measures the goodness of fit of the points to the least-squares plane and thus reflects the explanatory power of the independent variables taken together. It runs between zero and 1, with a coefficient of 1 indicating that all the points lie exactly on the least-squares plane. The greater the scatter of points about the least-squares plane, the lower the multiple correlation coefficient. It is perhaps helpful, in this context, to note that the multiple correlation coefficient can be thought of as representing the bivariate

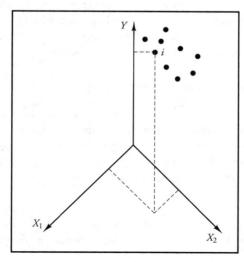

Figure 4.1 Scatter diagram for the three-variable case.

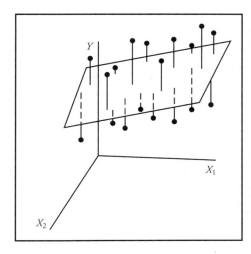

Figure 4.2 The least-squares plane minimizing the sum of the squared deviations in the vertical (*Y*) dimension. (From H. M. Blalock, Jr., *Social Statistics*, Rev. Ed., McGraw-Hill, New York, 1979, Fig. 19.2, p. 454. Reproduced with permission.)

correlation between the actual values of the dependent variable and those predicted by the least-squares equation (Blalock, 1979:483). The square of the multiple correlation coefficient is termed the *coefficient of multiple determination*, and this coefficient expresses the amount of variation in the dependent variable that is accounted for by the explanatory or independent variables.

Besides determining the goodness of fit of the whole equation, it is equally important to ascertain the relative importance of the individual explanatory variables. These individual contributions can be assessed in two different ways, the first of which involves computing *partial correlation coefficients*. Partial correlations refer to the relationship between each independent variable and the dependent variable, while controlling for one or more of the remaining independent variables. In the three-variable case, for example, one could calculate the partial correlation between Y and X_1, while controlling for the influence of X_2. This relationship can be expressed in symbolic form as $r_{12.3}$, where 1 is the dependent variable Y, 2 is the first independent variable X_1, and 3 is the second independent variable, X_2. When a single variable is controlled for, as in the present example, the resulting correlation is called a first-order partial; when two variables are controlled for, the resulting correlation is called a second-order partial; and so on. In keeping with this terminology, a simple bivariate correlation, with no controls, is sometimes referred to as a zero-order correlation.

The intuitive meaning of a partial correlation coefficient can best be grasped by first imagining that the variables Y and X_1 are, in turn, related to the control variable X_2. The residuals associated with these two relationships are then correlated in order to obtain a measure of the relationship between Y and X_1 that is independent of the effects of X_2. In other words, the partial correlation coefficient expressing the relationship between Y and X_1, while controlling for X_2, can be defined as the correlation between the residuals of the regressions of Y on X_2 and X_1 on X_2. As a result of this kind of analysis, the independent variables can be rank ordered on the basis of the strength of their relationship with the dependent variable Y, with those variables with the highest partial correlation coefficients being the most important and those variables with the lowest partial correlation coefficients being the least important.

A similar rank ordering, in terms of relative importance, can also be achieved by investigating the *regression coefficients*. If the independent variables are all measured in the same units, their relative importance can be immediately ascertained by reference to the regression coefficients b_1 and b_2 in equation (4.2). Recall that the coefficient b measures the change in Y associated with a unit change in X. A high value for b indicates that Y is very sensitive to changes in X, while a low value for b indicates that Y is not unduly sensitive to changes in X.

In those instances, however, where the independent variables are not measured in the same units, a direct comparison of the associated regression coefficients is unrewarding, as their magnitudes will not be independent of the units being used. For example, how does one compare the change in Y associated with a change of one dollar, on the one hand, and a change of one mile, on the other? In such situations it is necessary to calculate the *standardized regression coefficients*, or beta weights.

To understand what is meant by the term standardized regression coefficient, it is necessary to rewrite equation (4.2) a little more formally, as follows:

$$X_1 = a + b_{12.3}X_2 + b_{13.2}X_3 \qquad (4.3)$$

where X_1 is the dependent variable; X_2 and X_3 are independent variables; and a, $b_{12.3}$, and $b_{13.2}$ are constants. The regression coefficient $b_{12.3}$ indicates the relationship between X_1 and X_2, while controlling for X_3, while the regression coefficient $b_{13.2}$ indicates the relationship between X_1 and X_3, while controlling for X_2. Note that this subscript notation is the same as that used for partial correlations and that we have departed from the usual practice of denoting the dependent variable as Y merely in order to simplify the notation system.

The standardized regression coefficients associated with equation (4.3) can be obtained in either of two ways. First, each variable in the original data matrix can be converted to standardized form by expressing it in terms of standard deviations from the mean. Second, the unstandardized regression coefficients can be converted to standardized regression coefficients by simply multiplying the particular unstandardized coefficient by the ratio of the standard deviation of its associated variable to the standard deviation of the dependent variable. Either way, the resulting standardized regression coefficients associated with each independent variable are now directly comparable, thus allowing the independent variables to be rank ordered in terms of relative importance.

It should be remembered that the partial correlations and standardized regressions represent different measures of the relationships among variables and therefore give different results, although they usually rank variables in the same order of importance. The partial correlation coefficient indicates the amount of variation explained by one independent variable after the others have explained all they can. The standardized regression coefficient, on the other hand, indicates the change in the dependent variable that is associated with a standardized unit of change in one of the independent variables, while controlling for all the remaining independent variables.

A few final notes of *caution* are in order. First, the multiple correlation coefficient will ordinarily be larger than any of the zero-order or bivariate correlations, since it is impossible to explain less variation by adding further variables. Second, the magnitude of the multiple correlation coefficient, for a given set of variables, will usually be maximized in those situations where there are no intercorrelations between the independent variables. It is prudent, therefore, to search for independent variables that are relatively unrelated, with each accounting for a different portion of the total variation. Finally, where the independent variables are highly intercorrelated, both the partial correlations and standardized regressions will be sensitive to sampling and measurement errors. This problem is referred to as *multicollinearity* (Blalock, 1979:485).

A Multivariate Land Value Model for Chicago

One of the best illustrative examples of a multivariate land value model has been provided by Yeates and Garner (1971) for the city of Chicago, based on an earlier study by Yeates (1965), and it is instructive to summarize this model in terms of the model-building process outlined in Chapter 2. The *research problem* was to account for the spatial variation in land values within Chicago for the years 1910 and 1960, with two time periods being chosen to assess any significant change over time. Yeates and Garner expressed this research problem in visual form by producing land value surfaces for 1910 and 1960. A close inspection of these two maps suggested a number of variables to be included in the land value model. In particular, land values seem to decrease with increasing distance from the central business district, as predicted by Alonso's theory, and distance from Lake Michigan appeared to play a similar role.

In all, the model contains five variables and the interrelationships among these five variables can be represented by means of a *conceptual model* (Figure 4.3). The variable to be explained, land value, is placed on the right side of the diagram, and the explanatory variables are on the left side. All the explanatory variables—distance from the central business district, distance from Lake Michigan, distance from the nearest elevated-subway station, and distance from the nearest regional shopping center—are directly linked to the dependent variable.

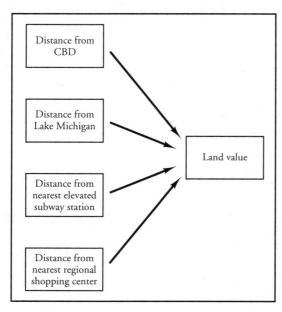

Figure 4.3 Conceptual model to explain the spatial distribution of land values in Chicago. (Based on M. H. Yeates, "Some factors affecting the spatial distribution of Chicago land values, 1910–1960," *Economic Geography,* 41, 1965, pp. 55–70.)

This conceptual model can be translated into the following *symbolic model*:

$$V_i = f(C_i, M_i, E_i, S_i) \tag{4.4}$$

where V_i is the land value at site i; C_i is the distance from the central business district; M_i is the distance from Lake Michigan; E_i is the distance from the nearest elevated-subway station; and S_i is distance from the nearest regional shopping center. More specifically, Yeates and Garner hypothesized the following relationships:

$$V_i = a - b_1 C_i - b_2 M_i - b_3 E_i - b_4 S_i \tag{4.5}$$

where the notation is the same as in equation (4.4), and a, b_1,..., b_4 are constants. In other words, land values were expected to be inversely related to each independent variable.

The model was *operationalized* by using data from Olcott's *Blue Book of Chicago Land Values*, which gives a front-foot land value for every block in the city. Rather than use all these data, however, a sample of 484 front-foot land values was chosen, using a systematic, stratified, random sample procedure. This procedure is designed to ensure an adequate spatial coverage of sample points, thus reducing the risk of having a cluster of sample points in one particular part of the city. For purposes of the statistical analysis, all the data were logarithmically transformed.

Multiple correlation and regression analysis were chosen as the appropriate techniques for *testing* the overall goodness of fit of the model and for assessing the relative importance of the explanatory variables (Table 4.1). For 1910 the coefficient of multiple determination was comparatively high, indicating that, in combination, the four explanatory variables accounted for over 75 percent of the variation in land values. Also, the signs of all the regression coefficients are negative, as hypothesized. These regression coefficients are direct indicators of the relative importance of the explanatory variables, because all those variables were measured in the same distance units. Distance from the central business district was the most important determinant of land values, followed, in order, by distance from Lake Michigan, distance from the nearest elevated-subway station, and distance from the nearest regional shopping center.

By 1960, however, the situation was different. First, the coefficient of multiple determination indicates that only just over 10 percent of the variation in land values was accounted for. In other words, the model was far more successful with respect to the 1910 data than with respect to the 1960 data. Second, the regression coefficient associated with distance from the nearest elevated-subway station is positive in 1960, reflecting the declining importance of this

Table 4.1 Chicago Land Values: Coefficients of Determination (R^2) and Regression Coefficients for 1910 and 1960

Year	R^2	C	M	E	S
1910	0.762	−0.935	−0.469	−0.300	−0.035
1960	0.112	−0.250	−0.120	+0.029	−0.124

Source: Table 9.4 (p. 256) from *The North American City,* Fourth Edition by Maurice H. Yeates. Copyright © 1990 by Harper & Row, Publishers, Inc. Reprinted by permission of HarperCollins Publishers, Inc.

form of rapid transit. Third, although the other regression coefficients remained negative, the individual contribution of each variable is considerably less than it was in 1910.

These results have since been extended and confirmed by McMillen and McDonald (1991). Using similar data, they estimated multivariate models for 1971 and 1981. These models contained a number of distance variables, including distance from downtown Chicago, distance from commuter railroad lines, and distance to O'Hare airport. The results indicated that residential land values have become less dependent on distance from Chicago and distance from commuter railroad lines. On the other hand, land values have become increasingly sensitive to distance from O'Hare airport, as the airport has emerged as a major employment center.

In general, then, because of highway improvements and the increased use of private automobiles, accessibility considerations have become less important as a determinant of land values. Other variables, besides these location factors, must obviously be included in contemporary land value models. In particular, site factors, representing the intrinsic attributes of the individual site, should be taken into account together with the more traditional situation or location factors.

A Multivariate Residential Land Value Model for Los Angeles

A multivariate residential land value model that incorporates both site and situation factors has been developed for Los Angeles by Brigham (1965). The conceptual model states that residential land values are a function of three major factors: accessibility to economic activities, amenities, and topography (Figure 4.4). This conceptual model can be expressed in symbolic form as follows:

$$V_i = f(P_i, A_i, T_i) \tag{4.6}$$

where V_i is the residential land value at site i; P_i is accessibility to economic activities; A_i is amenities; and T_i is topography.

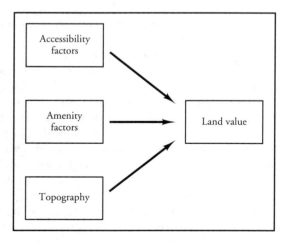

Figure 4.4 Conceptual model to explain the spatial distribution of residential land values in Los Angeles. (Based on E. F. Brigham, "The determinants of residential land values," *Land Economics*, 41, 1965, pp. 325–34.)

The *accessibility to economic activities* factor is operationalized in two different ways. First, it is measured simply by calculating the distance to the central business district. Such an operationalization assumes that the central business district is the major workplace and that it is equally accessible in all directions. Second, accessibility to economic activities is also measured by accessibility potential. This concept is especially valuable when there are a variety of major workplaces in the city, as the accessibility potential of any given site is a direct function of the magnitude of surrounding employment opportunities and an inverse function of the distance to those opportunities.

The *amenity value* associated with a particular site depends on the local neighborhood characteristics. In this instance, amenity value is measured by median family income, the percentage of dwelling units with more than 1.01 persons per room, and median single-family housing value. Income and housing value should be positively related to land value, while crowding should be negatively related to land value.

Finally, *topography* is treated as a binary variable, with relatively hilly sites being assigned the value 1 and relatively flat sites the value 0. Topography should have a bearing on land values through its effect on development costs. Generally, flat land can be comparatively easily subdivided into residential lots, whereas the costs associated with developing hilly land are far greater, due to the extra expenses incurred when putting in access roads, grading slopes, and providing utilities.

The *data* to calibrate the model and to determine how well these three factors account for the variation in land values were obtained from tax assessment appraisal records. Two arbitrary rays, or lines, were extended out from the Civic Center to the county boundaries. The first ray ran to the northwest of the Civic Center, through Hollywood, the Santa Monica Mountains, and out through the San Fernando Valley to the Ventura County boundary. The second ray ran to the northeast, through South Pasadena to the base of the San Gabriel Mountains. Each block that these rays passed through was treated as a sample point.

As some of the explanatory variables are highly intercorrelated, two equations are used to *test* the model for each ray (Table 4.2). For both rays, distance from the central business district is used in equation 1, while accessibility potential is used in equation 2, as these variables are too highly correlated to be used in unison. Overall, the coefficients of multiple determination suggest that the general model is very successful in accounting for the variation in residential land values.

Both accessibility variables behave as expected, as the partial correlation coefficients indicate that land values are negatively related to distance from the central business district and positively related to accessibility potential. The magnitudes of the partial correlation coefficients are different for the two rays, however. So, although there is clearly a relationship between land values and accessibility, the strength of that relationship varies from sample to sample. Housing value, one of the amenity variables, also behaves as expected in that it is in all cases positively correlated with land values. The median income variable, however, has an unexpected negative partial correlation for both equations, which is probably due to the high correlation between median income and housing value. Crowding is substituted for median income in the second ray, and here the result is exactly as expected, with the crowding variable exhibiting comparatively strong negative partial correlations. Finally, the binary level variable measuring topography also behaves as expected. For all four equations the partial correlation coefficient is negative, indicating that residential land values are relatively lower in the hilly areas.

Table 4.2 Los Angeles Residential Land Values: Coefficients of Determination (R^2) and Partial Correlation Coefficients

Ray 1	Distance to CBD	Accessi-bility Potential	Median Income	Housing Value	Topog-raphy	R^2
Equation 1	−0.89		−0.12	0.44	−0.64	0.87
Equation 2		0.53	−0.01	0.33	−0.53	0.79

Ray 2	Distance to CBD	Accessi-bility Potential	Crowding	Housing Value	Topog-raphy	R^2
Equation 1	−0.68		−0.83	0.17	−0.69	0.87
Equation 2		0.73	−0.87	0.20	−0.76	0.89

Source: E. F. Brigham, "The determinants of residential land values," *Land Economics*, 41 (1965), Table 1, p. 331.

In sum, Brigham's model successfully incorporates a variety of variables beyond the accessibility measures used by Yeates and Garner. The validity of the results, however, could be improved in at least two important ways. First, the sample could be made more representative by including more than two transects, and these transects should be chosen on the basis of some type of random sample design. Second, the stability of the relationships needs to be investigated by calibrating the model for a variety of different cities.

The two multivariate land value models that have just been described represent a rather different approach to the study of urban land values than that followed by Alonso. Alonso's model is based on a series of simplifying assumptions and is not directly testable, as no single city can meet all these assumptions. In contrast, although the models of Yeates and Garner, and Brigham, are not constrained by such simplifying assumptions, they are directly applicable only to Chicago and Los Angeles, whereas the level of abstraction involved in Alonso's model makes it generally appropriate for all cities.

A Regression Model for Industrial Land Values

Kowalski and Paraskevopoulos (1990) have estimated a multiple regression model for industrial land values in *Detroit*. The data consisted of 56 observations of industrial acreage sales drawn from five suburban communities. The model, with expected relationships, was as follows:

$$P = a - b_1 D + b_2 V + b_3 IP - b_4 A + b_5 FF + b_6 Y - b_7 R + b_8 L \qquad (4.7)$$

where P is price per acre; D is distance from the northern border of the submarket; V is a dummy variable that takes the value 1 if the parcel abuts a limited access expressway; IP is a dummy variable that takes the value 1 if the parcel is located in a platted industrial park; A is parcel size measured in acres; FF is front footage as measured by the number of feet of roadway that fronts a parcel; Y is the year in which the parcel was sold; R is a dummy variable that takes the value 1 if a parcel was sold in a recession period; L is a dummy variable that takes the value 1 if the grantee was a waste disposal operator; and a, b_1,..., b_8 are constants.

Equation (4.7) was transformed by taking the natural logarithms of both sides and then estimated using ordinary least-squares regression analysis. The signs of the coefficients were as expected, with a coefficient of multiple determination of 0.891 (Table 4.3). More specifically, distance and size had a negative impact on price, and prices were lower during recessions. The remaining coefficients were positive, indicating that exposure to an expressway, location within a platted industrial park, front footage, and landfill sites all have a positive impact on price. Finally, the positive coefficient for the year in which the parcel was sold demonstrates that prices were generally increasing over time.

4.2   The Urban Housing Market

As with land values, the intraurban variation in housing values is also amenable to investigation via multiple correlation and regression analysis. Before discussing examples of such housing value models, however, it is worthwhile considering the general nature of the urban housing market. In particular, we will focus our attention on the demand and supply of housing and the various kinds of individuals and institutions that comprise the major elements in the urban housing market.

Demand for Housing

At least three major factors determine the demand for housing. First, the overall *demographic* structure of the population plays a vital role in generating housing demand. The pres-

Table 4.3 Regression Results for the Industrial Land Value Model

Distance (D)	−0.093
Visibility (V)	0.448
Industrial park (IP)	0.511
Acres (A)	−0.453
Frontage feet (FF)	0.118
Year (Y)	0.097
Recession (R)	−0.519
Landfill (L)	0.593
Constant	10.06
R^2	0.891

Source: Kowalski and Paraskevopoulos, "The impact of location on urban industrial land prices," *Journal of Urban Economics*, 27 (1990), Table 1.

sure on housing resources is especially severe when a large proportion of the population is concentrated within the age group between 30 and 40. Similarly, the rate of family formation is a critical determinant of housing demand, and this rate tends to vary in concert with the demographic profile of households.

Second, the demand for housing is closely tied to family *income* levels. The rapid rise in real income that has occurred in the United States since World War II has greatly spurred the demand for new single-family housing. Rises in real income have also increased the replacement demand, which involves a demand for new housing based on changing tastes and rising aspirations. When replacement demand is high, and yet there is no growth in the number of households, certain portions of the housing stock are simply abandoned. The income elasticity of demand for housing is defined as the proportional change in the demand for housing that results from a unit change in the level of real income. So if, for example, a 20 percent increase in real income leads to a 10 percent increase in the quality of housing demanded, the income elasticity is 0.5. In the United States, estimates of the magnitude of this income elasticity have ranged from 0.55 to 1.63, with a figure of 1.0 often being quoted as the average (Bourne, 1981:127). More important, however, income elasticities tend to vary according to age, income, and ethnicity. Also, elasticities appeared to be higher among families with children.

Third, the demand for housing is also significantly influenced by the availability of *credit* and associated interest rates. Often the magnitude of the interest rate is of more immediate concern to a prospective homeowner than the price of the house itself, as monthly mortgage payments can vary substantially depending on the prevailing interest rates. When credit restrictions are comparatively severe, the number of new housing starts diminishes accordingly.

Supply of Housing

The supply of housing is also influenced by at least three major factors (Kinzy, 1992). First, the activity of private *developers* is of prime importance. Residential developers are especially responsive to vacancy rates as a means of monitoring the housing market and gauging its ability to absorb new dwelling units at appropriate prices. From the point of view of the buyer, of course, vacancies ensure a degree of choice and mobility, but housing suppliers hope for as low vacancy rates as possible.

Second, like the demand for housing, the supply of housing is also affected by the availability of cheap *credit*. Builders normally need to borrow large amounts of money in order to cover the initial construction costs. As a result, the money lenders, such as banks and savings and loan associations, can significantly influence the housing market, as they are able to regulate the flow of capital.

Finally, the *federal government* plays an important role in determining the supply of housing in the United States. The Housing Act of 1949 was of particular significance in this respect, as its stated goal of providing a decent home and suitable living environment for every American family led to programs of inner-city slum clearance and public housing construction and stimulated single-family housing construction in the suburbs through mortgage guarantees. Since that time, urban renewal and low-rent public housing schemes have been important instruments of direct government intervention in the urban housing market. The subsidies for new construction and slum clearance are administered by the Department of

Housing and Urban Development, and more recently subsidies on the demand side have also been instituted. The latter subsidies often take the form of vouchers that are intended to increase the purchasing power of low-income families.

Actors in the Housing Market

Besides considering the demand and supply of housing, the operation of the housing market can also be analyzed in terms of the actors involved, the behavior of some of which, such as residential developers and the federal government, has already been alluded to. Perhaps the most prevalent of these actors are the *occupiers* of housing, who mainly act on the basis of use value rather than exchange value. There are two major categories of occupiers, owners and tenants, and the latter do not ordinarily receive any exchange-value benefit when they improve the use value of a property by modernizing or altering it in some way. Most tenants wish to become owner–occupiers eventually, mainly because of the tax advantages and security of tenure associated with home ownership.

Landlords, on the other hand, tend to be involved primarily with considerations of exchange value. How much they will invest in improvements to their properties, however, depends on a variety of external factors, such as threats of highway clearance, urban renewal, vandalism, and general neighborhood deterioration. Also, the behavior of other, neighboring landlords is critical. For this reason, landlords are generally reluctant to make improvements unless they are sure that their peers will behave in a similar fashion. It is to the benefit of an individual landlord in the housing market to slightly underinvest relative to other landlords in the neighborhood. However, if all landlords react in this way, the housing stock will rapidly deteriorate. Note also that this kind of underinvestment is rational only for those landlords interested in the exchange value of housing and does not apply to owner–occupiers concerned with the use value of housing. As a consequence, the housing stock of an area tends to deteriorate more rapidly under the exchange-value system than under the use-value system, unless landlords cooperate with each other.

As already mentioned, *residential developers* and the construction industry in general play an important role in determining the overall supply of housing, with respect to both the type and quantity of that housing. The residential development process has three major stages. First, a potentially profitable site must be identified and purchased. This process requires establishing the general nature and scale of the project, examining large areas to determine the probability of finding suitable sites, and evaluating specific sites in the chosen areas (Baerwald, 1981). Second, the site must be prepared for development, which involves such things as clearance and the provision of water, road, and sewage facilities. In this context, it is usually cheaper to develop new land on the urban fringe than it is to clear and redevelop land toward the center of the city. The third stage involves the housing construction itself. Throughout these stages the developers are concerned with whether they will obtain permission to build, if financing will be available, and if they will sell for a profit.

Real estate agents play an important intermediary role in the housing market (Jud and Frew, 1986). They make a profit by charging for their services as coordinators between buyers and sellers, and for this reason they tend to encourage turnover in the housing market. Realtors also significantly influence migration patterns, as they are a major source of information about the availability of housing. In particular, of course, they tend to overrecommend areas in which they list homes and thus have the potential for creating localized imbalances between supply and demand (Palm, 1985).

The major role of *financial institutions*, such as banks, savings and loan associations, and insurance companies, is to regulate the flow of money into the housing market (Doling and Williams, 1983). These financial institutions tend to avoid areas of housing that are perceived to have high risks and low rates of return associated with them. The resulting lack of mortgage money in certain parts of a city seriously hampers the purchase and improvement of housing in those areas. Commercial institutions prefer to provide mortgage money to households involved in purchasing higher-priced homes, as the servicing costs on mortgage loans are constant, and therefore the larger the mortgage the greater the profit margin for the institution servicing it. State-chartered savings and loan associations are less profit oriented and more interested in financing housing in the lower price ranges. As a result, there is a clear relationship between different types of financial institutions and different segments of the housing market.

Finally, it is in those areas of the city that lack adequate mortgage money that the *federal and local government* tend to play an important role in the operation of the housing market. Generally, most governments have three major purposes in mind when intervening in the private market (Bourne, 1981:192). First, they try to ensure that the productive resources of society are allocated as effectively as possible. Second, they attempt to stabilize the economic system such that major fluctuations are minimized. Third, they encourage continuing economic growth and a reduction in the level of social inequality. Within this context, besides directly providing housing through various types of public housing programs, the government also influences the character of the current housing stock by means of zoning and housing code legislation. Also, some kinds of government agencies, such as the Veterans' Administration, are directly involved in the provision of mortgage money. Local governments influence the housing market by setting the rate for property taxes (McGibany, 1991).

Rent Control

The local government explicitly intervenes in the housing market when it enacts legislation concerning rent control. Approximately 200 cities and counties in the United States currently use some form of rent control to maintain the affordability of rental housing (Appelbaum and Gilderbloom, 1990). Conventional economic reasoning, however, suggests that rent control can ultimately lead to shortages in the apartment market. Following the analysis of Heertje et al. (1983:73–5), we can consider Figure 4.5, which shows hypothetical demand and supply curves for rental housing. The intersection of the demand and supply curves indicates an equilibrium rent of $320 per month, but the hashed line represents a $250 per month ceiling imposed by rent control. As can be seen, the ceiling of $250 creates a shortage of apartments; there is a demand for 50,000 units, but only 25,000 units are supplied.

This shortage will tend to have certain repercussions in terms of both demand and supply. On the demand side of the equation, a long waiting list for apartments will develop. These waiting lists might also encourage a black market in rental housing, whereby tenants

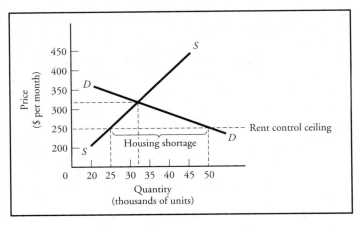

Figure 4.5 Rent control. (From A. Heertje, F. Rushing, and F. Skidmore, *Economics*, Dryden Press, Chicago, 1983, Fig. 4.3, p. 74.)

agree to pay higher rents than are legal. On the supply side, those units left on the market will be allowed to deteriorate. Landlords with property to rent that is worth more than $250 per month will either not maintain that property until its real value falls to $250 per month, or they will get out of the rental market by converting to condominiums. To argue thusly is not to deny that there is a problem in the rental housing market, but to suggest that rent controls set in motion a series of forces that might not always have beneficial consequences. In many cases, rent control has been used as a substitute for spending on public housing, especially by fiscally strained local governments (Ho, 1992).

4.3   Multivariate Housing Value Models

A Multivariate Housing Value Model for Chicago

Berry and Bednarz (1975) have constructed and tested a multivariate housing value model for the city of Chicago. Selling price of a sample of single-family homes in Chicago in 1971 was the dependent variable, while the *explanatory variables* were grouped into three general categories: location, housing characteristics, and neighborhood characteristics. The location variable represented an attempt to capture accessibility and was measured by the distance of the home from downtown Chicago. Housing characteristics were subdivided into housing characteristics per se, measured by floor space, lot size, and the age of the house, and housing improvements, measured by four dummy variables indicating the presence or absence of air conditioning, a garage, an improved attic, and an improved basement, plus the number of bathrooms. The neighborhood characteristics were also subdivided into two groups. First, median family income, the percentage of multiple-family dwelling units, and the amount of migration were grouped together to represent neighborhood socioeconomic status and mobility characteristics. Second, percent blacks, percent Cubans and Mexicans, and percent Irish were used as racial and ethnic variables.

Data on the housing characteristics and housing improvement variables were obtained from the Society of Real Estate Appraisers' Market Data Center for a sample of 275 single-family homes. All the neighborhood characteristics were generated from census data, with figures corresponding to the census tract in which the property was located. Finally, the accessibility variable was calculated in terms of straight-line distance.

The *results* of the multiple regression analysis are reported in Table 4.4, and as the variables were in logarithmic form the coefficients can be interpreted as elasticities (the expected percentage change in the dependent variable associated with a unit percentage change in the independent variable). Overall, the 15 explanatory variables account for just under 80 percent of the total variation in house prices, and nearly all the relationships are in the expected direc-

Table 4.4 Chicago Housing Values: Coefficient of Determination (R^2) and Regression Coefficients

Variable	Regression Coefficient
Housing characteristics	
Floor space	0.344
Age	−0.123
Lot size	0.123
Housing improvements	
Air conditioning	0.078
Garage	0.071
Improved attic	0.124
Improved basement	0.074
Number of baths	0.130
Neighborhood characteristics	
Median family income	0.720
Multiple family dwellings	0.061
Migration	−0.034
Race and ethnicity	
Percent blacks	−0.022
Percent Cubans and Mexicans	−0.013
Percent Irish	−0.016
Accessibility	
Distance to CBD	0.023
R^2	0.787

Source: Adapted from B. J. L. Berry and R. S. Bednarz, "A hedonic model of prices and assessments for single-family homes: Does the assessor follow the market or the market follow the assessor?" *Land Economics*, 51 (1975), Table 4, p. 31.

tions. With respect to the housing characteristics, floor space and lot size are both positively related to price, while age of the dwelling unit has a negative coefficient. Also, all the housing improvement variables exhibit the expected positive relationships.

In terms of the neighborhood characteristics, median family income has a positive influence on selling price, whereas migration, or the amount of population turnover, has a negative impact. A somewhat surprising result, however, is the positive relationship between the percentage of multiple-family dwelling units, or apartments, and house price. Ordinarily, one might expect proximity to apartments to bring down the housing value (Wang et al., 1991). In this context it should be noted that the rather unexpected behavior of the apartment variable might be due to its relatively high correlation with some of the other variables in the model. Such multicollinearity among the independent variables tends to result in rather unstable coefficients and makes it difficult to determine the true relationship between the independent and dependent variables involved.

The regression coefficients indicate negative relationships between house price and each of the three racial and ethnic variables: percent black, percent Cuban or Mexican, and percent Irish. These coefficients imply, then, that housing prices tend to be lower in those areas characterized by large racial or ethnic minorities. This result appears to be at odds with those obtained by some previous researchers, who have found, for example, that black households in U.S. cities tend to pay more than whites for comparable housing. It should be remembered, however, that the racial and ethnic variables are measured at the census tract level rather than for the individual houses. Finally, the coefficient for the accessibility variable reflects the fact that housing price tends to increase with increasing distance from the central business district. Table 4.5 reports the results obtained when successively entering the different groups of variables into the overall regression model. The housing characteristics and improvements were entered first, as it was felt that they are the most important variables in determining the value of housing. Similarly, the neighborhood characteristics were entered before the racial and ethnic variables in order to reflect their relative importance. The results indicate the increase in the coefficient of multiple determination (R^2) as each of the variable groups is added to the model.

Table 4.5 Increases in Explanatory Power with the Addition of Variable Groups

Variable Group Added	R^2
Housing characteristics	0.474
Housing improvements	0.568
Neighborhood characteristics	0.741
Race and ethnicity	0.772

Source: Adapted from B. J. L. Berry and R. S. Bednarz, "A hedonic model of prices and assessments for single-family homes: Does the assessor follow the market or the market follow the assessor?" *Land Economics*, 51 (1975), Table 3, p. 30.

Multivariate Housing Value Models for St. Louis

In a study similar to that of Berry and Bednarz, Mark (1977) attempted to identify the determinants of housing value in St. Louis, Missouri. He developed four different models, although some of the variables appear in more than one model. The first model is a housing characteristics model, containing such variables as housing age, number of rooms, housing size, and presence or absence of central air conditioning. The second model is termed a spatial model and contains a variable measuring distance from the central business district and a series of dummy variables representing location in terms of direction from the central business district. Third, an accessibility model was constructed, using such variables as distance from the central business district, distance from the airport, and the number of bus routes passing through the census tract. Finally, an area preference model was developed to measure characteristics of the local neighborhood, including median income, percent nonwhite, and expenditure per pupil in public schools.

The housing value data are actual sales prices for 6533 owner-occupied single-family dwelling units in St. Louis, and the ability of each of the four models to account for variation in these sales prices is reported in Table 4.6. Corrected coefficients of determination are calculated to take into account the different number of variables in each model; otherwise, it would be difficult to tell whether a particular model worked best due to having a greater number of explanatory variables or due to the inclusion of variables that are more important determinants of variation in housing values. These corrected coefficients indicate that the housing characteristics model works best, while the strictly spatial model has the least explanatory power.

A Multivariate Housing Value Model for Los Angeles

Heikkila et al. (1989) used regression methods to assess the impact of dwelling unit characteristics, neighborhood effects, and location on a sample of residential property sales in Los Angeles County. The sample consisted of almost 11,000 property sales, all of which were transacted in 1980. The dependent variable was house price per unit of lot area. There were 19 independent variables: 5 property or transaction characteristics, 4 neighborhood effects, and 10 locational variables.

Table 4.6 St. Louis Housing Values: Corrected Coefficients of Determination Associated with Four Different Models

Model	R^2
Housing characteristics	0.647
Spatial	0.345
Accessibility	0.407
Area preference	0.449

Source: J. Mark, "Determinants of urban house prices: A methodological comment," *Urban Studies*, 14 (1977), Table 2, p. 362.

In particular, the number of bathrooms, an index of dwelling-unit condition, square footage of living space, and age were used to measure property characteristics. Dwelling-unit condition was calibrated on a five-point scale as judged by property appraisers, and the age variable refers to the year in which the property was built. Month of sale was also included as a property-specific variable, as during 1980 the price of single-family homes increased by about 20 percent. Four census-tract variables were included to capture neighborhood effects: median household income, percentage of jobs held by professionals, percentage of black residents, and percentage of hispanic residents. Finally, 10 locational variables were used, including distance to the central business district, distance to the ocean, and distance to a number of subcenters in Los Angeles, such as Burbank and Santa Monica.

The multiple regression equation accounted for more than 93 percent of the variation in house prices (Table 4.7). All the coefficients were statistically significant at the 0.01 level, with the exception of property age, which was significant at the 0.05 level, and distance to the central business district, which was statistically insignificant. In addition, all the dwelling-unit and neighborhood variables have the expected signs, except for the percentage of hispanic residents. As the equation was estimated in logarithmic form, the regression coefficients can be interpreted as elasticities; the larger the coefficient, the greater the impact on house price. Thus, the most important variables were living area, neighborhood income, condition of the dwelling unit, distance to Santa Monica, and the occupational mix of the neighborhood. A later study by most of these same authors (Richardson et al., 1990) showed that distance to the CBD had been a statistically significant influence in 1970 and that its declining importance over the next 10 years was accompanied by the increasing importance of several subcenters. In other words, Los Angeles was becoming an increasingly polycentric city during the 1970s.

The previously described multiple regression models of housing value illustrate a number of *methodological issues*. First, one must be wary of specification errors with respect to the functional form of the model (Waddell et al., 1993). Second, it is assumed that there are no interaction effects between the independent variables and that multicollinearity is not a problem (Mark and Goldberg, 1988). Third, little explicit attention is focussed on the problem of spatial autocorrelation, which occurs when one encounters a clustering of similar or dissimilar values in geographical space (Can, 1990; 1992). Finally, it has been cogently argued that the role of neighborhood effects versus individual property effects is best investigated by using a series of multi-level models (Jones, 1991). Such models can be used to represent hierarchical structures, and they allow relationships to vary from place to place and also over time.

4.4 Causal Models of Housing Quality and Size

The multivariate models considered thus far have represented rather simplistic causal structures. In this section more sophisticated causal structures are investigated and then exemplified by two models of housing quality and housing size. These causal models allow one to distinguish between direct and indirect effects among the constituent variables (Cadwallader, 1986a).

Table 4.7 Determinants of Residential Property Values

Variable	
Constant	4.682
Number of bathrooms	0.077
Condition	0.225
Living area	0.645
Age	0.012
Month of sale	0.048
Income	0.293
Share of professional jobs	0.212
Share of blacks in population	−0.009
Share of hispanics in population	0.018
Distance to:	
Ocean	−0.099
Central business district	0.002
Santa Monica	−0.224
South Bay	0.110
Palos Verde	−0.047
Glendale	−0.029
Southwestern San Fernando Valley	0.016
Malibu	−0.028
West San Gabriel Valley	−0.048
Burbank	−0.029
Sample size 10,928	
$\overline{R}^2 = 0.933$	

Source: E. Heikkila et al., "What happened to the CBD-distance gradient?: land values in a policentric city," *Environment and Planning A*, 21 (1989), Table 1, p. 224.

Introduction to Causal Models

The land value and housing value models described previously are based on the assumption that only direct relationships exist between each of the explanatory variables and the associated dependent variable, as expressed in Figures 4.3 and 4.4. Such an assumption is overly simplistic, however, in two important ways. First, it does not allow for interrelationships between the explanatory variables. Second, it does not allow for feedback relationships, either between the dependent variable and the explanatory variables, or among the explanatory variables themselves.

In contrast, if it is hypothesized that each variable in a particular causal system is dependent on all the other variables in that system (Figure 4.6), a separate equation for each variable can be written as follows:

$$X_1 = a_1 + b_{12.3}X_2 + b_{13.2}X_3 + e_1 \tag{4.8}$$

$$X_2 = a_2 + b_{21.3}X_1 + b_{23.1}X_3 + e_2 \tag{4.9}$$

$$X_3 = a_3 + b_{31.2}X_1 + b_{32.1}X_1 + e_3 \tag{4.10}$$

where the error terms e_1, e_2, and e_3 represent the effect of all variables outside the system on X_1, X_2, and X_3, respectively. Note that all possible interrelationships among the three variables are represented and that each variable, in turn, is treated as the dependent variable.

When dealing with systems involving two-way causation, however, ordinary least-squares analysis should not be used to estimate the regression coefficients, as the error terms cannot be assumed to be uncorrelated with the independent variables in each equation. In these situations, when estimating *nonrecursive models*, other techniques such as two- or three-stage least-squares analysis should be adopted. As a result, we will postpone consideration of these more complex causal structures until later and concentrate instead on *recursive models*; models that involve only one-way causation and thus do not allow for feedback effects.

A simple three-variable recursive model is represented in Figure 4.7. All the arrows indicate one-way causality, and X_1 is an *exogenous* variable, determined by forces outside the system, while X_2 and X_3 are *endogenous* variables, as their values are at least partly determined by variables within the system itself. This causal structure can be mathematically described by the following three equations:

$$X_1 = e_1 \tag{4.11}$$

$$X_2 = b_{21}X_1 + e_2 \tag{4.12}$$

$$X_3 = b_{31.2}X_1 + b_{32.1}X_2 + e_3 \tag{4.13}$$

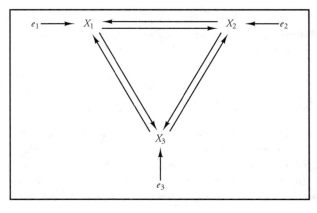

Figure 4.6 Three-variable nonrecursive model.

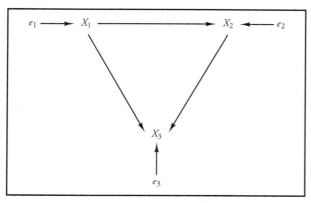

Figure 4.7 Three-variable recursive model.

Note that the a terms can be omitted by simply assuming that each variable is measured in standardized form, thus ensuring that the least-squares line passes through the origin and that the a terms, or intercepts, are therefore zero.

In general, any kind of causal structure can be described by an appropriate set of equations. For example, the four-variable recursive model represented by Figure 4.8a can be expressed as follows:

$$X_1 = e_1 \tag{4.14}$$

$$X_2 = b_{21}X_1 + e_2 \tag{4.15}$$

$$X_3 = b_{31.2}X_1 + b_{32.1}X_2 + e_3 \tag{4.16}$$

$$X_4 = b_{41.23}X_1 + b_{42.13}X_2 + b_{43.12}X_3 + e_4 \tag{4.17}$$

The empirical adequacy of this causal structure is best understood by successively removing individual causal arrows from Figure 4.8a (Blalock, 1964:66). If the arrow between X_1 and X_3 is eliminated (Figure 4.8b), it means that X_1 only has an indirect effect on X_3 via X_2. In other words, we are really saying that $r_{31.2} = 0$, meaning that when the intervening variable X_2 is controlled for the partial correlation between X_1 and X_3 should be approximately zero. Also, if $r_{31.2} = 0$, the associated standardized regression coefficient $b_{31.2} = 0$ and thus disappears from equation (4.16). If we next omit the arrow linking X_1 and X_4 (Figure 4.8c), we are saying that X_1 will only influence X_4 via X_2 and X_3. We are therefore predicting that $r_{41.23} = 0$ and that the associated standardized regression coefficient $b_{41.23}$ will disappear from equation (4.17). Finally, if the arrow linking X_2 and X_4 is removed (Figure 4.8d), we are hypothesizing that X_2 only influences X_4 via X_3, thus creating a simple causal chain going from X_1 to X_2 to X_3 to X_4. If this hypothesis is true, then $r_{42.13} = 0$ and $b_{42.13} = 0$.

Many verbal theories in social science can be usefully stated in terms of causal models and their associated equations. Such statements tend to highlight any ambiguities, or inconsistencies, in the original verbal theory and also cast that theory in a form that is empirically testable. An excellent example of this approach is represented by Mercer's (1975) application of causal modeling to metropolitan housing quality.

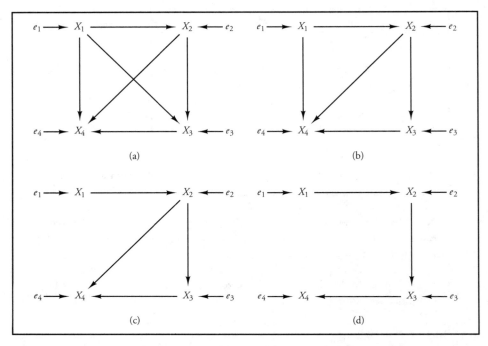

Figure 4.8 Examples of four-variable recursive models. (Based on H. M. Blalock, Jr., *Causal Inferences in Nonexperimental Research*, University of North Carolina Press, Chapel Hill, N.C., copyright © 1964, p. 66. Used by permission of the publisher.)

A Causal Model of Housing Quality

Mercer (1975) uses the causal modeling approach to evaluate a causal structure suggested by Muth (1969). Generally, Muth's thesis is as follows: given that poor people tend to live in poor-quality housing, and given that a disproportionate number of black Americans are poor, the observed relationship between blacks and poor-quality housing can be explained largely in terms of income. The associated causal model is represented in Figure 4.9. In particular, it is hypothesized that there is no direct link between X_1 and X_3, only an indirect link via X_2. This causal structure is of interest, as it implies, for example, that racial discrimination in the housing market is not a major factor.

Mercer tested the adequacy of this causal structure by using data for the *Chicago* Metropolitan Area, as reported in the 1950 and 1960 U.S. Censuses of Population and Housing. The data were collected at the census tract level, with 1060 observations for 1950 and 1216 observations for 1960. Housing quality was measured by determining the percentage of blighted dwelling units in each census tract, where blight is represented by the percent of dwelling units that had no private bath or were dilapidated in 1950, and by the percent of dwelling units dilapidated, deteriorating and lacking in plumbing facilities, or sound and lacking in plumbing facilities in 1960.

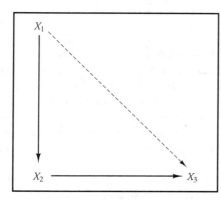

Figure 4.9 Hypothesized relationships between black Americans, income, and poor-quality housing. X_1, percent blacks in a census tract; X_2, median income; X_3, percent blighted dwelling units in a census tract. (Adapted by permission from John Mercer, "Metropolitan housing quality and an application of causal modeling," *Geographical Analysis*, 7, July 1975, pp. 295–302. Copyright © 1975 by the Ohio State University Press.)

The model predicts that there is no direct relationship between X_1 and X_3, meaning that $r_{31.2} = 0$. This partial correlation coefficient is computed as follows:

$$r_{31.2} = \frac{r_{31} - r_{21}r_{32}}{[(1 - r_{12}^2)(1 - r_{23}^2)]^{1/2}} \tag{4.18}$$

If, however, $r_{31.2} = 0$, then

$$r_{31} - r_{21}r_{32} = 0 \tag{4.19}$$

and therefore

$$r_{31} = r_{21}r_{32} \tag{4.20}$$

In other words, for the partial correlation between X_1 and X_3, while controlling for X_2, to be zero, the zero-order correlation between X_3 and X_1 should be equal to the product of the zero-order correlations between X_2 and X_1 and between X_3 and X_2. The results of this analysis are reported in Table 4.8. The differences between the values are sufficiently large to raise questions with respect to the acceptability of a Muth-type causal structure. The smaller gap between the values for 1960, however, suggests that income was beginning to play a more important role.

Table 4.8 Empirical Relationships among Black Americans, Income, and Poor-quality Housing in Chicago

Year	r_{31}	$r_{21}r_{32}$
1950	0.59	0.39
1960	0.45	0.33

Source: Reprinted by permission from John Mercer, "Metropolitan housing quality and an application of causal modeling," *Geographical Analysis*, 7 (July 1975), Table 1, p. 298. Copyright © 1975 by the Ohio State University Press.

Introduction to Path Analysis

Path analysis represents a straightforward extension of this causal modeling framework, as it involves estimating the magnitude of the linkages between variables, rather than merely focusing on their presence or absence. In addition, path analysis allows one to distinguish between the *direct* and *indirect* effects between variables and to decompose the correlation between any two variables into a sum of simple and compound paths. Some of these compound paths involve theoretically meaningful indirect effects, while others may represent noncausal components of association. Note that path analysis includes information provided by the previously described causal modeling technique, in that a nonexistent link is implied by a zero path coefficient.

Consider the path diagram represented by Figure 4.10. One-way arrows lead from each determining variable to each dependent variable, and two-headed arrows indicate unanalyzed correlations between variables that are not causally dependent on other variables in the system. Such two-headed arrows are drawn curved, rather than straight, to distinguish them from true causal links. The numerical values placed on the diagrams are the *path coefficients* or in the case of the two-headed arrows, simply the zero-order correlation coefficients.

If each variable is measured in standardized form, then the causal structure in Figure 4.10 can be represented by the following structural equations:

$$X_3 = P_{31}X_1 + P_{32}X_2 + P_{3u}R_u \tag{4.21}$$

$$X_4 = P_{41}X_1 + P_{42}X_2 + P_{43}X_3 + P_{4v}R_v \tag{4.22}$$

$$X_5 = P_{51}X_1 + P_{52}X_2 + P_{53}X_3 + P_{54}X_4 + P_{5w}R_w \tag{4.23}$$

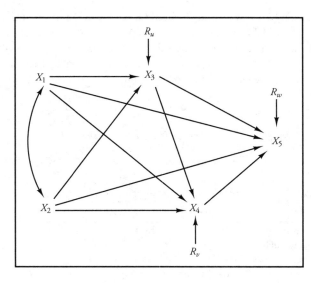

Figure 4.10 A path diagram. (Based on O. D. Duncan, "Path analysis: Sociological examples," *American Journal of Sociology,* 72, 1966, Fig. 1.)

where the P terms are path coefficients, and R_u, R_v, and R_w refer to the residual, or disturbance, terms. As can be seen from the diagram, these residual terms are uncorrelated with
any of the immediate determinants of the variable to which they pertain and are also uncorrelated with each other. In this context, the path diagram supplies important information
about the causal system that is not directly discernible from the structural equations themselves.

   In this simple recursive situation the path coefficients can be estimated from ordinary
least-squares regression analysis, as they represent standardized regression coefficients. Thus,
for example, P_{32} equals $b_{32.1}$, and P_{31} equals $b_{31.2}$. Similarly, P_{42} equals $b_{42.13}$, and P_{43} equals
$b_{43.12}$. Unlike the notation for partial regression and correlation coefficients, the notation for
path coefficients does not identify the other variables influencing the particular endogenous
variable under consideration. The path coefficient associated with the residual term is simply
the square root of the unexplained variation in the dependent variable, since standardized
variables have a variance of 1.

   One way to decompose the correlation coefficients between the constituent variables is
to use the following expression:

$$r_{ij} = P_{iq} r_{jq} \tag{4.24}$$

where i and j denote two variables in the system, and the index q runs over all variables from
which paths lead directly to X_i or, in other words, those variables having a direct impact on X_i.
This expression is the *basic theorem* of path analysis and states that the correlation between any
two variables can be decomposed into the sum of simple and compound paths, where a compound path is equal to the product of the simple paths comprising it.

   Using this formula in the context of Figure 4.10, it can be seen that

$$\begin{aligned}
r_{53} = {}& P_{53} + P_{51}P_{31} + P_{51}r_{12}P_{32} + P_{52}P_{32} + P_{52}r_{12}P_{31} \\
& + P_{54}P_{41}P_{31} + P_{54}P_{41}r_{12}P_{32} + P_{54}P_{42}P_{32} \\
& + P_{54}P_{42}r_{12}P_{31} + P_{54}P_{43}
\end{aligned} \tag{4.25}$$

This expression can be most easily obtained by referring to the path diagram and reading back
from variable i and then forward to variable j. When doing so, three instructions should be
observed. First, no path can pass through the same variable more than once. Second, no path
can go back along an arrow having started forward on a different arrow. Third, a path can go
in either direction along a two-headed arrow, but only one such two-headed arrow can be
used in any single path.

   In this way, correlations can be decomposed into four *components of association* (Alwin
and Hauser, 1981). As can be seen from Figure 4.11, these components involve a direct effect,
an indirect effect due to some intervening variable, an effect due to associated causes (or some
unanalyzed correlation), and an effect due to common causes (or spurious causation). Two
simple three-variable models can be used to describe these various components of association.
Looking at Figure 4.12a, we can decompose three different correlations. First,

$$r_{12} = P_{21} \tag{4.26}$$

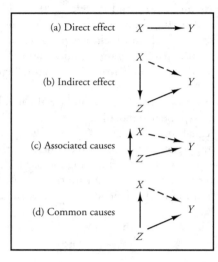

Figure 4.11 The components of association between X and Y. (From M. Cadwallader, "Structural equation models in human geography," *Progress in Human Geography*, 10, 1986, pp. 24–47.)

which means that the entire correlation between X_1 and X_2 is attributable to the direct effect P_{21}. Second,

$$r_{13} = P_{31} + P_{32}P_{21} \qquad (4.27)$$

which means that the correlation between X_1 and X_3 is generated by the direct effect P_{31} and the indirect effect $P_{32} P_{21}$. Third,

$$r_{23} = P_{32} + P_{31}P_{21} \qquad (4.28)$$

which means that the correlation between X_2 and X_3 is generated by the direct effect P_{32} plus correlation due to a common cause, $P_{31} P_{21}$.

Figure 4.12b illustrates the idea of *associated causes*. In this case,

$$r_{13} = P_{31} + P_{32}r_{12} \qquad (4.29)$$

where P_{31} is the direct effect and $P_{32} r_{12}$ represents that part of the correlation due to associated causes, or correlation with another cause. It should be noted that Figure 4.12b represents a single-equation regression model, where the correlation among the exogenous variables is indicated by the double-headed arrow. Such models correctly represent the direct effects of the two exogenous variables but supply no information about how any indirect effects are generated, as the causal structure of the relationship between the exogenous variables is unknown. In this sense, a major goal of theory construction should be to generate models that will make some of the exogenous variables endogenous (Duncan, 1975:43).

Of particular importance in this discussion is the distinction between total association and total effect. The *total association* between two variables is the sum of the four components of association noted above and is given by the zero-order correlation coefficient. The *total effect*, on the other hand, is the sum of the direct and indirect effects, where the indirect effects are transmitted by variables that intervene between the cause and effect of interest. Most

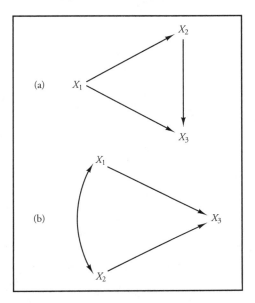

Figure 4.12 Two three-variable models that illustrate the four components of association. (From M. Cadwallader, "Structural equation models in human geography," *Progress in Human Geography*, 10, 1986, pp. 24–47.)

important, one cannot usually calculate the indirect effect by simply subtracting the direct effect from the zero-order correlation coefficient.

A Path Analysis of Housing Size

As an example of path analysis, we can look at a model of housing size, although the version presented here is a substantial simplification of the original formulation by Guest (1972). The model is represented in Figure 4.13, and it can be seen that there is one exogenous variable, distance from the central business district, while the other four variables are endogenously determined. In particular, size of housing is determined both directly and indirectly by all the other variables in the model.

The equations describing this recursive model are as follows:

$$X_1 = e_1 \tag{4.30}$$

$$X_2 = -b_{21}X_1 + e_2 \tag{4.31}$$

$$X_3 = -b_{31.2}X_1 + b_{32.1}X_2 + e_3 \tag{4.32}$$

$$X_4 = -b_{41}X_1 + e_4 \tag{4.33}$$

$$X_5 = b_{51.234}X_1 + b_{52.134}X_2 - b_{53.124}X_3 - b_{54.123}X_4 + e_5 \tag{4.34}$$

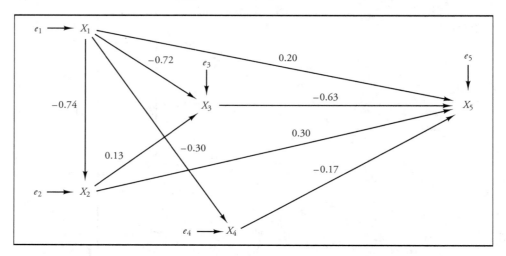

Figure 4.13 Causal model to explain variations in house size. X_1, distance from the central business district; X_2, housing age; X_3, housing density; X_4, industrial activity; X_5, housing size. (Based on A. M. Guest, "Patterns of family location," *Demography*, 9, 1972, Fig. 2, p. 165.)

where X_1 is distance from the central business district, X_2 is age of housing, X_3 is housing density, X_4 is industrial activity, and X_5 is housing size. Note that values for the intercept are not represented, as the data are in standardized form.

The model was tested using census tract data for the *Cleveland*, Ohio, Standard Metropolitan Statistical Area. Distance from the central business district, as defined by the U.S. Bureau of the Census, was measured in 1-mile concentric zones. Age of housing was measured by the proportion of household units built before 1940. Housing density was represented by the natural logarithm of the number of household units per acre of residential land. Industrial activity was the natural logarithm of the proportion of the land area in industrial use, and housing size was measured by the mean number of rooms per household unit.

All the standardized regression coefficients in this model, as expressed in equations (4.30) to (4.34), can be thought of as *path coefficients*, measuring the strength of the links in the causal structure. The values for these path coefficients have been included in Figure 4.13. They indicate, for example, that the link between distance from the central business district and housing age is fairly strong, whereas the link between industrial activity and housing size is comparatively weak.

As described above, the path coefficients also allow us to determine the *direct versus indirect* effect of variables on each other. For example, we can calculate the direct and indirect effect of distance from the central business district on each of the other variables in the system (Table 4.9). The zero-order correlation between distance from the CBD and housing age (r_{21}) can be decomposed into direct and indirect effects. However, in this particular model, there is only a direct path between these two variables (p_{21}), so the zero-order correlation is the same as the path coefficient. A similar situation occurs in the case of distance from the CBD and industrial activity, as again there is no indirect effect.

Table 4.9 Direct versus Indirect Effects of Distance from the CBD on Each Variable in the Housing Size Causal Model

Housing age:

$$r_{21} = p_{21}$$
$$-0.74 = -0.74$$

Housing density:

$$r_{31} = p_{31} + p_{32}p_{21}$$
$$-0.82 = -0.72 + (0.13)(-0.74)$$
$$= -0.72 + (-0.10)$$
$$= -0.82$$

Industrial activity:

$$r_{41} = p_{41}$$
$$-0.30 = -0.30$$

Housing size:

$$r_{51} = p_{51} + p_{52}p_{21} + p_{53}p_{31} + p_{53}p_{32}p_{21} + p_{54}p_{41}$$
$$0.55 = 0.20 + (0.30)(-0.74) + (-0.63)(-0.72) + (-0.63)(0.13)(-0.74) + (-0.17)(-0.30)$$
$$= 0.20 + (-0.22) + 0.45 + 0.06 + 0.05$$
$$= 0.20 + 0.34$$
$$= 0.54$$

Source: Based on A. M. Guest, "Patterns of family location," *Demography* 9 (1972), Table 2, p. 164, and Figure 2, p. 165.

The relationship between distance from the CBD and housing density, however, can be decomposed into direct and indirect effects. The direct effect is represented by the path coefficient p_{31}. There is also an indirect effect via the chain from housing density, to housing age, to distance from the CBD. In other words, housing age acts as an intervening variable. This indirect effect is calculated by obtaining the product of the two constituent path coefficients, p_{32} and p_{21}. In this way the zero-order correlation coefficient of -0.82 can be decomposed into a direct effect of -0.72 and an indirect effect of -0.10.

Finally, the relationship between distance from the CBD and housing size is more complex still, as there are a number of causal chains connecting these two variables. The direct effect is represented by the path coefficient p_{51}. In addition, there are four indirect effects, whose values are calculated by multiplying the constituent path coefficients. One of these indirect effects involves a three-link chain, going from housing size to housing density, from housing density to housing age, and from housing age to distance from the CBD. This particular indirect effect is measured by obtaining the product of the three path coefficients, p_{53},

p_{32}, and p_{21}. Overall, the zero-order correlation coefficient of 0.55 between housing size and distance from the CBD can be decomposed into a direct effect of 0.2 and an indirect effect of 0.34, although these two values do not exactly sum to 0.55 because of rounding errors.

In summary, the idea of being able to decompose relationships into direct and indirect effects, based on the existence of intervening variables, is a very seductive one. The complexity of both social and physical systems is such that unanticipated indirect effects are often produced when one tries to intervene or tamper with the system. Efforts to direct the national economy, for example, are often thwarted by unanticipated indirect effects. In the context of medicine, these indirect effects are commonly called side effects, although their impact on the patient can be substantial.

Chapter 5

Urban Retail Structure and Population Density Patterns

5.1   Elements of the Urban Retail Structure

Two of the spatial distributions most closely associated with that of land values are the patterns of retailing and population density. In this chapter we discuss the elements of the urban retail structure and the spatial organization of shopping centers. We then turn our attention to the pattern of population density and the idea of density surfaces. The retail structure of urban areas is composed of three main elements: specialized areas, ribbons, and retail nucleations (Berry and Parr, 1988; Jones and Simmons, 1990). After describing the major characteristics of these three elements, we will examine their relationship to the pattern of land values discussed in Chapters 3 and 4.

Specialized Areas

Specialized areas are concentrations of retail or service establishments that have something in common. Such concentrations make it easier for consumers to compare prices and quality and also allow the retailers to use each other's facilities. For example, major medical centers often have one central laboratory for analyzing blood samples or a central x-ray testing facility. One of the best examples of a specialized area is *"automobile row,"* where one finds new- and used-car showrooms, plus associated facilities such as body shops, service stations, and spare parts outlets. Buying a car involves a major outlay of money, so consumers wish to compare prices and quality carefully before coming to a final decision. This kind of comparative shopping is facilitated by the clustering of car showrooms.

Furniture districts are also a relatively common feature of larger urban areas. Such districts usually include a variety of different types of furniture stores, plus ancillary activities such as furniture repairs and rental furniture stores. As with automobiles, the purchase of furniture is a relatively expensive although infrequent event. Therefore, consumers generally wish to evaluate their options carefully and are willing to travel to some specialized area of the city in order to compare products.

Similarly, *medical districts* also represent a clustering of linked activities. Doctors, dentists, hospitals, and pharmacists are often located together in medical centers. Here the clustering tendency reflects not so much a desire to facilitate comparative shopping, but a desire to concentrate related activities and functions. For example, including a pharmacy in a medical center or clinic reduces the amount of time patients spend traveling.

In general, then, these specialized areas are associated with the provision of higher-order goods and services; that is, goods and services that are expensive but are not required on a daily basis. Because such areas often cater to the whole city, they require very accessible locations near freeways or major highways. Originally, these specialized areas arose from the informal clustering of similar establishments or related functions, but increasingly they are being developed according to a formal set of plans, especially in the case of complex medical centers.

Ribbons

Ribbon developments are composed of retail and service activities that are associated with automobile traffic. Such establishments include fast-food restaurants, drive-in banks, motels, and service stations. As the demand for these goods and services is generated by the flow of automobile traffic, it is those highways with the greatest traffic volume that tend to be the most densely developed. Freeways would not be included in this category, as their accessibility is restricted to a series of relatively widely spaced intersections and off-ramps.

The *traditional shopping street* is perhaps the most common form of ribbon development, especially in the older parts of cities. Such streets are generally associated with low-order convenience goods and services, such as groceries, laundromats, and drugstores. These goods and services are described as being low-order because they are relatively inexpensive and are demanded on a day-to-day basis. Often such streets are the scene of multipurpose trips, as one can be shopping for groceries while washing clothes in the laundromat.

The *urban arterial ribbons* are similar in many ways to the specialized areas, as they are composed of facilities associated with infrequent demand. They are different from the specialized areas, however, in that the individual establishments generally function independently of each other and are often large consumers of space. Lumberyards, funeral parlors, housing supplies, and television repairs and services exemplify the types of activities that are found along arterial ribbons.

Highway-oriented ribbons contain business types that are most closely related to traffic-generated demand. These ribbons represent the classic type of ribbon development, involving gas stations, motels, ice-cream parlors, drive-in banks, and other kinds of financial institutions. As with the arterial ribbons, but unlike the traditional shopping streets, there is usually very little functional linkage between the individual establishments. Highway-oriented ribbons are different from arterial ribbons, however, in that the goods and services they sell are less specialized and in more frequent demand.

Ribbon development in general is often accused of being a major contributor to visual blight in cities. This is because such ribbons are only haphazardly planned, especially in comparison with some of the large, modern shopping centers. They are usually adorned with large billboards, and most cities do not have effective policies or ordinances concerning the size and spacing of these billboards. The problem of visual blight has received especially critical attention when it is associated with major highways leading out to suburban communities.

Retail Nucleations

The third component of the urban retail structure is made up of various kinds of retail nucleations that are usually widely dispersed within cities. Some of these nucleations are unplanned, often at the intersection of two strips of ribbon development, while others are planned shopping centers. Whether planned or unplanned, however, these retail nucleations form a hierarchy of shopping centers, with the smaller centers providing low-order goods and services that are required on a daily basis, while the larger centers provide more specialized high-order goods and services that are required at more infrequent intervals.

The top of this hierarchy is represented by the *central business district* (CBD). As the name implies, the CBD is usually located toward the center of the city and attracts consumers from all parts of the city. In smaller cities, the CBD is still the major retailing, entertainment, and financial center, although in larger cities these functions have been progressively usurped by outlying regional shopping centers. This latter situation has created serious problems of commercial abandonment in central cities.

The CBD in large cities can be subdivided into smaller functional regions, creating a core area surrounded by a frame area (Herbert and Thomas, 1982:215). The *core area* is characterized by very intensive land use, as manifested by multistoried buildings of various kinds, and is usually the focus of a city's mass transit and freeway systems. It is dominated by (1) very specialized retailing activity; (2) financial institutions, such as savings and loan associations, banks, and insurance companies; and (3) the main offices of large companies.

By contrast, the *fringe area*, or frame, is characterized by less intensive land use, such as warehouses, transportation facilities, and light manufacturing. These are the activities that cannot afford the high-priced land at the core, partly because they are less sensitive to accessibility considerations and so have lower bid-rent curves, as discussed in Chapter 3, and also because they are space-consuming kinds of activities. Mixed in with these commercial activities are clusters of low-quality multifamily residences that are often in the process of becoming abandoned and being converted into more profitable uses.

The next level in the hierarchy of retail nucleations is represented by *regional shopping centers*, which contain such specialized functions as department stores, camera shops, and music stores. As mentioned previously, in some cities these outlying regional shopping centers have taken over the functions traditionally associated with the CBD, and they often contain cinemas, banks, legal offices, and real estate agents. As these shopping centers are very similar to each other in terms of level of specialization and the variety of goods and services provided, each tends to attract its consumers from a particular subregion within the city.

The *community centers* and *neighborhood centers* are responsible for providing less specialized goods and services that are required more frequently. Variety stores, clothing stores, and jewelers are often found in the community centers, while the neighborhood centers are occupied by even lower-order functions, such as supermarkets, bakeries, and laundromats. The neighborhood centers are responsible for serving the needs of very localized areas within the city, while the community centers serve somewhat larger subregions, although these are still smaller than the subregions served by the regional shopping centers. The exact nature of this hierarchy of retail nucleations will be examined in greater detail after we have considered the interrelationships between the overall urban retail structure and the pattern of land values.

The Urban Retail Structure and the Pattern of Land Values

As described in Section 3.3, the generalized urban land value surface has three major components (Figure 3.12). First, land values reach a peak toward the center of the city. Second, there are ridges of higher-valued land associated with major highways. Third, there are local peaks of higher land value at the intersections of those major highways. These three components are closely related to the elements of the urban retail structure that we have just discussed. Specifically, the CBD is located on the highest-valued land at the center of the city, the various outlying shopping centers are located at the local peaks, while the ribbons are associated with the ridges in the land value surface.

This close correspondence between the pattern of land values and the pattern of retailing should come as no surprise, as both land values and retailing are heavily dependent on accessibility considerations. We have already discussed how the various bid-rent curves associated with different types of land use respond to differing accessibility, thus ensuring that the land value pattern is similarly influenced. Retailing is equally sensitive to accessibility considerations, as retailing establishments depend on a steady flow of customers.

5.2   The Spatial Organization of Shopping Centers

The spatial pattern of shopping centers in cities can be explained by central place theory, although this theory was originally applied to settlement patterns (Jones and Simmons, 1990:153–61; O'Brien and Harris, 1991:71–7). Within the context of the intraurban structure of cities, central place theory attempts to account for the number, size, and spacing of shopping centers, or retail nucleations. The theory will be examined in three parts: a conceptual overview, the formal theory, and empirical tests of that theory.

Conceptual Overview

Imagine that we have a small bakery store located somewhere in the city. To stay in business the baker must earn normal profits. These normal profits are just sufficient to cover the operating costs associated with the bakery, including a reasonable wage for the baker, or entrepreneur. To realize these normal profits, the baker must reach a certain threshold sales volume. In other words, he or she must serve a certain minimum population or market area. This concept of a threshold, or minimum market area, is expressed diagrammatically in Figure 5.1. More formally, the *threshold* of a good or service is the minimum population required to

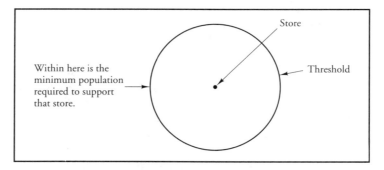

Figure 5.1   Threshold of a good or service.

support that good or service. If, in the long run, the entrepreneur fails to reach the threshold sales volume, he or she will go out of business. The size of the required threshold will vary from good to good, with jewellers, for example, needing a larger minimum population, or market area, than bakers. The calculation of these different threshold sizes will be discussed a little later.

A second concept that is of fundamental importance to central place theory is the *range* of a good or service. The price the consumer must pay for a particular good or service will vary according to how far away from the store he or she lives. In general, the farther away a person lives, the more the good will cost, as transportation costs must be added to the original price of the good. In other words, there will be a positive relationship between delivered price and distance from the store (Figure 5.2a).

According to microeconomic theory, as the price, or in this case delivered price, of a good or service increases, the demand for that good or service will generally decrease (Figure 5.2b). Just how quickly the demand decreases depends on the elasticity associated with the product in question. Where the decrease in demand is very small, as reflected by a gen-

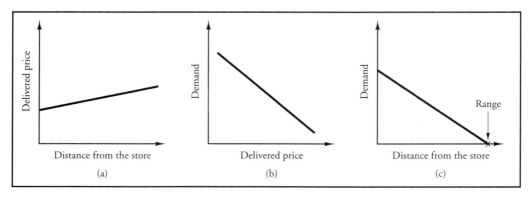

Figure 5.2   Range of a good or service.

tly sloping demand curve, it means that the demand for the good is relatively elastic and therefore comparatively insensitive to changes in price. On the other hand, where the decrease in demand is very dramatic, as reflected by a steeply sloping demand curve, it means that the demand for the good is relatively inelastic and therefore very sensitive to changes in price.

It follows from these two relationships that there must be a negative relationship between demand and distance from the store (Figure 5.2c). As delivered price increases with increasing distance and demand decreases with increasing delivered price, demand must also decrease with increasing distance. In fact, at a certain distance, the demand curve will intersect the horizontal axis, indicating that demand is effectively zero beyond that point. This particular distance corresponds to the range of a good or service, so we can state more formally that the range of a good or service is the maximum distance people will travel to purchase that good or service.

If the threshold and range are superimposed on each other, it is possible to see how bakers, or any other kind of entrepreneur, might earn greater than normal or excess profits (Figure 5.3a). In those situations where the range lies outside the threshold, the area between the threshold and the range represents excess profits that accrue to the entrepreneur. In other words, all the people living in the area between the threshold and the range are over and above the minimum population required to earn normal profits. Over the long run, we can expect other bakers to move into the area and thus soak up the excess profits (Figure 5.3b). Conversely, if stores are earning less than normal profits, we can expect some of them eventually to go out of business, thus increasing the market area of those remaining. In this sense the system always tends to be moving toward an equilibrium position in which all entrepreneurs earn normal profits.

As mentioned earlier, *threshold size* will vary across different types of goods and services. The threshold populations contained in Table 5.1 are from a classic study based on Snohomish County, Washington (Berry and Garrison, 1958). Specific values like these are most easily

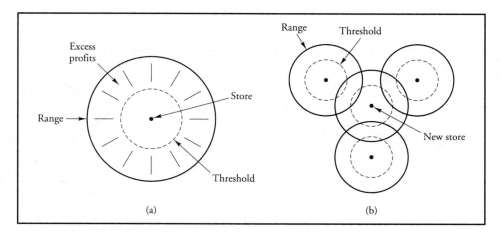

Figure 5.3 Threshold and range combined.

Table 5.1 Threshold Sizes for Selected Central Functions in Snohomish County, Washington

Central Functions	Threshold Size
Filling stations	196
Food stores	254
Restaurants	276
Appliance stores	385
Hardware stores	431
Drugstores	458
Furniture stores	546
Apparel stores	590
Florists	729
Jewelry stores	827
Sporting goods stores	928
Department stores	1083

Source: B. J. L. Berry and W. L. Garrison, "The functional bases of the central place hierarchy," *Economic Geography*, 34 (1958), Table 2, p. 150.

obtained by simply dividing the total population by the number of filling stations, for example. In this particular case, then, there are 196 people per filling station. Such a simple methodology for computing threshold sizes does involve certain problems, however. First, if some of the filling stations are earning less than normal profits, the true threshold is somewhat higher than 196. Conversely, if some filling stations are earning excess profits, the true threshold is somewhat lower. Second, we are not really dealing with a closed economic system. Some people from Snohomish County will travel outside the county to purchase their gasoline, while others living outside the county will purchase gasoline while traveling through it. If there are more people living within the county but purchasing gasoline outside the county than vice versa, the threshold size is overestimated. Finally, the threshold size will vary according to the size of the filling station and the income of the surrounding population. All other things being equal, larger filling stations will require larger threshold sales populations, or market areas. Also, those filling stations surrounded by low-income households will require a larger market area than those surrounded by high-income households, as the latter have greater disposable incomes and are therefore likely to use more gasoline.

In any event, despite these methodological shortcomings, Table 5.1 is worth close inspection, as it provides the basis for the central place *hierarchy* within cities. It should be remembered, however, given the previous discussion, that it is only the ordering of central place functions that will remain relatively stable from city to city, not the absolute threshold values themselves. The functions with the lowest thresholds, such as filling stations and food stores, are referred to as lower-order functions, while those with higher thresholds, such as jewelry stores and department stores, are referred to as higher-order functions. Following from this distinction, the smallest shopping centers, or neighborhood shopping centers, are characterized by lower-order functions, while the regional shopping centers are characterized by higher-order functions.

Furthermore, the size of the market area associated with each class of shopping center will be related to the kinds of functions contained within that center. More specifically, neighborhood centers will have comparatively small market areas, whereas regional centers will serve much larger areas. As a result, in any given city, we should expect to find fewer regional centers than neighborhood centers. The overall hierarchy of centers, according to the size and number of them at each level, comprises a series of steps (Figure 5.4). In a medium-sized city, for example, there might be one CBD, four regional shopping centers, and so on.

Finally, in terms of this conceptual overview of central place theory, it should be noted that the market areas associated with the different levels in the hierarchy of shopping centers will tend to nest inside each other, thus reflecting the *nesting principle* (Figure 5.5). Inside the market area for the CBD will be a series of regional center market areas, while inside each regional market area will be a series of community market areas, and so on. As a result, a person located at *A* in Figure 5.5 might go to the nearest community shopping center for lower-order goods, to the nearest regional shopping center for higher-order goods, and to the CBD for the most specialized goods.

The Formal Theory

Like many theories based on economic principles, central place theory contains a series of *simplifying assumptions* (King, 1984). After examining these simplifying assumptions and the associated spatial patterns, we will discuss what happens when the assumptions are relaxed. The first assumption is that we have a uniform distribution of population and a uniform distribution of purchasing power within our hypothetical city. That is, population density and income are everywhere the same. Second, it is assumed that the city is characterized by an isotropic transportation surface. There are no barriers to movement, such as rivers or mountains, and movement is equally possible in all directions. Furthermore, the transporta-

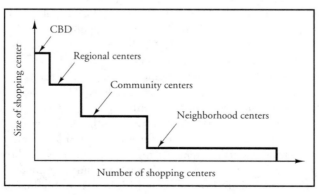

Figure 5.4 Hierarchy of shopping centers.

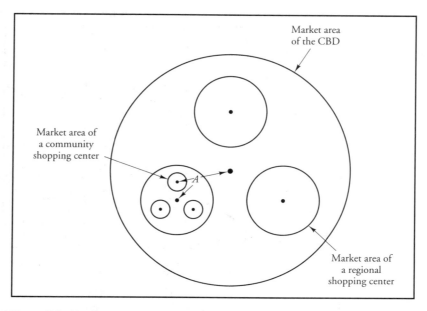

Figure 5.5 Nesting principle.

tion cost per unit of distance is the same in all directions. Third, it is assumed that consumers always patronize the nearest store offering the desired good. That is, consumers are interested only in minimizing transportation costs and are insensitive to variations in the price or quality of goods. Finally, it is also assumed that no excess profits are being earned by entrepreneurs. In other words, the thresholds and ranges associated with the various individual businesses will always coincide.

Given these assumptions and the previously discussed concepts of the threshold, range, and nesting principle, we would expect the *spatial pattern* of shopping centers and their associated market areas to be as represented in Figure 5.6. The market or trade areas form hexagons, as circular trade areas would either overlap or leave parts of the city unserved. These hexagonal trade areas preserve the assumption that each consumer patronizes the closest store offering the desired good or service.

Three major properties are associated with this abstract pattern of shopping centers. First, the frequency of occurrence of the different levels of shopping centers follows the progression, from large to small, 1, 2, 6. One can verify this property by examining the number of shopping centers dominated by each higher-order center. There is a community center at each corner of the regional center market area, and as each of these community centers is equidistant from three regional centers, we can say that the community centers are shared three ways. Because there are six such community centers on each regional center's trade area, there are six complete neighborhood centers, hence the progression 1, 2, 6.

Second, the progression of trade areas from largest to smallest is 1, 3, 9. This property can be verified by examining the number of trade areas dominated by each regional center.

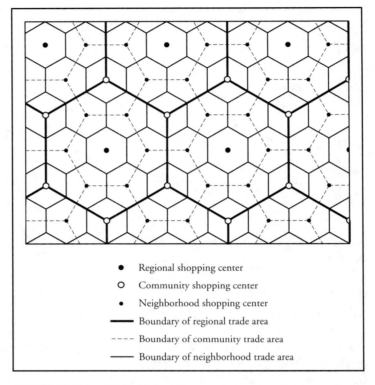

●	Regional shopping center
○	Community shopping center
•	Neighborhood shopping center
——	Boundary of regional trade area
----	Boundary of community trade area
——	Boundary of neighborhood trade area

Figure 5.6 The pattern of shopping centers according to central place theory. (Adapted from *The North American City,* Fourth Edition, by Maurice H. Yeates, p. 203. Copyright © 1990 by Harper & Row, Publishers, Inc. Reprinted by permission of HarperCollins, Inc.)

Each regional center dominates one regional trade area, and within that regional trade area there is one complete community trade area and portions of six others. Each of these portions represents a third of a community trade area, thus giving us one complete trade area plus six-thirds, or the equivalent of three. Finally, within each regional trade area there are seven complete neighborhood trade areas, plus portions of six others. These portions again represent thirds, thus giving us seven complete trade areas plus six-thirds, or the equivalent of nine; hence the progression 1, 3, 9.

Third, each shopping center is surrounded by a ring of six centers of the next lower level in the hierarchy, which are located at each corner of its hexagonal trade area. As a result, each lower-level center is equidistant between three centers of the next higher level in the hierarchy. This situation gives rise to a uniform spatial pattern in which shopping centers are distributed in the form of a triangular lattice.

Of course, this very regular pattern of shopping centers and associated market areas will become increasingly distorted as we *relax each of the initial simplifying assumptions.* First, it was assumed that population and income are uniformly distributed throughout the urban area. If population density is kept the same everywhere, but at a higher level than previously, the market areas will become smaller, as they need not be so large in order to capture the same threshold population. Conversely, if the population density is less everywhere, the market areas will become larger, and the shopping centers will be spaced farther apart. Similarly, if income remains constant, but at a higher level, the market areas will become smaller, as fewer people will be required to reach the same threshold sales volume, because each consumer will be spending more than previously. If income is reduced everywhere, the market areas will increase, and the shopping centers will be spaced farther apart.

Now, so far we have considered the consequences of changing the levels of population density and income, but we have still kept them constant throughout the city. If we allow population density and income to vary in magnitude, the trade area *size*, at any given level within the hierarchy, will also vary. In the more densely populated parts of the city the trade areas will be small and the shopping centers close together, whereas in the less densely populated areas the reverse will be true. Similarly, in those areas of the city characterized by high incomes, we would expect small trade areas and closely spaced shopping centers, whereas in the poorer areas we would expect large trade areas and widely spaced shopping centers.

Second, it was assumed that our hypothetical city has an isotropic transportation surface. If this assumption is relaxed, it means that accessibility is not the same in all directions, due to freeways, physical barriers, and the like. In any event, whatever the causes of variation in accessibility, the trade areas will respond by becoming elongated in those directions of greatest accessibility, and vice versa. In other words, the previously hexagonal trade areas will now be distorted into a variety of different *shapes*.

Third, it was assumed that consumers always minimize travel distance by going to the nearest store offering the desired good. If consumers no longer merely minimize distance, but also take into account other variables, such as the price of goods and services, the previously mutually exclusive trade areas will now *overlap*. For example, someone who lives very close to a particular neighborhood shopping center may travel across the city to another such center to take advantage of lower prices.

Finally, if the assumption that no excess profits are earned is relaxed, the threshold and range associated with each shopping center will no longer coincide. This situation indicates that the system is *dynamic*, because in those areas where the thresholds lie inside the ranges, and thus excess profits are being earned, new shopping centers will be developed. Conversely, in those areas where the thresholds lie outside the ranges, we would expect one or more shopping centers to go out of business in the long run.

In summary, the initial postulation of simplifying assumptions allows the spatial pattern of shopping centers to be understood more clearly. It is only when these assumptions are successively relaxed, however, that we move closer and closer to the actual pattern of shopping centers and trade areas in any given city. In particular, (1) relaxation of the uniform population and income assumption allows the trade areas to vary in size, (2) relaxation of the isotropic transportation surface assumption allows the trade areas to vary in shape, (3) relaxation of the distance minimization assumption allows the trade areas to overlap, and (4) relaxation of the excess profits assumption allows shopping centers to both enter and leave the system.

Empirical Testing

When applied to the intraurban retail structure, central place theory contains two major components that are amenable to empirical testing. First, data can be obtained to check the empirical validity of the suggested hierarchy of shopping centers. Second, one can investigate the spatial distributions associated with different kinds of retail establishments.

Berry and his associates (Berry, 1963), working in Chicago, suggested a four-tier *hierarchy*. They identified major regional centers, shopping goods centers, community centers, and neighborhood centers, and the general attributes of this hierarchy were consistent with the predictions of central place theory. First, the larger centers contained more ground-floor establishments and business types. Second, there were more low-level centers than high-level centers, although this generalization broke down at the lower end of the hierarchy, where there were fewer neighborhood centers than community centers. The main reason for the latter situation is that many of the smaller neighborhood centers were omitted from the analysis. Similar results have been obtained by Morrill (1987) for the Seattle metropolitan area (Table 5.2).

As expected, the *spatial distribution* of shopping centers in Chicago, at each level in the retail hierarchy, is fairly dispersed, with the various low-level centers filling in between the more widely spaced high-level centers. A more quantitative description of the degree of dispersion associated with retail establishments, however, can be achieved via the statistical technique known as *quadrat analysis* (Boots and Getis, 1988:17–35). Quadrat analysis is intended specifically for the investigation of point patterns, as it allows one to compare an observed point pattern with a theoretically expected point pattern in order to calibrate the goodness of fit. For our

Table 5.2 Hierarchy of Shopping Districts, Seattle Metropolitan Area, 1982

	Level	No. of Centers	No. of Establishments	Mean
A	Downtown Seattle	1	4,060	
B	Downtown Bellevue	1	1,115	
C	Major regional	5	3,550	710
D	Regional	9	4,180	464
E	District	18	4,220	234
F	Large neighborhood	38	4,725	124
G	Neighborhood	52	2,750	53
H	Small neighborhood	70	1,610	23
I	Suburban neighborhood	80	720	9
J	Corner	105	400	4
	Special	20	400	
	Metropolis	399	27,600	

Source: R. Morrill, "The structure of shopping in a metropolis," *Urban Geography*, 8 (1987), Table 1, p. 102.

purposes, a useful theoretical point pattern is a random pattern, as we can then judge whether the observed patterns are more dispersed than random or more clustered than random.

In the spatial context, a Poisson probability distribution will generate a random point pattern. Moreover, the mean of a Poisson distribution is equal to its variance, so a quick check for randomness in an observed point pattern merely involves comparing the mean and the variance. If, for the purpose of this comparison, we divide the variance by the mean, a value close to 1 implies that the pattern is approximately random, a value greater than 1 indicates that the pattern is more clustered than random, and a value less than 1 indicates that the pattern is more dispersed than random. Using this technique to analyze the distribution of various kinds of stores in Stockholm, Sweden, Artle (1965) demonstrated that the more specialized stores, such as antique stores and women's clothing stores, were especially clustered. The less specialized stores, however, such as grocery stores, were more ubiquitous.

The Changing Pattern of Retailing

One major change in the pattern of retailing since World War II has been the suburbanization of shopping centers and, in particular, the emergence of large, *planned shopping centers* (Lloyd, 1991; Simmons, 1991). At first these planned centers tended to be smaller than their unplanned counterparts, as developers tried to restrict competition between stores by controlling the level of functional duplication. More recently, however, the planned shopping centers have become incredibly large and have also incorporated other services besides retailing, such as movie theaters, skating rinks, and even miniature golf courses (Jackson and Johnson, 1991).

A second change in the pattern of retailing, somewhat related to the first, is that the planned, regional-level shopping centers have now become the catalysts for the overall suburbanization process, rather than merely the followers of that process. In other words, regional shopping centers often act as local *growth poles* (Baerwald, 1989; Williams, 1992). Cheap land is purchased at the periphery of the city, and the shopping center later becomes the focus for new apartment buildings, industrial parks, and so on. This kind of large-scale development will be less conspicuous in the future, however, if rising energy costs begin to seriously restrict consumer mobility. Also, some local planning boards have become increasingly disenchanted with both the various forms of pollution created by these large centers and their contribution to urban sprawl.

Changes in the scale of retailing have not only led to much larger shopping centers, but also to *larger individual stores*. These larger stores draw consumers from much wider trade areas and are therefore heavily dependent on automobile transportation. Associated with this change in scale has been the increasing importance of more specialized, higher-order goods. As a result, many of the individual stores and small neighborhood shopping centers that emphasize convenience goods have been losing trade and sometimes closing down.

5.3   The Pattern of Population Density

Like the urban retail structure, the pattern of population density in cities is very closely related to the underlying land value surface. In this section we consider some of the major population density models, the reason for the relationship between land value and population density, and how the pattern of population density has changed over time.

Population Density Models

Although a variety of population density functions have since been tested (Parr et al., 1988; McDonald, 1989), *Clark* (1951) first provided convincing empirical evidence to suggest that population density tends to decline in an exponential fashion with increasing distance from the CBD. That is, if population density is plotted against distance from the CBD, the resulting curve drops steeply at first and then more gradually (Figure 5.7a). This relationship is expressed symbolically as follows:

$$D_d = D_0 e^{-bd} \qquad (5.1)$$

where D_d is the population density at a given distance from the city center; D_0 is the density at the city center; e is the base of natural, or naperian, logarithms; b is the rate of change of density with distance; and d is distance. We have encountered a negative exponential function earlier in the book, represented by equation (3.8), and, as was the case then, the function can be linearly transformed by simply taking the natural logarithm of population density (Figure 5.7a). The equation corresponding to the linear form of the relationship is as follows:

$$\ln D_d = \ln D_0 - bd \qquad (5.2)$$

where ln is the natural logarithm and the remaining notation is the same as in equation (5.1).

The first major revision of Clark's model was independently suggested by *Tanner and Sherratt* (Thrall, 1988). They proposed that density declines exponentially with the square of distance (Figure 5.7b), as represented by the following equation:

$$D_d = D_0 e^{-bd^2} \qquad (5.3)$$

where the notation is the same as in equation (5.1). Again, this equation can be put in linear form by taking the natural logarithm of population density:

$$\ln D_d = \ln D_0 - bd^2 \qquad (5.4)$$

where the notation is the same as in equation (5.2).

Finally, *Newling* (1969) has suggested a quadratic exponential model (Figure 5.7c), which is represented as follows:

$$D_d = D_0 e^{bd - cd^2} \qquad (5.5)$$

where the notation is the same as in equation (5.1), except that we now have two parameters to be estimated, b and c. The b term is especially significant in Newling's model, as it measures the instantaneous rate of change of density with distance at the center of the city. In those situations where b is positive, it represents a density crater surrounding the CBD, where population density is comparatively low due to the presence of other kinds of land uses, especially those associated with commercial and retailing activity. Like the previous models, Newling's

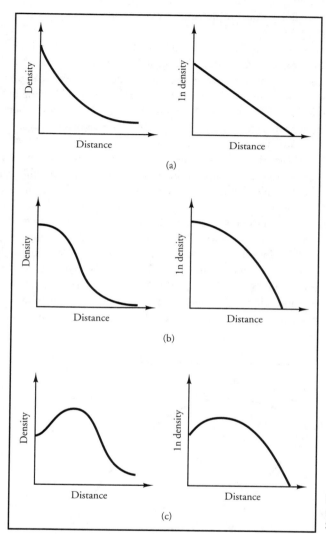

Figure 5.7 Population density models: (a) Clark; (b) Tanner and Sherratt; (c) Newling.

model can be linearly transformed by taking the natural logarithm of density, thus producing the following equation:

$$\ln D_d = \ln D_0 + bd - cd^2 \tag{5.6}$$

where the notation is the same as in equation (5.2).

Newling's model is important for two reasons. First, it incorporates the previous models of Clark and of Tanner and Sherratt. This particular property can be appreciated by considering equation (5.5). If c is zero and b is negative, Newling's model is the same as Clark's. On the other hand, if *b* is zero, Newling's model is the same as that proposed by Tanner and Sherratt.

The second major contribution of Newling's model is that it can be placed within a *dynamic framework* that allows for the emergence of a density crater in the CBD. Newling suggests that the density profile of a city evolves from the pattern suggested by Clark, through that suggested by Tanner and Sherratt, until a density crater emerges as in the Newling model. In other words, Clark's model is synonymous with a youthful stage, Tanner and Sherratt's model is synonymous with a mature stage, and Newling's model represents old age. When population density is in logarithmic form, this evolutionary sequence is expressed by Figure 5.8, where *t*, *t* + 1, ..., *t* + 4, indicate the different time periods. This diagram represents the "tidal wave of metropolitan expansion" and contains three major elements. First, the central density gets less over time. Second, the peak density is displaced outward. Third, the overall density gradient decreases.

This tidal wave of metropolitan expansion has been documented by Anderson (1985) for Detroit, and some empirical evidence of Newling's evolutionary schema has also been provided by Latham and Yeates (1970) in a study of population density change in Metropolitan Toronto. The linear form of Newling's model, equation (5.6), was calibrated and tested for a number of different time periods (Table 5.3). The first thing to note about the results is that the coefficient of multiple determination (R^2) is very stable across the different time periods and indicates that over 65 percent of the variation in population density within Toronto is accounted for by Newling's model.

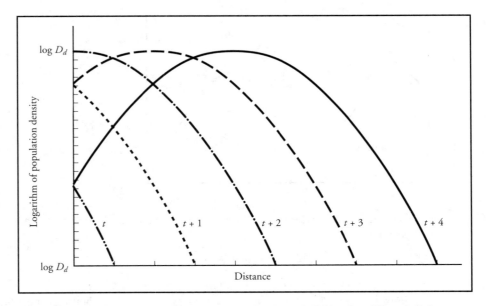

Figure 5.8 Population density curves over time. (From B. E. Newling, "The spatial variation of urban population densities," *Geographical Review*, 59, 1969, Fig. 6, p. 249, with the permission of the American Geographical Society.)

Table 5.3 Population Density Curves for Metropolitan Toronto

Year	Model Form	
1956	$\log D_d = 4.79 - 0.170d - 0.001d^2$	$R^2 = 67.12$
1961	$\log D_d = 4.45 - 0.052d - 0.006d^2$	$R^2 = 66.45$
1963	$\log D_d = 4.32 - 0.002d - 0.009d^2$	$R^2 = 68.43$

Source: Reprinted by permission from Robert F. Latham and Maurice H. Yeates, "Population density growth in Metropolitan Toronto," *Geographical Analysis*, 2 (April 1970), pp. 177–85. Copyright © 1970 by the Ohio State University Press.

Of greater importance, however, than the overall goodness of fit is the behavior of the individual parameters. For 1956, b is negative and c is close to zero, indicating that Clark's model is appropriate for this particular time period. In the following two time periods, however, b moves toward zero and c becomes increasingly negative, indicating a gradual shift from Clark's model toward the Tanner and Sherratt formulation. Over time, if b moves through zero to become positive, the full evolutionary cycle will be completed. In this context, it should be noted that the identification of a density crater, as implied by a positive value for b, is partly dependent on the kind of data being used. The density crater will emerge somewhat earlier if the population density values are for blocks rather than census tracts. Finally, it should also be noted that some researchers have argued for reformulating urban population density models in terms of the inverse power function, rather than the conventional negative exponential function (Batty and Kim, 1992).

The Relationship between Land Values and Population Density

In Chapter 3 it was demonstrated, via the use of bid-rent curves, why land value tends to decrease in a curvilinear fashion with increasing distance from the city center. In this chapter we have argued that population density behaves in a similar fashion. The reason for this close association between land value and population density is not hard to find. Quite simply, because the price of land decreases with increasing distance from the city center, regardless of other changes, land inputs will be substituted for other inputs, such as capital. Because the intensity of land use diminishes in this fashion, declining residential densities should also be expected. Note that this statement is consistent with the fact that high-intensity, multistory apartment buildings are usually found on high-valued land near the city center, while low-density, single-family dwelling units are usually found in more peripheral locations.

The Changing Pattern of Population Density

When analyzing change over time in the pattern of population density, two attributes of that pattern are usually monitored. First, we can examine changes in population density at the center of the city, or the *central density*. Second, we can examine changes in the *density gradi-*

ent, or the rate at which density decreases with increasing distance from the city center. These two attributes correspond to the two parameters D_0 and b in Clark's original model.

A comparative study of United States and Canadian metropolitan areas (Edmonston et al., 1985) demonstrated that the average central density of cities in both countries has been declining fairly consistently over time, although for each time period it is substantially higher in Canadian cities (Table 5.4). The values for the density gradient have also declined. Similar evidence has been provided by Bourne (1989), who looked at the changing density–distance relationship for 13 Canadian cities. He also notes that the declining coefficients of determination associated with this relationship reflect an increasing variability in the density surface.

As with land values, there is increasing evidence that the population density pattern in U.S. cities is best represented by a polycentric model. For example, Gordon et al. (1986) have identified 25 density peaks in Los Angeles County for 1970 and 1980. It is also noteworthy that these peaks and their associated gradients behave differently over time. Some peaks increase while others decrease, and the same is true for the gradients. In other words, the polycentric model suggests a much more complex pattern of change than does its monocentric counterpart.

5.4 Population Density Surfaces

It is also instructive to investigate the pattern of population density in terms of a three-dimensional surface. Thus far, by simply considering how density varies with distance from the city center, we have averaged out the directional variation. Although representing the density pattern in two dimensions is useful initially, especially since it facilitates comparison with our previous discussion of land values, less information is discarded if the pattern is analyzed in three dimensions. Within this context, we will first consider the statistical technique of trend surface analysis and then discuss some of the results of this analysis as applied to population density surfaces.

Table 5.4 Density Gradients and Central Densities in Canadian and American Metropolitan Areas, 1950/51 to 1975/76 (CMAs and SMSAs)

	Density Gradients		Mean Central Densities	
Year	Canada	USA	Canada	USA
1950/51	0.93	0.76	50,000	24,000
1960/61	0.67	0.60	33,000	17,000
1970/71	0.45	0.50	22,000	13,000
1975/76	0.42	0.45	20,000	11,000

Source: B. Edmonston, M. Goldberg, and J. Mercer, "Urban form in Canada and the United States: An examination of urban density gradients," *Urban Studies,* 22 (1985), Table 1, p. 213.

Trend Surface Analysis

Trend surface analysis is essentially a multiple regression technique whereby the general trend in a mapped variable, such as population density, can be identified (Griffith and Amrhein, 1991:406–16). As discussed in Chapter 2, a surface is created by drawing a series of *isolines* that join up points of equal value (Figure 5.9a). The most common type of isoline is a contour, which connects points of equal elevation, but we can also conceptualize land value surfaces, income surfaces, and so on. In the present instance, the isolines connect points of equal population density, thus creating a population density surface.

Such surfaces can be expressed symbolically as follows:

$$Z = f(X, Y) \tag{5.7}$$

where Z is the mapped variable, in this case population density, X is latitude, and Y is longitude. The trend surface analysis assumes that the mapped variable can be decomposed into

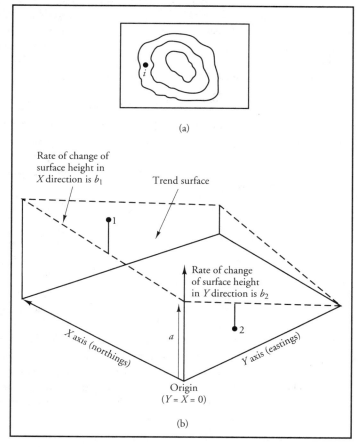

Figure 5.9 Trend surface analysis.

two components. First, there is a trend component, representing the large-scale systematic changes that extend across the map. Second, there is a local component, representing local fluctuations around the general trend.

So, for any particular location on the map, say i (Figure 5.9a), the observed height of the surface is made up of the trend component at that point, plus the local component, or residual. This statement can be expressed as follows:

$$Z_i = f(X_i, Y_i) \pm e_i \tag{5.8}$$

where Z_i is the height of the surface at location i; X_i and Y_i are the latitude and longitude associated with location i; and e_i is the local component, or residual, associated with location i.

The simplest form of surface is a *linear trend*, or inclined plane (Figure 5.9b). This surface can be described symbolically, for any particular location, as follows:

$$Z_1 = a + b_1 X_1 - b_2 Y_1 + e_1 \tag{5.9}$$

where Z_1 is the height of the surface at location 1; a is the height of the surface at the map origin, where X and Y are zero; b_1 is the rate of change of the surface height along the X axis; b_2 is the rate of change of the surface height along the Y axis; X and Y are latitude and longitude; and e_1 is the residual associated with location 1. Note that b_1 is positive, as the surface increases along the X axis, and b_2 is negative, as the surface decreases along the Y axis. Also, e_1 is positive, as the height of the surface at location 1 is above the general trend. Similarly, the surface can be expressed as follows for location 2 (Figure 5.9b):

$$Z_2 = a + b_1 X_2 - b_2 Y_2 - e_2 \tag{5.10}$$

where the notation is the same as in equation (5.9). The major difference is that the residual term is now negative, because the height of the surface at location 2 is beneath the general trend.

More *complex surfaces* can also be conceived of, generating a series of successively higher order polynomials of the following form:

$$Z = a + b_1 X + b_2 Y \tag{5.11}$$

$$Z = a + b_1 X + b_2 Y + b_3 X^2 + b_4 XY + b_5 Y^2 \tag{5.12}$$

$$Z = a + b_1 X + b_2 Y + b_3 X^2 + b_4 XY + b_5 Y^2 + b_6 X^3 + b_7 X^2 Y + b_8 XY^2 + b_9 Y^3 \tag{5.13}$$

where the notation is the same as in equation (5.9), and equations (5.11), (5.12), and (5.13) represent linear, quadratic, and cubic surfaces, respectively. A quadratic surface contains one inflection point, while a cubic surface contains two inflection points, and so on.

How well each of these surfaces fits an actual surface can be determined using the same kind of multiple correlation and regression analysis described in Chapter 4. In other words, each equation can be treated as a multiple regression equation, like the general case represented by equation (4.1). The overall goodness of fit is measured by the coefficient of multiple determination (R^2), and each individual regression coefficient can be estimated via least-squares analysis.

Results

Haggett and Bassett (1970) have used trend surface analysis to describe and compare the population density surfaces of 15 *U.S. cities*. They calculated the residential density for equally sized subareas, and locational coordinates were defined for the midpoints of those subareas. Linear, quadratic, and cubic surfaces were tested for each of the 15 cities, and the coefficients of multiple determination are reported in Table 5.5.

Note that, for all cities, the coefficients of multiple determination increase as one moves from the simple linear surface to progressively higher order surfaces. This situation is to be expected, as the higher-order surfaces contain more terms and thus account for more of the local variation in population density. Some cities are approximated quite well by even the linear surface, as in the cases of Milwaukee, Los Angeles, and New York, while other cities have such convoluted population density surfaces that they are not even approximated by a cubic surface. Perhaps the extreme example, in this context, is San Francisco, where the coefficient of multiple determination associated with the cubic surface is only 0.467.

It is also of interest to examine how population density surfaces have changed over time, and Hill's (1973) study of *Toronto* provides an interesting example. Hill tested up to a fifth-order surface for five different time periods, with the first-, second-, and third-order surfaces

Table 5.5 Coefficients of Multiple Determination for the Trend Surface Analysis of U.S. Cities

	Linear	Quadratic	Cubic
Atlanta	0.246	0.337	0.580
Cincinnati	0.097	0.661	0.857
Cleveland	0.329	0.689	0.996
Denver	0.386	0.788	0.831
Los Angeles	0.677	0.838	0.869
Miami	0.529	0.609	0.748
Milwaukee	0.719	0.944	0.961
Minneapolis–St. Paul	0.462	0.721	0.913
New Orleans	0.500	0.666	0.871
New York	0.629	0.795	0.875
Philadelphia	0.255	0.519	0.676
Pittsburgh	0.040	0.535	0.749
San Francisco	0.211	0.341	0.467
Seattle	0.207	0.558	0.651
St. Louis	0.002	0.591	0.688

Source: P. Haggett and K. A. Bassett, "The use of trend-surface parameters in inter-urban comparisons," *Environment and Planning*, 2 (1970), Table 1, p. 229.

representing linear, quadratic, and cubic surfaces, respectively (Table 5.6). As with Haggett and Bassett's results, for any given time period the higher-order surfaces always account for more variation than their lower-order counterparts.

Hill's major finding was that for each order surface the coefficient of multiple determination tends to get less over time. This phenomenon indicates that the population density surface is becoming less regular or, in other words, there are increasingly more inflection points in the surface. It is also noteworthy that Newling's model, when applied to Toronto (Table 5.3), accounted for approximately 67 percent of the variation in population density, thus emphasizing that curve fitting in a two-dimensional context smooths out the directional variation. Finally, the coefficients of multiple determination for Toronto are less than for most of the cities tested by Haggett and Bassett, partly because the latter used relatively large spatial data units and thus filtered out more of the original variation.

This last point serves as a reminder that the results of trend surface analysis should be interpreted with some caution. First, it is desirable to have a fairly uniform distribution of data points. A clustering of data points tends to inflate the R^2 value, and the area containing the cluster of points will have an undue influence on the shape of the surface. Second, higher-order surfaces are very demanding in terms of data. A perfect fit will always be obtained in those instances where the number of terms in the trend function equals the number of data points. Third, overall goodness of fit will be affected by the location of the boundary, while individual coefficients are difficult to interpret due to multicollinearity.

Table 5.6   Coefficients of Multiple Determination for Population Density Trend Surfaces, Metropolitan Toronto, 1941–1966

	Order of Surface				
Year	1	2	3	4	5
1941	0.228	0.355	0.486	0.540	0.597
1951	0.229	0.351	0.466	0.516	0.561
1956	0.245	0.379	0.486	0.522	0.567
1961	0.211	0.334	0.433	0.469	0.517
1966	0.180	0.298	0.373	0.403	0.462

Source: F. I. Hill, "Spatio-temporal trends in urban population density: A trend surface analysis," in L. S. Bourne, R. D. Mackinnon, and J. W. Simmons, eds., *The Form of Cities in Central Canada: Selected Papers,* Department of Geography Research Publication 11, University of Toronto, Toronto, 1973, Table 7.1, p. 110.

Chapter 6

Urban Social Areas

6.1   Residential Segregation

Having discussed the pattern of population density in cities, it is now of interest to examine the spatial distribution of different kinds of people. In particular, we wish to know the degree to which different kinds of people live in different parts of the city. That is, what is the extent of residential segregation in North American cities? After establishing the extent of segregation, it will be instructive to investigate whether this segregation has any particular spatial manifestation, thus leading to a discussion of the zonal and sector models. The general issue of social areas in cities will then be pursued via the topics of social area analysis and factorial ecology.

Index of Segregation

The extent of residential segregation associated with a particular subgroup, in a particular city, can be measured by the index of segregation (Peach et al., 1981). This index is expressed as follows:

$$S = \frac{\sum\limits_{i=1}^{N} |X_i - Y_i|}{2} \tag{6.1}$$

where S is the index of segregation, X_i is the percentage of a particular subgroup living in areal unit i, and Y_i is the percentage of the rest of the population living in areal unit i. The areal units are usually either blocks or census tracts.

123

A hypothetical example will clarify the use of this index. Imagine a city that is divided into five areal units (Figure 6.1a), and we wish to calculate the index of segregation associated with the distribution of students. Having listed the percentage of students and the percentage of the rest of the population in each subarea (Figure 6.1b), we then calculate the absolute differences. These differences are summed and divided by 2, yielding a segregation index of 40. The two extremes of the index are 0 and 100. If the students were distributed in exactly the same pattern as the rest of the population, all the absolute differences would be zero. If, on the other hand, all the students were in one areal unit, and the rest of the population was in the other areal units, the absolute differences would sum to 200, yielding an index of 100. Therefore, the closer the index is to 100, the more segregated is the particular subgroup under investigation.

Segregation According to Race and Ethnicity

From 1970 to 1980 the amount of segregation associated with blacks in U.S. metropolitan areas tended to decline (Massey and Denton, 1987), although New York was an exception to this general rule (see Table 6.1). Despite this decrease, however, the absolute levels of segregation associated with blacks were still considerably higher than those associated with Hispanics or Asians. For the five metropolitan areas reported in Table 6.1, the levels of segregation associated with Hispanics showed a slight increase, while those for Asians showed a

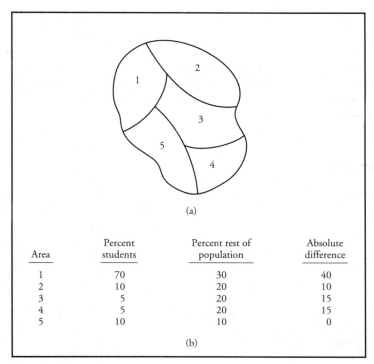

(a)

Area	Percent students	Percent rest of population	Absolute difference
1	70	30	40
2	10	20	10
3	5	20	15
4	5	20	15
5	10	10	0

(b)

Figure 6.1 Index of segregation.

Table 6.1   Residential Dissimilarity of Blacks, Hispanics, and Asians from Anglos in 60 U.S. Metropolitan Areas, 1970–1980

Metropolitan Area	Dissimilarity between Anglos and:								
	Blacks			Hispanics			Asians		
	1970	1980	Change	1970	1980	Change	1970	1980	Change
Chicago	0.919	0.878	−0.041	0.584	0.635	0.051	0.558	0.439	−0.120
Los Angeles	0.910	0.811	−0.099	0.468	0.570	0.102	0.531	0.431	−0.100
Miami	0.851	0.778	−0.073	0.504	0.519	0.015	0.392	0.298	−0.094
New York	0.810	0.820	0.010	0.649	0.656	0.007	0.561	0.481	−0.080
San Francisco	0.801	0.717	−0.084	0.347	0.402	0.055	0.486	0.444	−0.042

Source: D. Massey and N. Denton, "Trends in the residential segregation of Blacks, Hispanics, and Asians: 1970–1980," *American Sociological Review*, 52 (1987), Table 3, p. 815.

slight decrease. By the end of the decade the absolute levels of segregation were generally lower for Asians than for Hispanics, with the exception of San Francisco.

It is generally recognized that any satisfactory *explanation* of residential segregation will involve a number of causal factors (Clark, 1986a). There are, however, two extreme schools of thought as to why blacks and certain ethnic groups are concentrated in particular parts of the city. One viewpoint assumes the existence of free choice in the housing market and therefore attributes racial and ethnic differences in housing consumption to economic differentials. Specifically, it is argued that blacks and other minorities have lower incomes than whites and that they prefer to invest a smaller proportion of that income on housing. This kind of separation is also augmented by social preferences. For example, a recent study of the expressed preferences of four different ethnic groups in the Los Angeles metropolitan area showed that minorities prefer to live in neighborhoods primarily occupied by families like themselves (Clark, 1992a).

The alternative viewpoint emphasizes that the segregation of blacks and certain ethnic groups is not merely due to income differences and preferences, but also occurs as a result of discrimination in the housing market (Galster, 1988; Darden, 1989). For example, it has been argued that blacks have been discouraged from entering certain submarkets, especially by the activities of realtors, who are sometimes unwilling to help them locate suitable housing, and financial institutions, who may refuse to provide adequate mortgage money. According to this view, the self-segregation hypothesis associated with income and preferences is inappropriate.

The Process of Assimilation

The degree to which an ethnic or racial group remains a distinctive social and spatial entity depends on the process of assimilation. Assimilation is generally decomposed into behavioral assimilation, or acculturation, and structural assimilation. Behavioral assimilation occurs when members of the minority group acquire the sentiments and attitudes of other groups in the society and thus become involved in a common cultural life. Structural assimilation, on the other hand, refers to the distribution of minority members through the social systems of the society, especially its system of occupational stratification. Those ethnic groups that maintain a strong identity have usually created a set of ethnic institutions, relating to religion, education, politics, and business.

Boal (1976) has suggested that ethnic and racial residential clusters within cities serve four main *functions*. First, such clusters play a defensive role, whereby members of a particular minority can reduce their isolation and vulnerability. Second, ethnic and racial residential concentration serves an avoidance function and is often a "port of entry" before individual members are assimilated into the larger society. Third, these residential clusters allow ethnic and racial groups to preserve their own cultural heritage. Fourth, such clustering can also serve as the basis for "attack" functions against society in general, as spatial concentration may allow the group to elect its own political representatives.

Boal (1976) has also suggested a set of relationships among ethnic group distinctiveness, degree of assimilation, and the associated spatial pattern (Figure 6.2). If the entering group is very similar to the host society, rapid assimilation and few signs of spatial concentra-

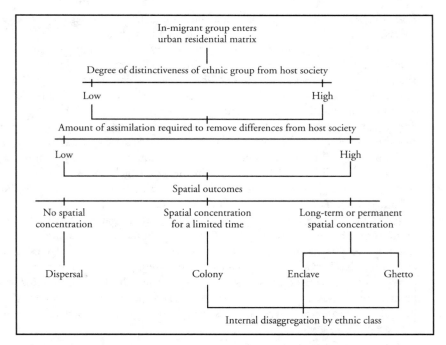

Figure 6.2 Ethnic groups, assimilation, and residential spatial outcomes. (From F. W. Boal, "Ethnic residential segregation," in D. T. Herbert and R. J. Johnston, eds., *Social Areas in Cities*, Vol. 1: *Spatial Processes and Forms*, Wiley, New York, 1976, Fig. 2.1, p. 57. Copyright © 1976 by John Wiley & Sons Ltd. Reprinted by permission of John Wiley & Sons Ltd.)

tion can be expected. In those cases where the ethnic group is very different from the host society, however, the assimilation process will take longer and residential clustering will occur.

Where such residential clustering is only a temporary stage in the assimilation process, the concentration is called a colony. A colony is normally short lived and primarily serves as a port of entry for a particular ethnic group. Once the influx of new ethnic group members begins to decline, such colonies tend to disappear. In other situations the residential clustering is long term due to a desire for cultural preservation on the part of the ethnic group and external pressure from the host society. Where the ethnic cluster represents a primarily voluntary phenomenon, it is called an enclave, as exemplified by Jewish communities in many cities. Those clusters that are more determined by external pressure and may be involuntary in nature are termed ghettos. Ghettos are often maintained by ethnic or racial prejudices within the local housing market.

6.2   The Classical Models

Having shown that U.S. cities are indeed segregated according to such attributes as occupation, ethnicity, and race, it is important to consider whether this segregation is manifested in terms of particular spatial patterns. Within this context, two very influential students of urban structure have suggested that intraurban residential variation can be conceptualized in terms of zones and sectors. It is to these models of Burgess and Hoyt, respectively, that we now turn our attention.

The Concentric Zone Model

The concentric zone model was formulated by Burgess (1925), a sociologist at the University of Chicago. After studying the land use and social characteristics of Chicago in the early 1920s, Burgess suggested that the urban pattern could be summarized in terms of five concentric land use zones (Figure 6.3). These land use zones not only described the pattern at a particular point in time, but also represented the successive zones of urban expansion.

The first, or innermost zone, was the central business district. This zone contained the downtown retailing district, plus major office buildings and banks. It was also a major recreational center, with theaters, museums, and social clubs. Surrounding this area was the wholesale business district, with its associated warehouses. Burgess called the second zone the zone in transition. This zone was characterized by an inner factory belt and an outer ring of deteriorating residential neighborhoods. The residential deterioration was caused by the encroachment of business and industry from the CBD, and these transitional neighborhoods were often the first homes of immigrants from Europe or black Americans from the rural South. The third zone was labeled the zone of independent workingmen's homes. Residents of this area had been able to move out of the second zone, although they still typically worked in the CBD. Many residents of these neighborhoods were second-generation immigrants. Burgess's fourth zone was the zone of better residences. It was here that the great mass of middle-class native-born Americans lived. Most of the housing consisted of single-family dwelling units with spacious yards. The fifth and outermost zone was the commuter's zone. This zone lay

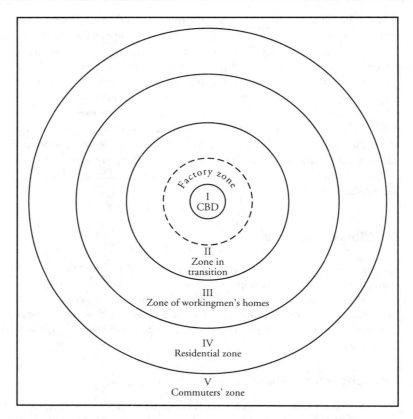

Figure 6.3 Concentric zone model. (Based on R. E. Park, E. W. Burgess, and R. D. McKenzie, *The City,* University of Chicago Press, Chicago, 1925, Fig. 1, p. 51. Copyright 1925 by the University of Chicago.)

outside the legal boundary of the city and consisted of a ring of small towns and villages. These settlements were mainly dormitory suburbs, with very little industry or employment of their own.

Burgess took care to emphasize that the zones should not be taken too literally, but rather that they represent an idealized pattern for the purposes of comparing deviations. The underlying mechanism that generated these distinctive zones was called the process of *invasion and succession*. Thus the prime characteristic of the classical ecological approach to urban structure, as represented by Burgess and his colleagues, was the utilization of a biological analogy, whereby different social groups, analogous to plant species, compete for space in the city (Timms, 1971:86). Different social groups came to dominate different parts of the city and form "natural areas." These natural areas and their associated pathologies were then subjected to further detailed examination.

The actual process of invasion and succession, by which a natural area came to be dominated by a new group, was divided into a series of stages. First, a small number of in-migrants from a different social group would penetrate a neighborhood. These in-migrants were usually upwardly mobile and sometimes even had higher incomes than the established population. Second, this initial penetration was followed by an invasion stage, in which large numbers of the new group replaced members of the old group. Third, there was a succession or consolidation stage in which the original minority group became the majority group. Fourth, and finally, there was a piling-up stage, which entailed a stabilization of the area in terms of its domination by the new group. The overall process can be conveniently represented by an S-shaped curve (Figure 6.4).

In order, however, for this process of invasion and succession to create the particular spatial pattern envisioned by Burgess, certain *preconditions* or assumptions about the city are necessary. First, the city needs to be growing in population and expanding at its edges. Second, there should be only one core to the city so that the zones are concentrically arranged around a single center. Third, new housing should be built primarily for the wealthy at the edge of the city so that the distinctive zones are created by the successive passing down of housing from the wealthy to the less wealthy. Fourth, there should be an efficient transportation system so that the wealthy are willing to live at the edge of the city. Fifth, the city's population should be relatively heterogeneous, with a wide variety of occupational, ethnic, and racial groups, so that different kinds of social areas can emerge. It is noteworthy, therefore, that the Chicago of the 1920s contained all these necessary ingredients.

As with every model, or idealized representation of reality, the concentric zone model has not been immune from *criticism*. First, it has been pointed out that the biological analogy is a potentially misleading one, especially since competition in the human world, unlike that

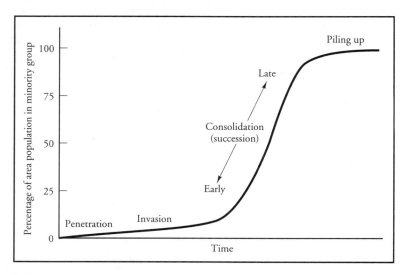

Figure 6.4 Process of invasion and succession, showing the typical stages according to the percentage of an area's population who are of the invading group. (From R. J. Johnston, *Urban Residential Patterns: An Introductory Review*, G. Bell, London, 1971, Fig. 6.3, p. 253.)

in the biological world, is restricted by conventions, laws, and institutions. Second, it has been argued that the model seriously underestimates the importance of sentimental attachments. In other words, residential neighborhoods are not merely the arenas of competition between social groups, as implied by the biological analogy, but they also represent certain symbolic qualities (Bassett and Short, 1980:15).

The Sector Model

Hoyt (1939) suggested that residential differentiation within cities could be summarized in terms of sectors rather that zones (Figure 6.5). Unlike Burgess, who had based his work on Chicago, Hoyt examined some 142 cities and concluded that socioeconomic status varied primarily in a sectoral fashion. On the basis of rental patterns, Hoyt made a number of general observations concerning these sectors, or pie-shaped wedges. First, the most highly valued residential areas were located in sectors on one side of the city and sometimes extended out continuously from the city center. Second, the middle-class or intermediate rental areas were often found on either side of the highest-rent areas. Third, the low-rent sectors were frequently found on the opposite side of the city to the high-rent sectors.

Hoyt studied the changing distribution of high-class areas in particular and suggested that the directional orientations and growth paths of these sectors were influenced by a variety of factors. First, the high-class areas tended to grow outward along major transportation routes. Second, they tended to grow toward high ground that was free from the risk of flooding. Third, they tended to extend toward the homes of the leaders of the community.

Like the concentric zone model, Hoyt's sector model has also been extensively *criticized* (Timms, 1971:227–9). One of the most ambiguous elements of the sector model is the definition of the sectors themselves. In Hoyt's original work the term sector is applied to areas that vary in size from single blocks to whole quadrants of the city. Also, the leaders of society are mentioned as being a major force in the expansion of high-socioeconomic-status sectors, but the identity of these leaders and the reasons for their appeal are never explicitly stated.

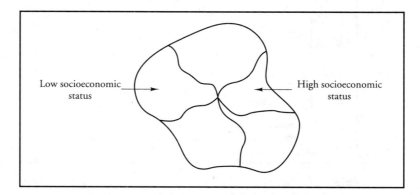

Figure 6.5 Sector model.

In any event, in purely geometric terms, the zonal and sector models are obviously at odds with each other, although it should be noted that Hoyt, while emphasizing the importance of sectoral differentiation, did include a zonal pattern within many of his sectors. Also, in terms of the underlying process responsible for producing these spatial patterns, Hoyt suggested a mechanism that has become known as the filtering of housing. The same filtering concept also played a role in the outward expansion of Burgess's zones.

The Filtering of Housing

The filtering-of-housing concept suggests that as a housing unit ages and deteriorates its relative market price will decrease, and it will become available to lower-income families (Weicher and Thibodeau, 1988). As a result, it is postulated that a number of moves at the periphery of a zone, or sector, will initiate a chain reaction of vacancies that eventually extends back to the city center. Johnston (1971:98) has conceptualized this chain reaction process as being composed of four stages (Figure 6.6). Thus, in terms of the classical zonal model, as the

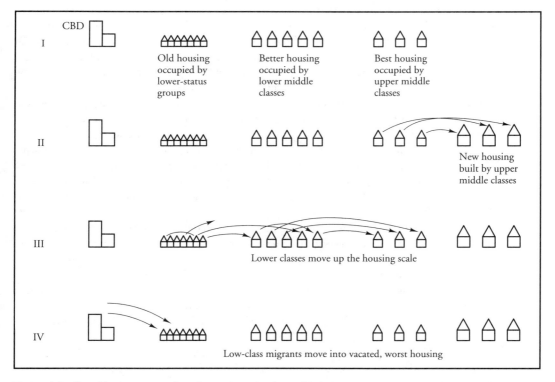

Figure 6.6 Simplified cross section through a city, from CBD to periphery, to illustrate the filtering process. (From R. J. Johnston, *Urban Residential Patterns: An Introductory Review*, G. Bell, London, 1971, Fig. 3.8, p. 98.)

housing stock in each zone ages it is occupied by a succession of lower-income groups. That is, the housing stock in each zone is filtered down over time. The effect of an influx of poor immigrants into the city center, therefore, was much like throwing a pebble into the middle of a pond and setting up a series of waves that eventually reach the edge of that pond.

Bourne (1981:149) has pointed out, however, that this argument contains a number of implicit *assumptions*. First, it is assumed that high-income families prefer to live in new rather than old housing. Second, these same high-income families are also assumed to prefer to live at the edge of the city rather than at the center. Third, it is assumed that housing inevitably depreciates with age, and that families prefer to move rather than renovate their existing homes.

The *rate of filtering* within a given housing market, or city, is influenced by two major factors: the rate of new housing construction and the rate of household formation. If new construction proceeds faster than the formation of new households, supply will exceed demand, and the market value of older housing will decrease rapidly, thus making it available to lower-income families. If, on the other hand, the rate of new construction is slow relative to the formation of new households, the rate of filtering will also be comparatively slow.

Transition probability matrices can be used to examine empirically the phenomenon of filtering. Table 6.2 contains data for the occupancy of 2495 homes in Kingston, Ontario, for the years 1958 and 1963. The head of household for each home is placed into one of seven occupational groupings, which have been ordered according to average income. The initial tally matrix indicates that of the 303 homes occupied by professional workers in 1958, by 1963, 6 were occupied by unskilled workers, 25 were occupied by service workers, 21 were occupied by craftsmen, and so on. In this way, the values to the left of the diagram represent homes that have filtered down to lower-socioeconomic groups, while the values to the right of the diagonal indicate homes that have filtered up to higher-socioeconomic groups.

The values in the tally matrix can be converted into transition probabilities, representing the probability that a home will be occupied by a particular type of family in 1963. These transition probabilities are obtained by calculating each entry as a proportion of the total number of homes in each row. Thus the proportion of homes occupied by professionals in 1958 that are occupied by unskilled workers in 1963, is 6 divided by 303, giving a transition probability of only 0.02. The values along the diagonal represent homes that are still occupied by the same occupational category, and these are the largest transition probabilities. In other words, during this particular 5-year period, a great many homes did not filter either up or down the socioeconomic hierarchy.

The filtering concept is of particular importance because it has become an underlying principle of *housing policy* in many countries. It is argued that an emphasis on the construction of middle- and upper-income housing will lead, as a result of the filtering process, to improved housing for all. There are a number of possible objections to this policy, however. First, rather than filtering down to lower-income families, older housing might be converted to other uses such as offices, or it might even be demolished. Second, lack of mortgage availability often restricts the movement of poorer families into available dwelling units. Third, as has already been mentioned, in order for there to be a significant amount of downward filtering, the supply of housing must exceed the demand for housing. Fourth, use of the filtering principle implies that newer housing should be for the rich, and older housing should be

Table 6.2 Tally Matrix and Transition Probability Matrix for 2495 Homes in Kingston, Ontario, 1958–1963

1958 State	| 1963 State							Total
	1	2	3	4	5	6	7	Total
				Tally Matrix				
Unskilled	179	62	39	7	18	10	5	320
Service	116	359	103	31	32	12	18	671
Craft	64	95	337	13	22	29	8	568
Clerical	11	23	12	66	7	6	13	138
Sales	27	40	25	15	96	18	23	244
Managerial	10	24	19	12	23	146	17	251
Professional	6	25	21	9	20	15	207	303
Total	413	628	556	153	218	236	291	2495
			Transition Probability Matrix					
Unskilled	0.559	0.194	0.122	0.022	0.056	0.031	0.016	1.0
Service	0.173	0.534	0.154	0.046	0.048	0.018	0.027	1.0
Craft	0.113	0.167	0.593	0.023	0.039	0.051	0.014	1.0
Clerical	0.080	0.167	0.087	0.478	0.051	0.043	0.094	1.0
Sales	0.111	0.164	0.102	0.061	0.394	0.074	0.094	1.0
Managerial	0.040	0.095	0.076	0.048	0.092	0.581	0.068	1.0
Professional	0.020	0.083	0.069	0.030	0.066	0.050	0.682	1.0

Source: M. H. Yeates, *An Introduction to Quantitative Analysis in Human Geography*, McGraw-Hill, New York, 1974, Tables 7–4 and 7–5, pp. 178–9. Reproduced with permission.

occupied by the poor. Fifth, even when older housing does filter down, it often burdens low-income families with prohibitive maintenance costs. In sum, it can be argued that filtering is neither an efficient nor a humane means of providing housing for low-income families (Bourne, 1981:154). To maximize the benefits to the poor, a certain amount of new housing should probably be constructed specifically for that income level.

6.3   Social Area Analysis

Social area analysis attempts to provide a broader framework for the analysis of residential differentiation within cities by examining the underlying dimensions of urban society (Davies and Herbert, 1993). This particular approach was first developed by Shevky and Williams (1949) in a study of Los Angeles and was later elaborated on by Shevky and Bell (1955) in a study of San Francisco. Essentially, social area analysis classifies census tracts according to three basic constructs that supposedly summarize the important social differences among those census tracts. The three constructs are economic status, or social rank; family status, or urbanization; and ethnic status, or segregation. We will discuss the meaning of these constructs, how they were operationalized, how they were used to generate a typology of social areas, and the various ways in which social area analysis has since been criticized.

The Constructs

According to Shevky and Bell, the basic premise underlying social area analysis is that the city cannot be understood in isolation from the overall society of which it is a part. In other words, the social characteristics of urban life must be investigated within the context of the social characteristics of society at large. Thus, according to this thesis, residential differentiation within cities has its origins in the changing social differentiation of society.

The increasing modernization or development of society results in increasing *societal scale*, whereby the social and economic interchanges within the society enlarge in terms of both range and intensity (LaGory and Pipkin, 1981:99). In social area analysis, increasing societal scale is also held to be synonymous with the growth of an urban–industrial society. People become more economically dependent on each other due to specialization of the labor force and improvements in transportation technology, and these changes result in new patterns of social differentiation (Figure 6.7).

First, because of the increasing division of labor, an occupational status system develops, and this system becomes the basic element of social stratification in industrial society. Second, the family becomes less important as an individual economic unit, and there is a weakening of the traditional organization of the family. Third, the improved transportation technology results in greater mobility, and the associated increased freedom of choice in terms of where to live in the city leads to a greater sorting of the population and the segregation of various ethnic and racial groups. In summary, an increase in societal scale tends to socially differentiate

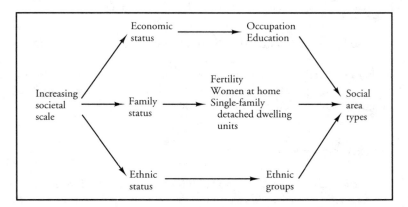

Figure 6.7 Shevky–Bell schema of social constructs and social area types. (Based on E. Shevky and W. Bell, *Social Area Analysis*, Stanford University Press, Stanford, Calif., 1955, p. 4.)

people within cities according to economic status, family status, and ethnicity, and these three social constructs provide the criteria for differentiating social areas within cities.

Operationalization

The three constructs were originally operationalized by the use of six different variables, or indexes (Figure 6.7). Economic status was measured by occupation, described as the total number of operatives, craftsmen, and laborers per 1000 employed persons, and by education, described as the number of persons who had completed no more than 8 years of schooling per 1000 persons aged 25 years and over. Family status was represented by three different variables: fertility, the number of children 4 years old or less per 1000 women aged 15 to 44; women at home, the number of females not employed in the labor force in relation to the total number females aged 15 years and over; and single-family dwelling units, the number of single-family homes as a proportion of all dwelling units. Finally, the ethnic status construct was simply measured as the number of people in specific minority groups as a proportion of the total population.

Data for these variables were obtained for each census tract in a city and then combined according to the following procedure, to provide an overall value for each construct. First, the scores on the original variables were standardized, so that each variable ran between 0 and 100. For example, if there were values of 22, 62, and 72 percent for a particular variable, then 22 percent became 0, 72 percent became 100, and 62 percent became 80, as it is four-fifths of the distance between 22 and 72 percent. This transformation lent greater validity to the comparison of scores across different variables. The overall construct scores were simply obtained by computing the average of the transformed scores for the variables associated with each of the three constructs.

Typology of Social Areas

On the basis of their scores on each of the social area constructs, the census tracts were classified into a series of social area types, and Figure 6.8 depicts the *social space* diagram for Winnipeg, Canada. The vertical and horizontal axes in the social space diagram are urbanization and economic status, respectively, where the urbanization dimension is merely the reverse of the previously described family status construct, with high values for urbanization indicating low values for family status. Each census tract has a unique location within this two-dimensional space, as represented by a circle, according to its overall score on each construct.

The social space is usually partitioned into 16 cells, or social areas, by dividing the urbanization and economic status constructs into four intervals. The four intervals on each axis correspond with the ranges of 0 to 24, 25 to 50, 51 to 74, and 75 to 100 within the two major constructs. Shaded circles indicate those census tracts that have a higher-than-average percentage of some particular minority group. As one would expect, there tend to be fewer shaded circles in the right-hand cells, as these cells are associated with the highest levels of economic status.

The distance between census tracts within this abstract social space can be thought of as representing *social distance*, rather than physical distance. For example, those census tracts that

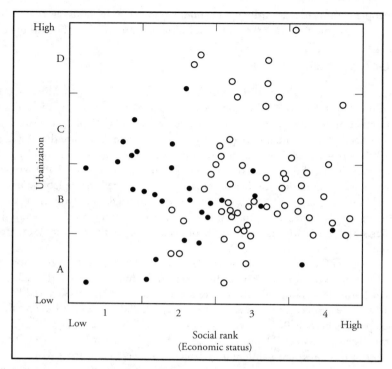

Figure 6.8 Social space diagram for Winnipeg. (From D. T. Herbert, *Urban Geography: A Social Perspective*, David & Charles, Newton Abbot, England, 1972, Fig. 40, p. 142.)

fall within the upper-right cells, synonymous with high values for both urbanization and economic status, are obviously very similar with respect to these two dimensions and thus separated by comparatively short social distances. Such a situation can occur even when the census tracts are located in different parts of the city and are therefore separated by relatively large physical distances. In this respect, it is interesting to note that the term social area was originally used to describe a cluster of census tracts in social rather than geographical space. The later use of this term, however, implied a contiguous territorial unit.

Criticisms

Although it is generally accepted that social area analysis represents a promising attempt to erect a logical framework for the analysis of urban residential differentiation, it has been criticized by a number of authors. The theoretical framework fails to explain why residential areas should be relatively homogeneous or why they should differ from each other. Also, the theory does not provide strong justification for using the three particular constructs of economic status, family status, and ethnic status to differentiate among social areas (Ley,

1983:76). In fact, it can be argued that the theoretical framework erected by Shevky and Bell is merely an a posteriori rationalization for their choice of constructs.

Social area analysis really involves two distinct parts, and there is no real explanation as to how a theory of social change can be translated into a static typology of residential differentiation. In other words, it requires no small inferential leap to move from a broad theory of social change to empirical regularities in the residential location of different kinds of households (Bassett and Short, 1980:18). Also, in this context, the concept of increasing societal scale is probably too ambiguous to provide the key element in an overall theory of urban social structure, as it appears to be used synonymously with such general terms as industrialization, urbanization, and modernization.

Besides being criticized on theoretical grounds, social area analysis has also been criticized for empirical reasons. In particular, the variables used to measure the three constructs are open to question (Carter, 1981:258), and the methodology used for operationalizing the three constructs is statistically unsophisticated. The variables employed simply assume the three constructs to be correct, without providing any test of their empirical validity. Such tests were later carried out using the technique known as factor analysis.

6.4   Social Area Analysis via Factor Analysis

The phrase "social area analysis via factor analysis" refers to the empirical testing of the three social area constructs using the multivariate statistical technique of factor analysis. After gaining an intuitive understanding of factor analysis, we shall look at some of the results obtained when this technique is applied to North American cities.

Factor Analysis

Factor analysis was originally developed by psychologists as a means of reducing a large number of variables to a smaller number of underlying factors, dimensions, or components (Shaw and Wheeler, 1985:273-295). These underlying factors summarize the original variables, although as one moves from, say, 20 variables to only 3 factors, some of the original variation is obviously lost. Just how much of the original is lost, however, can be readily computed. It should also be noted that there are a variety of factor analysis techniques and that, because of its widespread use in this particular context, we will be emphasizing the technique known as *principal components analysis*.

It is convenient to think of principal components analysis in terms of four matrices. The first matrix (Figure 6.9a) is simply a *data matrix* of the kind represented previously in Figure 2.2. The cases are the rows, and the variables are the columns. As we have N cases and M variables, the overall size of the matrix is $N \times M$. The data in this matrix are standardized so that each variable is expressed in terms of standard deviations from its mean, which is the same kind of standardization procedure as that discussed in Chapter 4. The data matrix is then transformed into a *correlation matrix* (Figure 6.9b). Each variable is correlated with every other variable, using the correlation coefficient described in Chapter 3, thus creating an

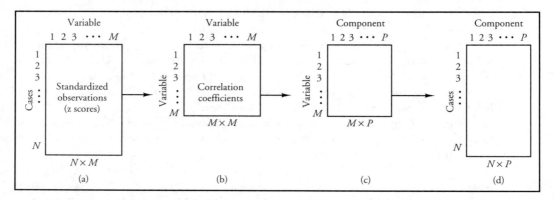

Figure 6.9 Major steps in principal components analysis: (a) data matrix; (b) correlation matrix; (c) components matrix; (d) matrix of component scores.

$M \times M$ matrix of correlation coefficients. The diagonal cells are given the value 1, as they represent the correlation of each variable with itself.

The correlation matrix is then transformed into the *components matrix* (Figure 6.9c). Each column in the matrix is a component that represents a group of interrelated variables. The variables are the rows, so if we have P components, the size of this matrix is $M \times P$. Each cell in the matrix contains a component loading and, as we shall see later, these component loadings indicate the strength of the relationships between the variables and the underlying components. Thus the component loadings are used to interpret which group of variables is summarized by each particular component. Finally, by multiplying the original data matrix by the components matrix, we obtain the *matrix of component scores* (Figure 6.9d). In this matrix the cases are once again the rows, while the components are the columns. Each cell contains a component score, and these scores indicate the value for each case on each component. In other words, after starting with values for a large number of variables for each case, we finish with values for a smaller number of components for each case.

This rather abstract framework will be more readily understood if we reexamine it within the specific context of social area analysis. In this situation the cases are census tracts within a city, while the variables are those used by Shevky and Bell. That is, occupation, education, fertility, women at home, single-family dwelling units, and the percentage of people in a specified minority group, such as blacks. Each cell in the data matrix contains a value for each census tract for each of these six variables.

Ideally, according to social area analysis, the resulting *correlation matrix* should look like the one represented in Figure 6.10a. If the crosses refer to relatively strong correlations and the circles to relatively weak correlations, it can be seen that three groups of variables emerge: first, education and occupation are highly intercorrelated; second, women at home and single-family dwelling units are highly intercorrelated; and, third, the percent black variable is relatively independent of the other five.

These three groups of variables will then form three distinct components in the *components matrix* (Figure 6.10b), where the crosses represent relatively high component loadings,

	(1)	(2)	(3)	(4)	(5)	(6)
Occupation (1)	—					
Education (2)	X	—				
Fertility (3)	O	O	—			
Women at home (4)	O	O	X	—		
SFDU (5)	O	O	X	X	—	
Percent black (6)	O	O	O	O	O	—

(a)

	Economic status	Family status	Ethnic status
Occupation	X	O	O
Education	X	O	O
Fertility	O	X	O
Women at home	O	X	O
SFDU	O	X	O
Percent black	O	O	X

(b)

Figure 6.10 Idealized (a) correlation and (b) components matrices associated with social area analysis.

while the circles represent relatively low component loadings. Occupation and education both have high loadings on the first component, so this is labeled "economic status." Fertility, women at home, and single-family dwelling units all have high loadings on the second component, so this is labeled "family status," while percent black comprises the "ethnic status" component. The final matrix, containing the component scores, simply indicates the value for each census tract for each of these three components.

Before testing the empirical validity of social area analysis by examining whether these three components really do emerge from a principal components analysis of particular cities, it is helpful to visualize the statistical procedure in geometric terms. First, imagine that we have two census tracts and six variables (Figure 6.11), although in reality we would need many more census tracts for this type of analysis. Both census tracts have above average values on the first three variables, while census tract 1 is below average and census tract 2 is above average on the remaining three variables.

If lines are drawn from the origin to each of the six points in this two-dimensional census tract space, we obtain six vectors, one for each variable (Figure 6.11b). The cosine of the

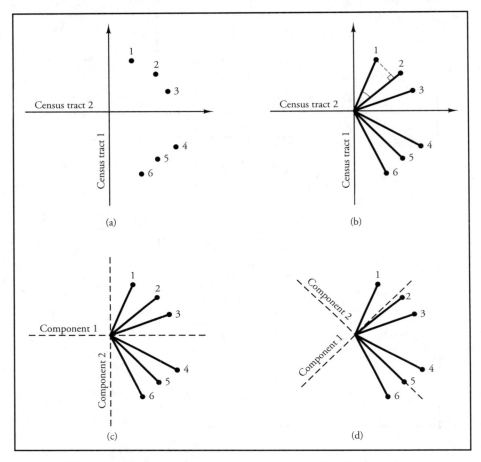

Figure 6.11 Geometric interpretation of principal components analysis: (a) location of variables in census tract space; (b) variables represented as vectors; (c) location of the principal axes; (d) location of the principal axes after rotation.

angle between each pair of vectors, or variables, denotes the *correlation coefficient* between those variables. An angle between 0 and 90 degrees indicates a positive correlation, with 0 representing a perfect linear relationship. Negative correlations are indicated by angles between 90 and 180 degrees. If the vectors are of unit length, the same information is represented by the projection of one vector on another. The longer the projection is, the greater the correlation coefficient. In other words, the closer together the vectors in this two-dimensional space, the more strongly are the variables related.

Having discussed correlation coefficients in geometric terms, we can now consider a geometrical representation of the *component loadings* (Figure 6.11c). The first component is placed through the six vectors, or variables, in such a fashion that it maximizes the sum of the squared component loadings. The component loadings are analogous to the correlation coefficients in Figure 6.11b, but in this case each vector is projected onto the components. In other words, variable 3 is closely related to the first component and thus has a relatively high component loading, while variable 1 has a relatively low component loading. If the loadings for all six variables are squared and summed, the location of the first component maximizes the sum. In technical terms, this sum of the squared component loadings is called an *eigenvalue*, and is used to determine the proportion of the total variation that is summarized by each component. The second component is introduced at right angles to the first, meaning that the two components are statistically independent of each other. The projections of each variable on this second axis provide the loadings for component two. Additional axes, or components, can be introduced at right angles to each other, although these axes cannot be represented in a two-dimensional diagram. In fact, as many axes can be fitted as there are variables, in this case six.

To interpret and label the different components, it is obviously advantageous to have some very high loadings, while the rest are very low, as in Figure 6.10b. The situation represented in Figure 6.11c is not ideal in this sense, as the loadings associated with the first component are neither exceptionally high nor exceptionally low. In other words, the component is not easily interpreted, or labeled, as a simple structure has not been achieved. This simple structure can often be obtained, however, by *rotating the axes* (Figure 6.11d). Now component 1 goes through the middle of the first three variables, while component 2 goes through the middle of the other three variables. As a result, both components have three high loadings and three low loadings, and thus each component summarizes three variables. If the axes, or components, are rotated while maintaining a right angle between them, the rotation is termed an orthogonal rotation. If, on the other hand, a right angle is not maintained, we have an oblique rotation in which the components are no longer statistically independent of each other.

Empirical Tests

A major test of the empirical validity of the three constructs hypothesized by Shevky and Bell has been carried out by Van Arsdol et al. (1958). Using 1950 census tract data, they undertook factor analyses of several large U.S. cities. We will concentrate on the results they obtained for *Minneapolis* and *Seattle*. The factor loadings for Minneapolis and Seattle, which are analogous to the previously described component loadings, are provided in Table 6.3. As expected, three main factors emerge for both cities. The first factor represents economic status, labeled "social rank" by Van Arsdol et al., and is composed of occupation and education, as the highest factor loadings for these two variables are in the social rank column. The second factor represents family status, labeled "urbanization" by Van Arsdol et al., and is composed of fertility, women in the labor force, and single-family dwelling units. The third factor represents ethnic status, here labeled "segregation," and is made up of percent black. Note that ordinarily one would expect women in the labor force to have a negative factor loading due to its inverse correlation with fertility and single-family dwelling units. In this particular instance, however, the values for the latter two variables have been subtracted from 1000.

Oblique rotations were used to obtain these factor loadings, so the factors are not necessarily orthogonal, or statistically independent of each other. The correlations between the three possible pairs of factors are reported in Table 6.4, where SR denotes social rank, or economic status; U denotes urbanization, or family status; and Seg. denotes segregation, or eth-

Table 6.3 Factor Matrices for Minneapolis and Seattle

	Social Rank	Urbanization	Segregation
Minneapolis			
Occupation	0.762	0.112	0.121
Education	0.670	−0.094	−0.026
Fertility	0.152	0.698	−0.205
WLF	−0.050	0.913	−0.061
SFDU	−0.084	0.757	0.224
Black	0.103	−0.050	0.423
Seattle			
Occupation	0.848	0.110	0.213
Education	0.631	−0.181	−0.257
Fertility	0.030	0.859	−0.159
WLF	0.040	0.907	−0.085
SFDU	−0.132	0.743	0.162
Black	−0.039	−0.080	0.592

Source: M. D. Van Arsdol, Jr., S. F. Camilleri, and C. F. Schmid, "The generality of urban social area indexes," *American Sociological Review*, 23 (1958), Table 2, p. 281.
Note: Underscores indicate highest factor loading for each variable. WLF, women in the labor force; SFDU, single-family dwelling units.

Table 6.4 Correlations between the Three Factors for Minneapolis and Seattle

	SR-U	SR-Seg.	U-Seg.
Minneapolis	0.002	−0.675	0.047
Seattle	0.122	−0.454	0.237

Source: M. D. Van Arsdol, Jr., S. F. Camilleri, and C. F. Schmid, "The generality of urban social area indexes," *American Sociological Review*, 23 (1958), Table 5, p. 283.
Note: SR, social rank; U, urbanization; Seg., segregation.

nic status. For both cities the correlations between social rank and urbanization and between urbanization and segregation are comparatively small. There are, however, fairly strong relationships between the social rank and segregation factors. More specifically, these relationships are negative, indicating that those census tracts with high scores for social rank tend to have low scores for segregation, and vice versa.

In summary, the factor analyses for Minneapolis and Seattle suggest that the three constructs derived from social area analysis are indeed empirically valid. This is not to say, however, that all cities exhibit exactly the same social structure. In fact, some of the other cities investigated by Van Arsdol et al. have slightly different factors, although the modifications are only minor.

6.5   Factorial Ecology

The more recent factorial ecological approach to studying the residential structure of cities became popular during the 1960s (Davies, 1984). This approach differs from social area analysis via factor analysis in two important ways. First, a larger number of variables are used in the factor analysis in order to test the robustness of the three major underlying dimensions of residential differentiation. Second, factorial ecology also places much greater emphasis on the spatial patterns associated with those dimensions. This focus on spatial patterns necessitates an interest in the component scores, as well as the component loadings.

In this section we consider first the underlying dimensions of residential differentiation uncovered by factorial ecologies of Chicago, Baltimore, and Montreal. Then the spatial patterns associated with those dimensions, the ecological structures of four medium-sized U.S. cities, and various criticisms of the factorial ecological approach to understanding residential differentiation are considered.

A Factorial Ecology of Chicago

The geographer Rees (1970) has undertaken an exhaustive study of the factorial ecology of Chicago. We will focus on the results he obtained from a factor analysis of 12 variables across 1324 census tracts in the Chicago metropolitan area. These 12 variables include one measure of education, two measures of occupation, three measures of income, two measures of age, and one measure each of family size, race, housing age, and housing quality.

Three factors were extracted, and the highest factor loading for each variable has been underlined (Table 6.5). Based on the pattern of factor or component loadings, the three factors have been labeled "socioeconomic status," "stage in the life cycle," and "race and resources," and it is immediately apparent that these three factors are very similar to the social area constructs described previously. Socioeconomic status contains the education, income, employment, and housing variables. As one would expect, percent families with annual incomes under $3000, percent substandard housing, and percent unemployed workers all have negative loadings, as they are inversely related to socioeconomic status. Stage in the life cycle is composed of family size and age variables. Note that percent population over 65 has a negative loading, as this variable is inversely related to population per household and percent population under 18. Finally, the race and resources factor is dominated by percent population black, although it also has some fairly high positive loadings on low-income families, substandard housing, and unemployment.

The row labeled "explained variance" in Table 6.5 indicates the percentage of the total variation that is accounted for by each factor. Recall that the sum of the squared factor loadings in each column gives us the eigenvalue associated with each factor. Using this information, the percent explained variance is simply computed by dividing the respective eigenvalues

Table 6.5 Factor Structure of the Chicago Metropolitan Area

	Socioeco-nomic Status	Stage in the Life Cycle	Race and Resources	Commu-nality
Median school years completed	0.920	−0.011	−0.048	0.850
Percent white-collar workers	0.846	−0.220	−0.203	0.805
Percent families with incomes over $10,000	0.771	−0.096	−0.484	0.837
Median annual income	0.746	−0.059	−0.510	0.820
Percent housing built after 1950	0.697	0.434	−0.168	0.702
Percent families with incomes under $3000	−0.646	−0.167	0.597	0.802
Percent substandard housing	−0.627	−0.197	0.488	0.670
Percent unemployed workers	−0.618	0.035	0.566	0.705
Population per household	0.032	0.928	−0.045	0.864
Percent population under 18	−0.133	0.867	−0.064	0.733
Percent population over 65	−0.102	−0.847	−0.241	0.786
Percent population black	−0.277	0.172	0.876	0.848
Explained variance	37.3	22.3	19.3	

Source: P. H. Rees, "Concepts of social space: Toward an urban social geography," in B. J. L. Berry and F. E. Horton, eds., *Geographic Perspectives on Urban Systems*, Prentice Hall, Englewood Cliffs, N.J., 1970, Table 10.16, p. 356.
Note: Underscores indicate highest factor for each variable.

by the total number of variables and multiplying by 100. In this particular instance, the three factors together account for 78.9 percent of the original variation. In other words, the 12 variables can be conveniently summarized by just three factors, as only 20 percent of the original variation has been lost.

The communalities indicate from which particular variables this information has been lost. A communality represents the sum of the squared factor loadings for an individual variable, so the communality for median school years completed is 0.85. This value tells us that 85 percent of the original variation in this variable has been preserved by the three factors, with only 15 percent being lost. By contrast, over 30 percent of the variation in percent substandard housing has been lost by reducing 12 variables to three factors.

A Factorial Ecology of Baltimore

Knox (1987) has undertaken a factorial ecology of the city of Baltimore using census tract data for 1980. Twenty-one variables were used as input and the principal components analysis, with a varimax rotation, generated four major underlying dimensions, or components (Table 6.6). Together, these components accounted for 72.2 percent of the variation

Table 6.6 Baltimore City: Factor Structure in 1980

(A) Explanatory Power of Each Factor			
Factor	Percent Variance Explained	Cumulative (%)	Eigenvalue
I	32.5	32.5	6.8
II	18.2	50.7	3.8
III	14.4	65.1	3.0
IV	7.1	72.2	1.5

(B) Nature of the Factors		
Factor	Loadings	
I.	Underclass	Rented housing	0.88
		Extreme poverty	0.79
		Vacant dwellings	0.70
		Inadequate kitchens	0.68
		Single persons	0.68
		Unemployment	0.65
II.	Socioeconomic status	Two or more bathrooms	0.85
		Family income	0.84
		College degree	0.78
		Managers, administrators and professionals	0.75
III.	Youth/migrants	Migrants	0.88
		Age 19–30	0.88
		Sex ratio	0.84
		Spanish speaking	0.57
IV.	Black poverty	Blacks	0.82
		Italian origin	−0.68
		Single-parent families	0.55
		Poverty	0.54

Source: P. Knox, *Urban Social Geography: An Introduction*, 2d ed., Longman, New York, 1987, Table 4.5, p. 129.

in the initial data set. The first component was labeled "underclass" and is associated with variables like rented housing, extreme poverty, and unemployment. The second component represents the more traditional socioeconomic status dimension, with high loadings for income, education, and occupation. The third component is associated with those census tracts exhibiting high proportions of in-migrants, young people, and Spanish-speaking people, while the fourth component represents black households, single-parent families, and poverty.

A Factorial Ecology of Montreal

Similarly, Le Bourdais and Beaudry (1988) have factor analyzed 1971 and 1981 census tract data for metropolitan Montreal. In particular, the factor analyses involved 59 variables and 561 census tracts for 1971 and 60 variables and 654 census tracts for 1981. They identified six major factors for both time periods, which accounted for 80.2 percent of the original variation in 1971 and 78.0 percent of the original variation in 1981. Also, for both time periods, three of the six factors represented each of the three classical dimensions: family status, socioeconomic status, and ethnicity.

Of particular interest is the fact that the dimensions for 1981 are remarkably similar to those for 1971. In fact, a correlation analysis applied to the two matrices of factor loadings reveals that the composition of each factor is virtually equivalent for the two time periods. As can be seen in Table 6.7, the coefficients along the diagonal are all equal to or greater than 0.96. Thus, at least at the aggregate level, there is evidence of substantial temporal stability in the overall residential structure. This temporal stability is matched by a persistent similarity across different cities at one point in time, as demonstrated by Davies and Murdie's (1991) study of 24 Canadian metropolitan areas using 1981 census data.

Table 6.7 Montreal: Interdimension Correlations, 1971 and 1981

1971 Dimension		Family		Socioeconomic		Ethnic	
		1[a]	5	2	4	3	6
Family	1	0.998	0.045	−0.015	−0.019	−0.013	0.022
	5	0.037	−0.963	−0.097	−0.020	0.069	0.241
Socioeconomic	2	0.021	−0.109	0.993	−0.012	−0.016	−0.035
	4	0.019	−0.015	0.010	0.999	−0.026	0.031
Ethnic	3	0.011	0.070	0.024	0.026	0.997	0.004
	6	−0.032	0.235	0.060	−0.027	−0.022	0.970

Source: C. Le Bourdais and M. Beaudry, "The changing residential structure of Montreal 1971–81," *Canadian Geographer*, 32 (1988), Table 4, p. 105.
[a] The numbers refer to the order in which the factors emerge.

Spatial Patterns

The spatial patterns are generated by mapping the factor or component scores for each of the three factors or components. A variety of studies have shown that significant regularities occur in these patterns across a large number of cities, including Chicago. First, the socioeconomic status factor tends to form sectors (Figure 6.12a). The high-status sectors contain census tracts with high values for variables such as income and education, while the reverse is true of the low-status sectors. Second, the family status factor tends to be distributed according to zones rather than sectors (Figure 6.12b). The zones toward the center of the city are dominated by census tracts containing small families, often single-person families, where the head of households tends to be either very young or very old. As one moves out toward the peripheral zone, the factor scores for family status increase, indicating a greater preponderance of large families with middle-aged heads of household. Third, the ethnic status factor tends to form clusters (Figure 6.12c). Each cluster represents a group of census tracts that is dominated by a particular racial or ethnic group, such as blacks, Italians, or Chinese. Such areas are known as black ghettos, Little Italy, or Chinatown, and they often represent a microcosm of the whole city, with their own internal differentiation according to economic status and family status.

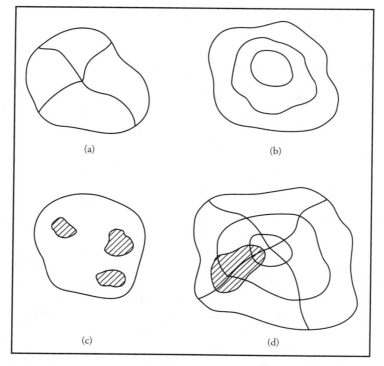

Figure 6.12 Idealized spatial patterns: (a) socioeconomic status; (b) family status; (c) ethnic status; (d) the urban mosaic.

Finally, each of these three spatial patterns can be superimposed to create the *urban mosaic*. Each segment or areal unit in Figure 6.12d represents a region that is characterized by particular kinds of households, and these internally homogenous social areas are differentiated according to economic status, family status, and ethnic status. Of course, social area boundaries are more convoluted than those depicted in Figure 6.12d, and it is helpful to think of the social areas as forming the individual pieces of a large-scale jigsaw puzzle.

From this spatial perspective it becomes clear that the classical models of Burgess and Hoyt are complementary rather than competitive. The spatial structure of cities is both zonal and sectoral, with family status being primarily zonal and economic status primarily sectoral. Analysis of variance is the statistical technique most often used empirically to validate this last statement, as it allows one to compare, for example, the variation between the zones with the variation within those zones. If the variation between the zones is significantly greater than the variation within the zones for a particular factor, we can conclude that the factor is distributed zonally. As we shall see in the next section, however, the spatial patterns can also be tested by multiple correlation and regression analysis using dummy variables.

In summary, factorial ecological studies have produced two major sets of conclusions. First, as predicted by social area analysis, residential differentiation can be summarized in terms of economic status, family status, and ethnicity. Second, these three dimensions of residential differentiation are spatially distributed in terms of sectors, zones, and clusters, respectively.

The Factorial Ecology of Medium-sized Cities

As most factorial ecological studies have focused on large cities, it is of interest to see if the results obtained at that level also hold for smaller cities. With this in mind, and as part of a larger study concerned with residential mobility, the present author (Cadwallader, 1981a) chose four U.S. cities: Canton (Ohio), Des Moines (Iowa), Knoxville (Tennessee), and Portland (Oregon). These particular cities were chosen because of their intermediate size, although Portland is somewhat larger than the other three, and because they are located in different parts of the country.

Census tract data from the 1970 U.S. Census were used, and the choice of variables was circumscribed by the desire to maintain comparability with the classic social area studies, although it was decided not to include the ethnic status variables, as their role in determining patterns of social variation is less important in smaller cities. Six variables were chosen to represent the two factors, socioeconomic status and family status. The socioeconomic status variables consisted of median income, median school years completed by persons 25 years or over, and the number of professional, technical, and kindred workers as a percentage of the total employed. The family status variables were persons per household, percentage females 16 years old and over in the labor force, and persons under 18 as a percentage of the total population.

The underlying dimensions produced by these six variables were identified, as in most factorial ecological studies, by the use of principal components analysis with an orthogonal rotation. The resulting two dimensions, or components, correspond closely to the economic status and family status dimensions found in previous studies (Table 6.8). The only variable

Table 6.8 Factor Matrices for Canton, Des Moines, Knoxville, and Portland

	SES	FS	Comm.	SES	FS	Comm.
		Canton			Des Moines	
Education	0.95	−0.08	0.91	0.96	−0.03	0.92
Income	0.94	0.07	0.89	0.91	0.34	0.94
Occupation	0.94	−0.07	0.89	0.88	−0.27	0.85
Persons per household	0.21	0.95	0.95	0.05	0.95	0.91
Percent under 18	−0.04	0.93	0.87	−0.01	0.95	0.90
Women in labor force	0.20	−0.74	0.59	0.03	−0.61	0.37
Percent total variance	46.0	38.8		42.2	39.5	
		Knoxville			Portland	
Education	0.94	−0.14	0.91	0.95	0.04	0.90
Income	0.91	0.21	0.86	0.90	0.33	0.92
Occupation	0.87	−0.09	0.76	0.91	−0.06	0.83
Persons per household	0.02	0.97	0.93	0.12	0.95	0.92
Percent under 18	−0.17	0.95	0.93	−0.05	0.95	0.91
Women in labor force	−0.03	−0.29	0.09	−0.10	−0.58	0.35
Percent total variance	41.7	33.3		42.8	37.6	

Source: M. T. Cadwallader, "A unified model of urban housing patterns, social patterns, and residential mobility," *Urban Geography*, 2 (1981), Table 1, p. 122.
Note: SES, socioeconomic status; FS, family status; Comm., communality.

that is inadequately represented in this structure is the percentage of females 16 years old and over in the labor force, which has communalities ranging from 0.59 in the case of Canton to only 0.09 in the case of Knoxville. Despite these low communalities, however, the variable behaves as expected in that it is always negatively associated with the family status component. Its declining importance as an indicator of social differentiation, since the pioneering studies of the 1950s, is undoubtedly due to the increased labor force participation of females from all levels of the socioeconomic hierarchy.

To verify the apparent similarity among the components obtained for the four cities, *coefficients of congruence* were calculated for each pair of cities. The coefficient of congruence is a measure of similarity between component loadings, which ranges in value from +1 for perfect agreement, through 0 for no agreement, to -1 for perfect inverse agreement. In the present instance the coefficients for both socioeconomic status and family status are all extremely high, indicating that the components are almost identical for all four cities (Table 6.9). For example, the coefficients of congruence for Canton and Des Moines are 0.99 in the case of the socioeconomic status component and 0.97 in the case of the family status component.

Table 6.9 Coefficients of Congruence

	(1)	(2)	(3)	(4)
Canton (1)	—	0.97	0.95	0.98
Des Moines (2)	0.99	—	0.96	0.99
Knoxville (3)	0.98	0.99	—	0.97
Portland (4)	0.98	0.99	0.99	—

Source: M. T. Cadwallader, "A unified model of urban housing patterns, social patterns, and residential mobility," *Urban Geography*, 2 (1981), Table 2, p. 122.
Note: Coefficients for family status above the diagonal; coefficients for socioeconomic status below the diagonal.

Multiple correlation and regression analysis was used to test for the *zones* and *sectors* that are supposedly associated with family status and socioeconomic status, respectively. For this purpose, each city was partitioned into six sectors and six zones, and the extent to which these sectors or zones account for the spatial variation in either of the two dimensions, as measured by their component scores, was determined by calculating the multiple correlation coefficient for the following equation:

$$Y_i = b_1 X_{i1} + b_2 X_{i2} + b_3 X_{i3} + b_4 X_{i4} + b_5 X_{i5} \qquad (6.2)$$

where Y_i is the value for census tract i of the component being tested, and $X_{i1},..., X_{i5}$ are dummy variables denoting the zone or sector in which any census tract i is located. This procedure is slightly different from analysis of variance, which is normally used when searching for zonal or sectoral variation, and it is important to note that when using dummy variables there should be one term fewer than there are zones or sectors. Also, the multiple correlation coefficients derived in this manner should be made directly comparable by using exactly the same number of zones and sectors for each city.

In general, the multiple correlation coefficients indicate that the variation in both socioeconomic status and family status can be equally well accounted for by either zones or sectors (Table 6.10). For example, in the case of Canton, approximately 45 percent of the variation in the socioeconomic status dimension is accounted for by the zones, and approximately 53 per-

Table 6.10 Multiple Correlation Coefficients for the Zones and Sectors

	Zones	Sectors	Zones	Sectors
	Canton		Des Moines	
Socioeconomic status	0.669	0.730	0.629	0.581
Family status	0.537	0.577	0.522	0.705
	Knoxville		Portland	
Socioeconomic status	0.605	0.536	0.214	0.734
Family status	0.556	0.292	0.737	0.230

cent is accounted for by the sectors. An important exception to this rule, however, is Portland, where, as postulated by other factorial ecological studies, the variation in socioeconomic status is primarily sectoral, while the variation in family status is primarily zonal. This finding suggests that the degree of spatial sorting associated with residential structure might be related to spatial scale, with larger cities being likely to exhibit higher degrees of spatial sorting than their smaller counterparts.

To verify further these results concerning the spatial patterns, the same regression equations were calibrated using the original variables, rather than the components, as the dependent variables. This second test was performed as a response to the argument that component scores might be misleading as indicators of spatial pattern in that their values depend, to a certain extent at least, on variables that are associated only slightly with the particular component under investigation. This problem is especially apparent when there are a large number of variables involved in the study, in which case the minor variables, in concert, can have a major influence on the component scores. The multiple correlation coefficients for the variables used here, however, indicated conclusions similar to those arrived at by using the component score themselves.

Criticisms

Criticisms of the factorial ecological approach to understanding residential differentiation within cities can be grouped into three main categories. First, it is unclear to what extent the results are dependent on the particular *research design* employed. It is possible, for example, that the underlying factors and their associated spatial patterns would be somewhat different if blocks were used rather than census tracts. Similarly, the results tend to vary according to the type of factor analysis performed and the different kinds of orthogonal and oblique rotations. Even within a given type of factor analysis, various problems are associated with the interpretation and labeling of those factors.

Second, it can be argued that factorial ecology is a purely *descriptive* form of analysis, as it fails to identify the processes that result in the urban mosaic. Attempts to address these processes generally involve investigating residential mobility. At the aggregate or macro level, the classical ecologists invoked such concepts as invasion and succession to help understand residential mobility (Section 6.2). At the individual or micro level, on the other hand, attempts have been made to model residential mobility within the framework of individual choice behavior (Chapter 11). In addition, the operation of the urban housing market has been scrutinized, with particular emphasis being given to the role of intermediaries in the housing market, such as real estate agents and financial institutions (Chapter 4).

Finally, the social areas identified by factorial ecology do not always constitute cohesive *communities*. Using the terminology discussed in Chapter 2, these social areas are uniform regions, but not necessarily functional regions. That is, they are relatively homogeneous with respect to certain specified variables, such as income, education, and family size, but they might not be characterized by a high degree of internal interaction. Attempts to measure the amount of internal interaction have focused on activity patterns associated with the workplace, friends, and clubs (Everitt, 1976), and social areas are not always related to behavioral and attitudinal characteristics.

Despite such criticisms, however, there is no doubt that these socially uniform residential neighborhoods play a distinctive role in the creation of urban society (Scott, 1980:124–7). First, by providing an environment that reinforces the ideological orientations of the dominant neighborhood group, they help to socialize the children in mutually acceptable ways. Second, they tend to sustain cultural homogeneity, as exemplified by certain ethnic neighborhoods. Third, they symbolize the social status of their inhabitants.

Chapter 7

Urban Industrial Structure

7.1   Principles of Industrial Location

When deciding where to locate their factories, industrialists consider a variety of factors (Chapman and Walker, 1991; Webber, 1984). We first examine these factors in general and then assess how they influence the spatial distribution of industrial activity within cities, commenting especially on the suburbanization of industry and the associated phenomenon of planned industrial parks. The economic base concept is then introduced as a technique for describing an individual city's economy and also for explaining urban growth. Finally, input–output analysis is discussed to provide a more detailed accounting of the flows within an economic system, and to derive the multiplier effect for individual sectors of the economy.

Attributes of the Raw Materials

If the raw materials used by a particular industry lose a great deal of *weight* during the manufacturing process, that industry tends to be located close to its raw materials, as it is cheaper to transport the finished product. The amount of weight loss is measured by the material index, which is calculated by dividing the weight of the nonubiquitous raw materials by the weight of the finished product. If the index is greater than 1, it is best to locate near the raw materials. The iron and steel industry is a good example of an industry that has traditionally been raw material oriented, due to the weight loss involved in the production of pig iron.

Perishable raw materials also encourage the associated industry to be raw material oriented in terms of location. For example, the fruit canning and preserving industry is located where the fruit is grown, in Florida and California, while the milk products industry tends to be located in those states where dairy farming is important, such as Wisconsin. In general, however, the improvement of refrigeration techniques has made the perishability factor increasingly less significant.

The *value per unit weight* of the raw materials is also an important determinant of industrial location. A raw material that has a comparatively high value per unit weight can bear the cost of transportation much better than can a raw material that has a comparatively low value per unit weight. As a result, industries that are dependent on raw materials with low values per unit weight tend to locate near those raw materials. Finally, the *substitutability* and *number* of raw materials must be considered. The locational pull of any particular raw material is less in those situations where the industry can substitute one raw material for another. Similarly, the more raw materials an industry uses, the less the locational influence of any one of those raw materials.

Attributes of the Finished Product

Some products undergo a *weight* gain during the manufacturing process, as reflected by a material index of less than 1. The soft-drink industry provides an excellent example of this phenomenon, as the flavors are transported in liquid or syrup form to local bottlers, who add water and bottle. As the weight of the final product is largely composed of the water and bottle, transportation costs are minimized by locating near the products' market. Consequently, the soft-drink industry tends to be concentrated in large cities. A second consideration, in terms of the attributes of the finished product, is the perishability of that product. A *perishable* product tends to make an industry market oriented. The bakery industry is a good example here, as it requires the prompt delivery of fresh goods.

The *value per unit weight* is a third important attribute of the finished product. Those industries with high-valued products per unit weight, such as watches, are more able to absorb transportation costs and are less tied to a market location. On the other hand, those with low-valued products per unit weight, such as cement, are more sensitive to transportation costs and tend to be market oriented in terms of location. Finally, a *bulky* finished product also encourages an industry to be market oriented. For example, the furniture industry tends to be located near the major markets because of the transportation costs associated with moving such bulky items as tables and sofas. Automobile assembly plants provide a similar example. Although many of the parts are produced in Detroit, the automobiles themselves are assembled in plants that are dispersed throughout the United States.

Transportation Costs

Transportation costs are perhaps the single most important determinant of plant location. In general, transportation costs increase with increasing *distance*, although in a curvilinear fashion (Figure 7.1a). The cost per mile tends to decrease with increasing distance because the same fixed costs, associated primarily with loading and unloading, have to be paid regardless of distance. More specifically, freight costs are often stepped, with the steps being superimposed on the general curvilinear relationship (Figure 7.1b). These steps indicate that the same rate is charged for a given range of distances, thus obviating the necessity to compute a separate rate for each individual journey. If there is a break of bulk point, or discontinuity, the

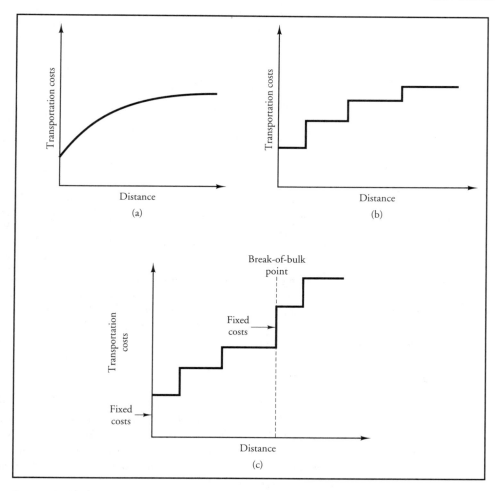

Figure 7.1 Relationship between transportation costs and distance: (a) general relationship; (b) stepped transportation costs; (c) discontinuity in transportation costs.

freight rate will increase at that distance, before beginning to taper off again (Figure 7.1c). Such break of bulk points occur, for example, when transferring from land to water, thus creating a new set of fixed costs.

Transportation costs are also influenced by factors other than distance (Healey and Ilberry, 1990). First, the form of the relationship between transportation costs and distance tends to vary according to the *mode of transportation* (Figure 7.2), with road transportation being cheapest for short distances, rail being cheapest for intermediate distances, and water being cheapest for long distances. Second, transportation costs will vary according to the *size of shipment*, because large shipments permit economies in terms of both administrative and terminal costs. Third, the *perishability* of the product makes a difference, with such goods as

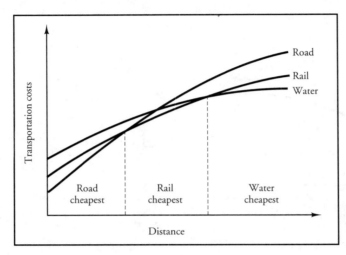

Figure 7.2 Variation in transportation costs for different modes of transportation.

fruits and vegetables generating higher charges because of the special handling involved with refrigeration or rapid transportation. Fourth, it is usually cheaper to transport *lower-valued products* because of the special packing and insurance costs associated with high-valued products. Finally, transportation rates will be cheaper in those situations where there is active *competition* among the various modes of transportation.

The Factors of Production

The factors of production— land, labor, and capital—are the three essential elements of the manufacturing process once all the raw materials have been assembled. These factors are combined in a variety of ways by different firms and industries, creating labor-intensive industries, where labor is a major input; capital-intensive industries, where capital is a major input; and so on. In recent years, there has been a move toward more capital-intensive industries, which are highly mechanized and require relatively little labor.

All industries need a certain amount of *land* on which to locate their factories (Watts, 1987:106–11). Some firms, however, have special requirements concerning the physical attributes of the site. Many firms, for example, use large volumes of water in the manufacturing process or require rivers or lakes into which effluence can be deposited. Other firms might require an extensive area underlain by solid bedrock to support special equipment. As discussed in Chapter 3, the cost of land in cities varies systematically. In general, land values decline with increasing distance from the city center, although ridges of higher-valued land also extend along the major traffic arteries, and certain sites might be unavailable because of zoning ordinances prohibiting industrial activity in that part of the city. Overall, however, the cost of land is probably less important in the long run, when considering potential locations, than are labor and capital costs. The cost of land is a major expenditure when starting a factory, but it becomes far less significant when costed over a long period or when rent is computed as a proportion of the total production costs.

Firms consider two major characteristics of the *labor* supply when deciding where to locate (Storper and Walker, 1983, 1984). First, the quality of labor, in terms of certain specialized skills, might be important. Some industries, such as the aerospace industry, need a large pool of highly skilled technical labor and so tend to be attracted to regions like southern California. The other main consideration is the cost of that labor. This factor is especially important for those industries that are comparatively labor intensive, such as the clothing industry. Berry et al. (1976:149) have identified four major kinds of regions in which labor costs tend to be low: first, where the supply of labor is increasing at a faster rate than the demand for labor; second, where economic opportunities are decreasing relative to a large local labor force; third, where employment opportunities are available only for a subset of the population, such as males; and, fourth, where the cost of living is exceptionally low so that real wages are high relative to money wages. However, although these regional differences in labor costs are undoubtedly important in the short run, in the long run such differences tend to be equalized by labor migration to high-wage areas. Also, the spread of unionism has tended to decrease the regional variation in labor costs.

Finally, *capital* costs can be divided into those associated with financial capital on the one hand and fixed capital equipment on the other. Financial capital refers to the money required to start up a firm, and although financial capital is generally mobile, it is easier to secure loans in rich areas such as southern California than in poor areas such as Appalachia (Walker and Storper, 1981; Gertler, 1984). Also, especially for small firms, loans are more easily obtained from local banks and credit institutions, where the entrepreneur is well known. Fixed capital equipment is much less mobile than financial capital and is thus similar to land as a factor of production, as buildings and machinery are comparatively fixed in space. This situation leads to industrial inertia, as it is usually cheaper to occupy existing industrial facilities than it is to construct a new factory. When a completely new factory is contemplated, however, industries consider the regional variation in construction costs. The costs of machinery and its repair are also subject to a certain amount of regional variation.

Economies of Scale

Production costs may also differ according to various economies of scale. These economies are usually divided into two types: those that are internal to the individual firm and those that are external to it. Internal economies of scale imply that, up to a certain point, the cost of producing a unit of output decreases with increasing plant size. Our major interest, however, is in the external economies of scale, as these are an important component of the overall location decision (Scott, 1983).

External economies of scale can be subdivided into urbanization economies and localization economies. *Urbanization economies* refer to the benefits that accrue to industries by locating in large cities, where there are already a large number of other industries. First, a large city can provide a varied labor force, with a whole range of specialized skills. Second, large cities tend to minimize transportation costs because they provide a local market and superior transportation facilities. Third, the larger the city, the greater the number of ancillary services that it can supply, such as fire protection, police protection, gas, electricity, and waste disposal. Often, however, some of these same economies of scale become diseconomies of scale if the city grows too large. Transportation facilities become overloaded and congested, pollution and crime become major problems, and competition for space increases the cost of land.

Localization economies occur when the firms in the same industry, or closely related industries, cluster in the same location. First, a pool of skilled labor becomes available to the industrial firms. Second, local services tend to adapt to the needs of a particular type of industry. Third, research facilities and marketing organizations can be shared by a group of similar industries. Fourth, similar industries can also share parts suppliers and machine repairers. Often, these localization economies result in cities that are dominated by a particular industry or group of industries.

Other Factors

Energy considerations also influence the location of industry. Most modern industries use some form of powered machinery, so fuel costs may contribute significantly to the overall production costs. There are large regional variations in the cost of fuel, with coal being comparatively cheaper in the North and East and oil being less expensive in the South and West. In addition, most firms are subject to some form of *taxes* on their revenue. These taxes may be paid to a local authority, such as the city or county, to the state, and to the federal government. As local taxes can vary considerably from place to place, industries consider this factor when choosing a suitable location (Charney, 1983). In particular, tax rates appear to be highest toward the city center and then decrease as one moves toward the suburbs.

The *government* itself can also profoundly influence the spatial distribution of industrial activity. First, direct expenditures by the federal government, such as defense spending, are not evenly distributed throughout the county. Second, government land use zoning ordinances prohibit industrial activity in certain areas. Third, the government uses financial inducements, such as low-interest loans, to attract industry to economically depressed regions such as Appalachia. Finally, *historical accidents* have influenced the location of industry. For example, some cities have certain industries simply because the founder of the industry happened to be born there. Also, the local invention of new machinery has often resulted in the growth of a particular industry in a particular city.

7.2   The Location of Industry within Cities

Having examined the factors that influence industrial location in general, we can now turn our attention to the locational pattern of industries within cities (Webber, 1982; Scott and Paul, 1991). In particular, we will discuss industrial types, the suburbanization of industry, the spatial mismatch hypothesis, and the development of industrial parks.

Industrial Types

In a classic study of industrial location within cities, Pred (1964) identified seven types of industry. First, *ubiquitous industries*, whose markets comprise the entire metropolitan area, tend to concentrate toward the edge of the central business district. Food processing, such as bread and cake plants, is illustrative of this type of industry, especially since the basic raw materials are nonlocal in origin. There is often a strong link with wholesaling activity, and the many warehouses and multistory buildings near the central business district are attractive to large-scale food manufacturers.

The second type of industry is the *centrally located communication economy industry,* in which face-to-face contact between buyer and seller is a distinct advantage and so a central location is highly desirable. Job printing is a good example of this kind of industry, because a downtown location maximizes accessibility to such consumers of printed material as lawyers, theaters, and advertising agencies. A second example is provided by the fashion clothing industry in New York's garment district. Note that both job printing and the garment industry tend to be characterized by small plants, so the high rents associated with central locations are not a major problem.

The third type of industry involves *local market industries with local raw material sources.* That is, those industries that use locally produced raw materials to manufacture goods that are also sold within the same urban area. Ice manufacturing plants are a good example of this type of industry, as are industries using raw materials that are by-products of local firms involved in iron and steel production. These industries are essentially randomly distributed throughout the city, although there is often some concentration toward the central business district.

Nonlocal market industries with *high-value products* comprise Pred's fourth type of industry. As transportation costs are relatively insignificant to these industries because of the high value per unit weight of the products, they are not constrained to locate near transportation arteries or railroad terminals. As a result, the spatial distribution of firms is generally random, although, again, there is sometimes a tendency to cluster in downtown locations. The manufacture of computer equipment is an excellent example of this kind of industry.

The fifth type of industry involves noncentrally located communication economy industries. These industries also have high-value products and serve national markets, but they realize communication economies by clustering together. Good examples of these industries are electronics and the aerospace industry, which locate together partly to keep abreast of the latest innovations. Unlike other industries to whom communication economies are important, however, these industries are not usually located downtown, as they are independent of the various kinds of services associated with the central business district.

The sixth type of industry is the *nonlocal market industry on the waterfront.* This industry is highly dependent on adequate transportation facilities and often involves foreign materials and markets. These industries are usually heavy industries, such as petroleum refining and ship building, whose raw materials and finished products are most easily moved via water transportation. Finally, there are those *industries oriented toward national markets.* These industries have bulky finished products, with relatively low value per unit weight, so they are strongly influenced by transportation considerations. Such industries include the iron and steel industry and motor vehicle assembly plants.

Northam (1979:413–17) has suggested an alternative classificatory schema that focuses on the identification of manufacturing zones within cities. He suggests that there are generally three major zones. The *central zone,* surrounding the central business district, is dominated by industries that have relatively low space needs per worker, but relatively high-value raw materials and finished products. In other words, land is used intensively in order to offset the high cost of land and taxes. Food products, the garment industry, and printing are the type of industries in this central zone. It is also in this zone, however, that the best examples of industrial blight are found. Many of the older firms and buildings are located here, often intermixed with nonindustrial land use, and there is a high incidence of dilapidation and abandonment. Congestion is also a problem, as most of the industries are labor intensive, but lack adequate parking facilities.

The next zone, or *intermediate zone*, as one moves away from the central business district, is characterized by industries engaged in the production of household appliances, automobiles, and various kinds of machinery. Such industries often require large amounts of storage space, extensive horizontal assembly lines, and considerable parking space for employees. Usually, there are a number of informal industrial districts within this zone, dominated by firms engaged in similar activities, as exemplified by the food-processing firms in the stockyards district of Chicago.

Beyond this intermediate zone is an *outer zone*, which also has a number of clustered industrial districts. The firms occupy large sites and are attracted by the lower land costs in suburban areas. Industries involved with national rather than local markets tend to predominate, and the industrial districts are usually located close to major traffic arteries, such as interstate highways. The number of industries in this zone has been steadily growing in recent years, and in the following section we consider some of the factors responsible for the suburbanization of industry.

The Suburbanization of Industry

Since World War II, industrial activity within U.S. cities has become increasingly decentralized (Erickson, 1983; Waddell and Shukla, 1993), and the factors that have contributed to this situation are often categorized into two groups (Scott, 1982). On the one hand, a series of factors has made the central city increasingly unattractive for industry. First, especially in older cities, many of the industrial plants have become obsolete, and new industries have constructed modern facilities in the suburbs rather than renovate old buildings in the city center. Second, planning restrictions on industry and urban renewal in central areas have often combined to reduce the amount of land available for industrial use. Third, traffic congestion has contributed to the decreasing attractiveness of the central city. Fourth, the high price of land in central cities, combined with the high tax rates on industry, have encouraged centrally located firms to vacate their present locations in order to capitalize site values. Fifth, the interaction of labor shortages, high levels of unionization, and comparatively high wages have helped detract from the attractiveness of central-city locations.

On the other hand, there are a variety of reasons why peripheral locations have become more attractive. First, the development of modern freeway systems has made the suburbs accessible for the movement of raw materials and finished products. Second, the development of horizontal plant layouts has drawn industry toward the comparatively cheap land to be found in the suburbs. Third, the decentralization of the labor force and the fact that the managerial staff in particular reside in the suburbs have added to the attractiveness of peripheral locations. Fourth, most major airports are located in the suburbs.

Such a listing of often disparate factors, however, does not necessarily lead to the development of broader, integrative theories that can address the process of industrial decentralization. An attempt to develop a more cohesive perspective can be conveniently labeled the incubation, product cycle, and hierarchical filtering theory. As the name implies, this theory has three major parts to it, involving the locational requirements of small firms entering the industrial system, the changing pattern as these firms grow and mature, and, finally, the filtering of firms down through an urban hierarchy.

Small, new firms tend to be attracted to the center of the city, where they can participate in economies of scale. In this sense the central city acts as an incubator for immature firms. Those firms that remain in business will eventually grow and expand and ultimately begin to generate their own internal economies of scale. It is at this stage that they will move to the suburbs to take advantage of the cheaper land and lower taxes on the periphery. The incubator thesis complements the idea of a product cycle in that, as the market for a new product begins to expand, the production process is standardized, firms grow larger, and peripheral locations become increasingly attractive (Taylor, 1986; Johnson, 1991). Finally, with respect to the hierarchical filtering process, the largest firms tend to establish branch plants, often using the cheaper and relatively unskilled labor found in smaller towns. So, as a particular industrial process develops and matures, it often leaves the largest metropolitan areas and is filtered down through the urban hierarchy.

Scott (1982) has sketched out an even broader theoretical perspective, emphasizing the essentially interdependent nature of locational choice and choice of production technique. He argues that, within the context of the capitalist commodity producing process, the locational patterns of urban industry can be divided into two main categories. First, labor-intensive firms tend to locate toward the center of the urban labor market. Second, capital-intensive firms are attracted by the relatively cheap land inputs at the peripheral locations (Blackley and Greytak, 1986). Historically, the manufacturing industry has increasingly substituted capital for labor, thus generating a more decentralized locational pattern.

Whatever the explanation, the various components of this *shift in industrial activity* have been well documented for Vancouver, Canada (Steed, 1973). Steed used the following accounting equation to measure the elements of industrial change:

$$X = b - d + m - e \tag{7.1}$$

where X is the net change in the number of industrial plants in any particular area of the city; b is the number of new plants, or births within the area; d is the number of plants closing down, or deaths within the area; m is the number of plants moving, or migrating into the area; and e is the number of plants leaving, or emigrating from the area.

This equation was calibrated for 13 subareas within Vancouver. Between 1955 and 1965, by far the greatest decrease in the number of industrial plants was experienced by the most central area, while the greatest increases were experienced in the more peripheral locations. Steed's equation indicates that the declining importance of the downtown area was due primarily to the out-migration of plants. On the other hand, the peripheral areas grew mainly because of the generation of new plants, or births, and only secondarily because of plant immigration.

The Spatial Mismatch Hypothesis

This suburbanization of industry has had a number of serious *consequences for central cities*. Perhaps the most important of these, as discussed in Chapter 2, has been the reduction in the central city's tax base. Also, however, as identified by De Vise (1976) for Chicago, there have been wide-ranging repercussions in terms of the spatial distribution of employment opportunities and the associated length of the journey to work. Between 1960 and 1970 Chicago lost 211,000 jobs, while its suburbs gained 548,000 jobs. Moreover, employment opportunities suburbanized twice as fast as the labor force. Blacks have been especially hurt by this shift of employment opportunities (Holzer, 1991). First, a disproportionate number of blacks are in the suburbanizing blue-collar occupations. Second, blacks often cannot afford suburban housing and are sometimes excluded on racial as well as economic grounds. The suburbanization of jobs has also increased the length of the journey to work. De Vise (1976) indicates that in the Chicago metropolitan area the average work trip was 5.5 miles in 1960, compared to 7 miles in 1970. Finally, as a corollary of the locational shift in jobs, there has been increased freeway congestion and a general underutilization of public transportation facilities.

Yeates and Garner (1980:371–5) have suggested three possible *policy responses* for alleviating the problems generated by the changing distribution of industry within cities. First, a conservative philosophy would simply let the problems be solved by forces operating in the marketplace. The vacated properties in the central city will be taken over by other types of land use that are more capable of earning a profit in that particular location. Meanwhile, the increased demand for labor in the suburbs will have a tendency to raise wages, while the labor surplus in the inner city will tend to reduce wages. In the long run, industry will return to the inner city to take advantage of these lower wages, thus increasing employment opportunities. In other words, it is contended that natural market forces will always act to remove any short-term disequilibrium in the distribution of industrial employment.

A second way to reduce industrial unemployment in the inner city would be to encourage the return of industrial plants to that part of the city, perhaps through the use of urban enterprise zones (Erickson, 1992). This approach exemplifies the liberal interventionist perspective, which argues that market forces are insufficient to redress the balance and that they should be augmented by various kinds of government programs. Such programs might include labor retraining, to upgrade certain specialized skills, and low-interest loans for firms willing to locate in the central city. More radical interventionists would advocate the state ownership of industry so that its location would be simply a matter of government policy.

A third major approach involves moving potential employees closer to the jobs rather than trying to move the jobs back to the employees. This approach also represents a liberal interventionist perspective, as it would require programs to help relocate the inner-city poor, who would ordinarily be unable to rent or own homes in the suburbs. These programs might include direct housing supplements for low-income families, mortgage schemes to encourage the construction of low-cost housing in the suburbs, and government construction of low-income apartment buildings in the suburbs.

Planned Industrial Parks

The increasing suburbanization of industry has been paralleled by the emergence of planned industrial parks, which are usually located toward the edge of the city. These industrial parks have been developed according to comprehensive plans that include the provision of streets, warehouses, sewers, utilities, and other ancillary services. The aim has been to generate a community of industries that are compatible and often functionally linked. As such, planned industrial parks are very different from informal concentrations of industrial activity, where such factors as the overall spatial organization of the area and the compatibility of individual industries are given little explicit consideration.

Although the first planned industrial parks were developed during the early part of the twentieth century, it was not until after World War II that the concept became really popular. During the period 1940–1959, over 1000 planned industrial districts were developed in the United States, whereas only 33 had been developed prior to that time (Buck, 1980). Most industrial parks in the United States have been developed by private corporations, such as railroads, industrial firms, and real estate brokers. Some industrial parks, however, were originated by nonprofit organizations, such as chambers of commerce and development commissions, while a few have been developed by local government, port, and airport authorities.

Public ownership is generally at the municipal level and reflects an effort to attract industry in those situations where private initiative is lacking. The local government receives direct benefits in terms of increased tax revenue and indirect benefits through increased employment, which enlarges the payroll of the community and thus leads to a greater demand for various goods and services. Public ownership can be a disadvantage if frequent administrative and political changes hamper the process of long-term development management, however, and often the most successful parks are those owned and managed by industrial developers. Such developers tend to retain ownership of many of the buildings, and the park is well maintained in order to attract tenants.

There are a variety of reasons why these planned industrial parks have become so popular. First, the lack of industrially zoned land in central cities has forced firms to pursue alternative locations. Second, the newer single-story plants, which facilitate horizontally organized production lines, require more space than is easily acquired in the central city. Third, the new industrial parks are generally free of traffic congestion and provide ample parking space for employees. Fourth, many industrial entrepreneurs appreciate the administrative help provided by the park management organizations. Fifth, as discussed previously, certain important economies of scale are derived when industries cluster together. Sixth, most industrial parks maximize accessibility considerations by being located near major highways, railroads, or airports. Seventh, services such as police and fire protection tend to be much more efficient in planned industrial parks.

Although all planned industrial parks have many characteristics in common, Hartshorn, (1992:430–6) has identified three major types of industrial parks. The first he calls *fabrication, distribution, and warehouse parks*. These parks are the most numerous, and they involve mainly light manufacturing and market-oriented industries. The individual firms are often very small, and although peripheral locations are the most popular, some of these parks have been developed in downtown areas as part of urban renewal projects. When the overall development becomes very large, hotel complexes, shopping centers, and even golf courses are included in these predominantly industrial facilities, and great care is taken to create a pleasing landscape.

Hartshorn's second category involves technology parks, or *research and science parks,* which are often located next to major universities. An excellent example of this type of park is the Research Triangle Park, which is located toward the center of a triangle formed by Duke University in Durham, North Carolina State University in Raleigh, and the University of North Carolina at Chapel Hill. These universities provide a pool of research consultants and also ancillary services such as computers, libraries, and research laboratories. Purely research parks are fairly rare, however; science parks, incorporating various kinds of light manufacturing, are more common. The electronics and aerospace industries are especially attracted to these parks, as their products involve large inputs of research and development.

The third and most recent kind of industrial park is the *business park.* These parks, although dominated by office buildings, often contain some light manufacturing, hotels, restaurants, and professional services, including doctors and lawyers. In fact, this variety of functionally compatible activities is perhaps the major characteristic of business parks, and even residential functions are sometimes successfully integrated into the overall development scheme.

7.3   The Economic Base Concept

The economic base concept has been a very popular technique for describing an individual city's economy and for explaining differential urban growth rates. In this section we examine the basic/nonbasic ratio, the economic multiplier, and a model of urban growth that is based on the economic multiplier.

The Basic/Nonbasic Ratio

The economic activity in any city can be divided into two sectors: basic and nonbasic (Stabler and St. Louis, 1990). *Basic activity,* or the basic sector, refers to those goods and services that are produced within a settlement but sold outside that settlement. In contrast, *nonbasic activity,* or the nonbasic sector, refers to goods and services that are produced within a settlement and sold within that settlement. The basic sector has been variously described as the export, surplus, or city-forming sector, while the nonbasic sector has been referred to as the local, residentiary, or city-serving sector. According to export or staple theory, the basic sector is the most important sector in terms of influencing urban growth, especially in the short run. Fluctuations in exports are of paramount importance to any urban area, as it is these exports that generate money for the community. As a result, forecasts of economic activity in cities are often based on the relationships between basic and nonbasic activity.

Symbolically, the total economic activity in a given city can be represented as follows:

$$TA = BA + NBA \qquad (7.2)$$

where *TA* is total activity, *BA* is basic activity, and *NBA* is nonbasic activity. The basic/nonbasic ratio is then simply expressed as the ratio of the number of employees in the basic sector to the number of employees in the nonbasic sector. If the labor force is equally divided between the two sectors, the basic/nonbasic ratio is 1:1. For every person employed in basic activity there is also one person employed in nonbasic activity. If, on the other hand, 25 percent of the labor force is in the basic sector, leaving 75 percent in the nonbasic sector, the ratio is 1:3. For every person involved in basic activity there are three people involved in nonbasic activity. In this situation, assuming that the ratio is stable, an increase of 20 new employees in basic activity will be matched by an increase of 60 new employees in nonbasic activity, resulting in an overall increase of 80 employees.

Obviously, there is some variation in the basic/nonbasic ratio across different cities, but this variation appears to be primarily related to size. Such a relationship is to be expected, as large cities perform many specialized functions for themselves that are not found in smaller cities. In addition to population size, two other factors help explain the variation in basic/nonbasic ratios. First, a city's location can make a difference. If a city is located next to a much larger metropolis, its nonbasic sector might be comparatively undeveloped, as many of the goods and services usually provided by that sector will be provided by the larger metropolis. On the other hand, a city of similar size that is relatively isolated has to be more self-sufficient and thus generates a larger nonbasic sector. Second, variation in the basic/nonbasic ratio is sometimes attributable to the population characteristics of cities. For example, cities with a small proportion of younger people will require fewer schools, thus diminishing nonbasic activity. Similar kinds of differences can be related to income variation.

The Economic Multiplier

Consideration of the basic/nonbasic ratio leads naturally to an examination of the economic multiplier (Frey, 1989; Mulligan and Kim, 1991). In essence, the economic multiplier relates growth in basic activity to growth in total activity. Consequently, given estimates of the future magnitude of basic or export activity, the economic multiplier allows us to forecast changes in total activity.

If the level of nonbasic activity is assumed to be a constant proportion of total activity, then

$$NBA = kTA \qquad (7.3)$$

where *k* represents the proportion of activity that is nonbasic and the remaining terms are the same as in equation (7.2). Equation (7.3) can be rewritten as follows:

$$TA = \frac{1}{k}NBA \qquad (7.4)$$

simply by dividing both sides by *k*. Therefore,

$$TA = \frac{1}{1-k}BA \qquad (7.5)$$

where the terms are the same as in equations (7.2) and (7.3). For example, if the proportion of activity in the nonbasic sector is three-fifths, the proportion in the basic sector must be two-fifths. Finally, equation (7.5) can be rewritten as follows:

$$TA = mBA \qquad (7.6)$$

where m is the economic multiplier relating total activity to basic activity.

In summary, the economic structure of any city can be characterized by three terms, all derived from simply dividing the economy into two sectors, as represented by equation (7.2). First, the basic/nonbasic ratio relates basic activity to nonbasic activity. Second, the constant k relates nonbasic activity to total activity. Third, the economic multiplier, m, relates basic activity to total activity.

An arithmetic example will help to clarify these relationships. If we return to our previous example of an increase in 20 employees in basic activity being matched by 60 new employees in nonbasic activity, for an overall increase of 80 new employees in total activity, the following values are generated. First, the basic/nonbasic ratio is 20:60, or 1:3. Second, rewriting equation (7.3) to solve for k, we get

$$k = \frac{NBA}{TA} \qquad (7.7)$$

which in this example is 60 divided by 80, or three-fourths. Third, rewriting equation (7.6) to solve for m, we get

$$m = \frac{TA}{BA} \qquad (7.8)$$

which in this example is 80 divided by 20, or an economic multiplier of 4. In other words, each new job in the basic sector generates four new jobs overall.

Despite its computational simplicity, however, the economic base concept suffers from a number of *methodological and philosophical problems*. First, there is the problem of choosing an appropriate unit of measurement to determine the relative magnitudes of the basic and nonbasic sectors. Employment figures are usually used, both because these data are comparatively easy to obtain and because the multiplier is often used to predict population changes. A more sensitive measure, however, might be wage rates, as changes in nonbasic employment are not always simply a function of changes in basic employment. For example, just a salary increase in basic activity leads to increased demand for nonbasic activity, thus increasing jobs in the nonbasic sector.

Second, it is often difficult to decide whether an establishment's function is basic or nonbasic. In many cases part of a firm's product is exported outside the city, while the rest is sold locally. In such situations it is necessary to determine the proportion of the firm's labor force that is associated with its basic activity and the proportion that is associated with its nonbasic activity. Similar problems arise with respect to service functions. For example, in the context of hotels, some guests come from outside the city and are therefore part of the basic sector, whereas other guests come from within the city and so contribute to the nonbasic sector.

Third, the values obtained for the basic/nonbasic ratio depend partly on how the spatial extent of the city is defined. The size of the base area determines what activities are treated as basic, because it determines which transactions involve exports. All other things being equal, the larger the base area is, the greater the number of activities that are nonbasic. For example, if only the central city is included in the base area, goods and services sold to the suburbs are regarded as exports. If, however, the suburbs are included in the base area, those same goods and services now represent nonbasic activity.

Fourth, in a more philosophical context, it is not clear that the basic or export sector is the only determinant of growth in a city. It has sometimes been argued that the nonbasic sector is equally important in determining urban growth. For example, in the southwest United States much of the recent urban growth appears to have been generated by the increasing importance of service activity. Finally, it should be noted that values for the economic multiplier are obtained by averaging over the whole city. In reality, however, the economic multiplier varies from industry to industry, reflecting the fact that some industries have greater growth-generating potential than do others. A more detailed analysis of the urban economy, at the level of individual industries, is provided by input–output analysis, which will be discussed shortly. First, however, we will consider a more detailed conceptualization of how the economic multiplier, or multiplier effect, is actually related to urban economic growth.

Urban Economic Growth

The relationship between the multiplier and urban economic growth has been discussed at some length by Pred (1977:88–93). His conceptual framework involved a simple causal chain (Figure 7.3). Imagine an isolated city in which a new industry is located or an existing industry is enlarged. This new industry will initiate a chain reaction effect. Purchasing power in the city will rise, due to the increased number of jobs, thus increasing the local demand for goods and services, such as housing, shops, restaurants, and banks. This increased demand will in turn create more jobs in the construction industry, the retailing industry, and so on. Thus, for every new job in the factory, a number of other jobs are created in the local economy. Exactly how many more jobs are created is measured by the economic multiplier. Finally, the city eventually reaches a new threshold population and is large enough to support another new industry, thus initiating another round of growth.

The different effects that the introduction of a new industry will have on a city's economy are summarized by the following equation:

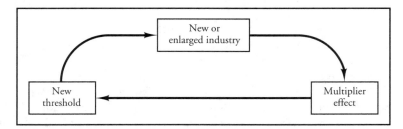

Figure 7.3 Economic multiplier and urban growth. (Based on Allan R. Pred, "Industrialization, initial advantage, and American metropolitan growth," *Geographical Review,* 55, 1965, Fig. 1, p. 165.)

$$M_i = E_i + \sum_{j=1}^{N} E_{ij} + \sum_{k=1}^{M} E_{ik} \tag{7.9}$$

where M_i is the total employment generated by the new industry i; E_i is the employment *directly* created by industry i; E_{ij} is the employment *indirectly* created by the demands of industry i on other industries ($j = 1$ through N); and E_{ik} is the *induced* employment created by the impact of industry i on other sectors of the economy ($k = 1$ through M). In this schema, the employment directly created by the new industry refers to the new employees actually working in that industry. The employment indirectly created by the new industry refers to the jobs that are created by industry i demanding certain inputs from other industries. These inputs are summed over the N other industries involved. Finally, the induced employment created by the new industry refers to the jobs that are generated by the increased local demand for housing, banks, doctors, and so on.

This scenario assumes that there are no major leakages in terms of the multiplier effect. Such leakages can occur in two ways. First, the new industry might buy some of its materials from outside the local area, thus reducing its indirect effect. Second, the new employees of that industry might travel outside the local area to buy certain goods and services, thus reducing the induced effect.

This particular chain reaction process described by Pred is one example of a more general process called the principle of *circular and cumulative causation*, whereby any change in a social system tends to set up forces that both increase and reinforce that change. In other words, within the urban context, growth tends to breed more growth. The process of circular and cumulative causation can also work in reverse, however, with a contraction in industrial activity leading to further contraction via the same multiplier effect. In this case, when an industry leaves a city, the labor force and overall purchasing power of that city decline accordingly. As a result, there is less demand for local goods and services, thus decreasing still further the number of jobs in the city. Eventually, the urban population will fall below a certain threshold size, and another industry will leave.

It is of interest, within this context, to examine why for some cities the principle of circular and cumulative causation leads to increasing growth, whereas for other cities it results in decline. Thompson (1965:22) investigated the growth rates of different-sized cities and suggested the concept of an *urban size ratchet*. By this concept he meant that once, a city reaches a certain size, a kind of ratchet mechanism comes into effect, locking in past growth and preventing future contraction.

There are a number of reasons why such a ratchet effect might operate. First, as cities grow, their industrial structure tends to become more diversified. This notion of industrial diversification can be visualized in terms of Lorenz curves (Figure 7.4). The various industries in a particular city are placed along the horizontal axis, from largest to smallest in terms of the number of employees, while the vertical axis represents the cumulative percentage of the total labor force. Curve A indicates a completely diversified economy in which each industry has an equal share of the total labor force, whereas curve B indicates a city that has an overwhelming majority of its labor force employed in its largest industry. Those cities that are closer to the situation represented by curve A, usually the larger ones, tend to be less vulnerable to economic depressions than those represented by curve B. In the latter case, a decline in demand associ-

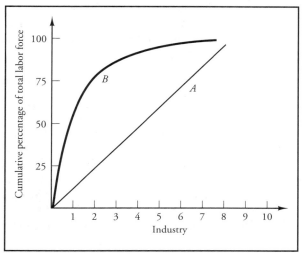

Figure 7.4 Lorenz curves showing different degrees of industrial diversification.

ated with the major industry could be devastating, as other industries would be unable to pick up the slack. Second, large cities can generate more political support, and therefore more federal and state assistance, than can their smaller counterparts. Third, large cities tend to generate more economies of scale, as discussed in Section 7.1. Fourth, as industry becomes increasingly consumer oriented, big cities provided immediate access to a large potential market.

Despite these apparent advantages, however, a number of large U.S. cities have recently begun to decline in population. This decline is due at least partly to the onset of diseconomies of scale (Burnell and Galster, 1992). Such diseconomies are represented hypothetically in Figure 7.5, which suggests how the cost per person for various kinds of services varies with city size. The cost curve for all services combined suggests that medium-sized cities are more effi-

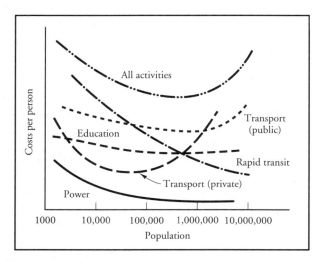

Figure 7.5 Economies and diseconomies of scale. (From *The Spatial Organization of Society* by Richard L. Morrill, © 1970 by Wadsworth Publishing Company, Inc., Belmont, California 94002. Reprinted by permission of the publisher.)

cient than either very small or very large cities, but there is remarkably little agreement with respect to what constitutes an optimal city size. One problem, as the individual cost curves indicate, is that diseconomies of scale emerge at different city sizes, depending on the particular service that is being considered.

7.4   Input–Output Analysis

Input–output analysis is a more detailed technique than the economic base concept for characterizing the economic structure of cities (Hewings, 1985; Thomas, 1990). Unlike the economic base concept, multiplier effects are not averaged for the whole city, but are computed for each individual sector of the economy. As a result, input–output analysis allows us to trace the effects of an increase or decrease in production in one sector of the economy on all the other sectors (Holland and Cooke, 1992). The input–output approach has become a major ingredient of economic planning in the industrially developed nations of the world.

Input–Output Coefficients

We will discuss input–output analysis using an example adapted from Abler et al. (1971). Suppose we have a simple urban economy that is composed of two major industries, automobile manufacturing and the iron and steel industry. The money flows in this two-industry economy, during a given time period, can be represented in terms of a flow matrix (Table 7.1). This flow matrix indicates that, of the $20 worth of total output produced by the automobile industry, $5 worth is purchased by the iron and steel industry, while $15 worth goes to final demand, or individual consumers. Similarly, of the $10 worth of total output produced by the iron and steel industry, $8 worth is purchased by the automobile industry, and $2 worth goes to consumers. The output for the city as a whole, during this time period, is $30.

As the name implies, the flow matrix simply summarizes the flow of money within an urban economy and so highlights the interconnection, or linkages, among the different sectors of that economy. The exact nature of these linkages is specified by the input–output coefficient (Szyrmer, 1992), which is measured as follows:

$$a_{ij} = \frac{x_{ij}}{X_j} \qquad (7.10)$$

where a_{ij} is the input–output coefficient expressing the input from industry i required to pro-

Table 7.1 Flow Matrix for Input–Output Analysis

Producing Industry	Purchasing Industry		Final Demand	Total Output
	Automobile	Iron and Steel		
Automobile	$0	$5	$15	$20
Iron and steel	8	0	2	10
			Total output for the economy = $30	

duce a unit of output in industry j; x_{ij} is the input, or amount of goods, flowing from industry i to industry j; and X_j is the total output of industry j.

The input–output coefficients for our particular example are represented in Table 7.2. First, we can consider the case where automobile manufacturing is the input, or producing sector, and the iron and steel industry is the output, or purchasing sector. In this situation, substituting in equation (7.10), we have

$$a_{AI} = \frac{x_{AI}}{X_I} \tag{7.11}$$

where A represents the automobile industry and I, the iron and steel industry. Using the values given in the flow matrix (Table 7.1), we get the following:

$$a_{AI} = \frac{\$5}{\$10} = \$0.5 \tag{7.12}$$

This input–output coefficient indicates that, to produce each dollar's worth of output, the iron and steel industry must purchase one half-dollar or 50 cents worth of input from the automobile industry.

Similarly, we can consider the case where the iron and steel industry is the input sector, and automobile manufacturing is the output sector. Again substituting in equation (7.10), we have

$$a_{IA} = \frac{x_{IA}}{X_A} \tag{7.13}$$

where the notation is the same as in equation (7.11). Using the values given in the flow matrix (Table 7.1), we get the following:

$$a_{IA} = \frac{\$8}{\$20} = \$0.4 \tag{7.14}$$

In other words, to produce each dollar's worth of output, the automobile industry must purchase 40 cents' worth, of input from the iron and steel industry.

Once the input–output coefficents have been obtained, the information contained in the original flow matrix can be specified by two simple equations, as follows:

$$X_A = 0.5X_I + 15 \tag{7.15}$$

and

$$X_I = 0.4X_A + 2 \tag{7.16}$$

Table 7.2 Input-Output Coefficients

Producing Industry (input)	Purchasing Industry (output)	
	Automobile	Iron and Steel
Automobile	—	$0.5
Iron and steel	$0.4	—

where the notation is the same as in equation (7.11). Equation (7.15) indicates that the total output of the automobile industry equals half the total output of the iron and steel industry, because for every dollar's worth of output by the iron and steel industry 50 cents' worth must be purchased from the automobile industry, plus $15 of final demand. Similarly, equation (7.16) indicates that the total output of the iron and steel industry equals four-tenths the total output of the automobile industry, plus the final demand of $2.

Equations (7.15) and (7.16) are simultaneous equations that can be solved to find the total output for each industry, as we have two equations and two unknowns, X_A and X_I. First, to find X_A, we can take equation (7.15) and multiply both sides by 2, giving

$$2X_A = X_I + 30 \tag{7.17}$$

Then, subtracting 30 from both sides, we have

$$X_I = 2X_A - 30 \tag{7.18}$$

Therefore, from equations (7.16) and (7.18),

$$0.4X_A + 2 = 2X_A - 30 \tag{7.19}$$

We now have one equation with one unknown and so can easily solve for X_A. If both sides of equation (7.19) are multiplied by 5, we obtain

$$2X_A + 10 = 10X_A - 150 \tag{7.20}$$

Then, subtracting $2X_A$ from both sides and adding 150 to both sides, we have

$$8X_A = 160 \tag{7.21}$$

Therefore,

$$X_A = 20 \tag{7.22}$$

Then, substituting the value for X_A in equation (7.16), we have

$$X_I = 0.4\,(20) + 2 \tag{7.23}$$

so

$$X_I = 10 \tag{7.24}$$

Note that these two values, $X_A = 20$ and $X_I = 10$, match the total outputs in the original flow matrix, so the two simultaneous equations, equations (7.15) and (7.16), are an accurate reflection of the flows, or interactions, within the economy.

Measuring the Impact of Change

One great advantage of calculating input–output coefficients and being able to represent the economy in terms of equations is that it becomes relatively easy to measure the overall impact of a change in any particular sector of that economy. For example, suppose that the final demand associated with the iron and steel industry increases from $2 to $6 so that we now have the following pair of simultaneous equations:

$$X_A = 0.5X_I + 15 \tag{7.25}$$

and

$$X_I = 0.4X_A + 6 \tag{7.26}$$

Equation (7.25) is exactly the same as the original equation (7.15), while equation (7.26) is the same as equation (7.16), except that the final demand has been changed from 2 to 6.

By solving equations (7.25) and (7.26) in the same way that we solved equations (7.15) and (7.16), we can measure the overall impact of changing the final demand for the iron and steel industry. As before, both sides of equation (7.25) are multiplied by 2, giving

$$2X_A = X_I + 30 \tag{7.27}$$

and 30 is subtracted from both sides to give

$$X_I = 2X_A - 30 \tag{7.28}$$

Therefore, from equations (7.26) and (7.28)

$$0.4X_A + 6 = 2X_A - 30 \tag{7.29}$$

If both sides of equation (7.29) are multiplied by 5, we obtain

$$2X_A + 30 = 10X_A - 150 \tag{7.30}$$

Then, subtracting $2X_A$ from both sides and adding 150 to both sides, we have

$$8X_A = 180 \tag{7.31}$$

Therefore,

$$X_A = 22.5 \tag{7.32}$$

Then, substituting the value for X_A in equation (7.26), we have

$$X_I = 0.4\,(22.5) + 6 \tag{7.33}$$

so

$$X_I = 15 \tag{7.34}$$

Thus the new output for the automobile industry is \$22.5, while that for the iron and steel industry is \$15.

By inspecting the revised flow matrix (Table 7.3) we can see that a fairly minor change in the final demand for the iron and steel industry has had widespread repercussions for the whole economy. Because the total output of the iron and steel industry has increased by \$5, another \$2.50 worth of input must be purchased from the automobile industry, thus giving the automobile industry a new total output of \$22.5. This same increase of \$2.50 in the total output of the automobile industry, however, means that another \$1 worth of input must be

Table 7.3 Revised Flow Matrix for Input–Output Analysis

Producing Industry	Purchasing Industry		Final Demand	Total Output
	Automobile	Iron and Steel		
Automobile	$0	$7.5	$15	$22.5
Iron and steel	9	0	6	15
			Total output for the economy = $37.5	

purchased from the iron and steel industry, as for each dollar's worth of output in the automobile industry 40 cents' worth of input has to be purchased from the iron and steel industry. So, because of the interrelationships between the two industries, as measured by the input–output coefficients, an increase of $4 in the final demand for the iron and steel industry has increased the total output of the overall economy from $30 to $37.5. In other words, the multiplier effect associated with the iron and steel industry is $7.5 divided by $4, or approximately 1.8.

We have only been considering a two-industry economy, but it is a fairly simple matter to add other industries. For example, the equations for a three-industry economy would be as follows:

$$X_1 = a_{11}X_1 + a_{12}X_2 + a_{13}X_3 + d_1 \tag{7.35}$$

$$X_2 = a_{21}X_1 + a_{22}X_2 + a_{23}X_3 + d_2 \tag{7.36}$$

$$X_3 = a_{31}X_1 + a_{32}X_2 + a_{33}X_3 + d_3 \tag{7.37}$$

where X is the total output for industries 1, 2, and 3; a is the input–output coefficient representing the interrelationship between each pair of industries; and d is the final demand for each industry. Four industries would generate four equations, five industries would generate five equations, and so on. Such equations can be used to measure the overall impact of a change in consumer demand associated with one or more of the individual industries. Also, the input–output coefficients can be expected to change over time due to technological innovations that allow alternative inputs to be substituted in the production process. For example, if the automobile industry begins to build cars out of wood rather than metal, the flow of inputs from the iron and steel industry to the automobile industry will obviously decrease, resulting in a lower input–output coefficient.

In summary, input–output analysis records a great deal of information about the economy in the form of flows between individual industries or sectors. Moreover, the calculation of input–output coefficients allows us to predict the impact of an increase or decrease in activity in any one industry on the economy as a whole. The higher the input–output coefficients, the greater the impact, while if all the input–output coefficients are zero, the individual industries are independent of each other. In the latter situation, the economic health of one sector of the economy would not influence the economic health of any other sector of that economy, as there would be no multiplier effects.

In practice, input–output analysis involves certain problems (Richardson, 1985; Dewhurst et al., 1991). First, it is often extremely difficult to obtain all the necessary flow data to undertake a very disaggregated examination of a city's economy. Second, the input–output coefficients are partly dependent on how the industries, or economic sectors, are classified in the first place. Nevertheless, despite these problems, input–output analysis provides great insight into the economic structure of urban areas, and is an extremely important tool of urban and regional planning.

Chapter 8

The City of the Mind

8.1   The Behavioral Environment

The *city of the mind* refers to the structure of the city as it is perceived by individual residents. People can call forth mental or cognitive maps of the cities in which they live, and these cognitive maps influence their behavior within those cities (Duncan, 1987). We begin our discussion by considering the concept of a behavioral environment, as this concept underlies the behavioral approach to urban geography. Spatial representations of the behavioral environment, as expressed in cognitive maps, are then considered, followed by a discussion of the geometric and appraisive information contained within such maps.

We can make a distinction between the phenomenal and the behavioral environments. The *phenomenal environment* refers to the objective, physical environment in which behavior takes place, whereas the *behavioral environment* is the subjective, psychological environment in which decisions are made that are then translated into overt action in the phenomenal environment. A fundamental axiom of the behavioral approach is that decision making has its roots in the behavioral environment rather than in the phenomenal environment. In other words, a person's behavior is based on his or her perception of the environment, not on the environment as it actually exists.

Social and physical facts of the phenomenal environment are incorporated into the behavioral environment, and two major processes are at work in this transformation of information: selectivity and distortion. The notion of *selectivity* indicates that each of us selects certain items of information to be incorporated into our behavioral environment, as it would be impossible to maintain complete information. Even the information selected, however, tends to be *distorted* in various ways (Lloyd and Heivly, 1987; Tversky, 1992), so the behavioral environment is both an incomplete and an inaccurate representation of its phenomenal counterpart. As we shall discuss later, variables such as socioeconomic status and length of residence help to explain the various kinds of selectivity and distortion that take place.

While the spatial structure of the phenomenal environment is portrayed in traditional maps, based on longitude and latitude, the behavioral environment is represented in terms of a cognitive or mental map. Similarly, physical or objective distance has as its counterpart cognitive or subjective distance (Coshall, 1985). Before discussing the concepts of cognitive maps and cognitive distance more fully, however, we consider first two studies that verify the previously mentioned axiom that spatial behavior can best be understood within the context of the behavioral environment.

In a study based on a sample of households located in West Los Angeles, the present author (Cadwallader, 1975) investigated the spatial rationality of *consumer behavior* in order to demonstrate that more consumers think they patronize the closest supermarket than actually do. In other words, the hypothesis was that consumers are more rational with respect to cognitive distance than they are with respect to physical distance. To test this hypothesis, the sampled group of consumers all lived within three blocks of each other and so possessed almost identical opportunity sets in terms of the available supermarkets. Each consumer was asked to estimate the distance from his or her home to each of five local supermarkets. Time estimates were used to provide a measure of cognitive time distance, and the method of direct magnitude estimation, which will be described later, was used to obtain a measure of cognitive mileage distance. The shortest road distances to the supermarkets varied between 0.43 miles and 1.75 miles.

The data were analyzed to determine the proportion of consumers who thought they were using the closest supermarket, as opposed to the proportion of consumers who were actually using the closest supermarket. To accomplish this comparison, the consumers were also asked to indicate which supermarkets they patronized most often. Note that they were not asked to identify which one they had used on their last shopping expedition, as that might have represented only a minor fluctuation in their overall shopping strategy. The results indicate that the hypothesis was well founded (Table 8.1). That is, for this particular group of consumers, more people thought they went to the nearest supermarket than actually did. The evidence suggests that the consumers are at least intendedly rational with respect to distance and that cognitive distance, or distance as it occurs in the behavioral environment, is a better predictor of consumer behavior than is physical distance, or distance as it occurs in the phenomenal environment.

Table 8.1 Spatial Rationality of Consumer Behavior

Patronizing Closest in Terms of:	Percent
Physical distance	53
Cognitive mileage distance	70
Cognitive time distance	72

Source: M. T. Cadwallader, "A behavioral model of consumer spatial decision making," *Economic Geography*, 51 (1975), Table 2, p. 342.

On a more general level, these results pose an interesting problem of interpretation. Consumers might choose between supermarkets partly on the basis of a priori distance estimates. Alternatively, the distance distortions might be attributable to some kind of cognitive dissonance process that occurs after a store has been chosen. In general, cognitive dissonance takes place after any decision in which a person has chosen between two fairly attractive alternatives. The person tends to reduce the dissonance by exaggerating the attractive features of the chosen alternative and the unattractive features of the rejected alternative. In the context of cognitive distance and consumer behavior, this argument would suggest that, after patronizing a particular supermarket, the consumer will begin to rationalize his or her decision and imagine that the chosen supermarket is relatively closer than the alternatives.

An examination of *pedestrian behavior* in the Monroe district of Philadelphia (Ley, 1974:219–26) also testifies to the importance of the behavioral environment as a means of understanding spatial behavior. A sample of residents were asked to specify what routes they would take when walking from their home at night to particular locations within Monroe. Only 27 percent of the responses represented a direct route between the origin and destination, while the largest group of responses indicated an indirect route, adding one-fourth or even one-half or more to the length of the journey.

To explain this apparently irrational behavior, Ley computed a stress surface, based on answers to the question, "Are there any blocks around here which have a bad reputation?" Such a stress surface can be represented by isolines joining points of equal perceived stress (Figure 8.1), and in the case of Monroe there were two notable peaks of higher-than-average stress, associated with the turfs of local gangs. When considered in the light of this perceived stress surface, the pedestrian behavior appeared to be much more rational. The area was traversed by following the valleys in the stress surface, just as a hiker might traverse a mountainous area by following the river valleys and avoiding the higher elevations. In other words, pedestrian behavior is governed by a kind of "invisible landscape," which is incorporated within the behavioral environment but is not a part of the phenomenal environment.

8.2   Cognitive Maps

Early in the present century, Trowbridge (1913), in a paper concerned with the perception of direction, used the term *imaginary map*. This paper specifically addressed the cognitive representation of large-scale environments and implied that people possess spatial images of those environments. Trowbridge investigated these spatial images by asking subjects to indicate the directions of such cities as New York, London, and San Francisco on a circular piece of paper. Although he did not indicate the scoring procedure, he concluded that more than half the subjects revealed adequate imaginary maps.

The term *cognitive map* was apparently coined by Tolman (1948). His experimental data consisted of observations on rat behavior, but he believed the results had significance for human behavior. In general, Tolman contended that during the learning process something akin to a field map of routes and paths becomes established in the rat's brain, and it is this map that ultimately determines the rat's behavior.

A cognitive map represents, therefore, an individual's model of objective reality (Golledge and Stimson, 1987:70). The use of the term map tends to imply that our internal representations of the physical world are in map form. This is not necessarily the case, how-

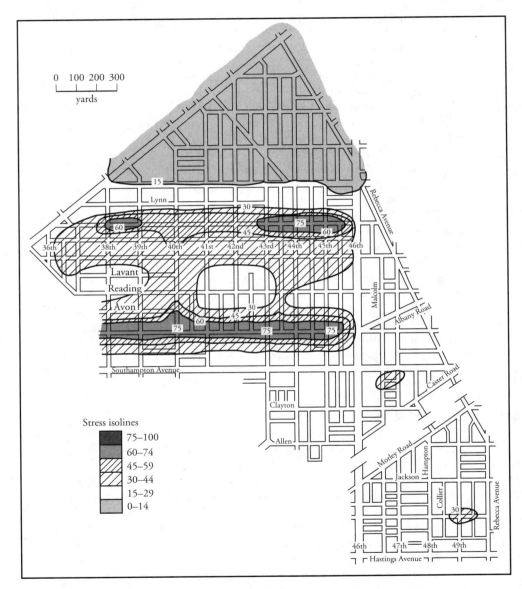

Figure 8.1 Monroe stress surface. (From D. Ley, *The Black Inner City as Frontier Outpost: Images and Behavior of a Philadelphia Neighborhood*, Monograph 7, Association of American Geographers, Washington, D.C., 1974, Fig. 36, p. 221.)

ever. In the present context, the term map is being used to indicate a functional analog, as we are interested in a cognitive representation that has the functions of a cartographic map, but not necessarily the physical properties of such a map (Downs, 1981). In this respect, *cognitive mapping* refers to the process by which information about the spatial environment is organized, stored, recalled, and manipulated, while a *cognitive map* is simply the product of this process at any particular point in time (Lloyd, 1989a).

The Image of the City

One of the first, and certainly most influential investigations of cognitive maps was undertaken by Lynch (1960). He selected a small sample of residents from three U.S. cities and analyzed their perception of the downtown area of those cities. A variety of methodologies were used, such as verbal lists of distinctive features and detailed descriptions of trips through the city, but the most important technique involved the use of sketch maps: subjects were asked to make a quick map of the central city, covering all the main features.

Five major features, or elements as Lynch called them, were abstracted from these maps (Lynch, 1960:47–8). First, *paths* represented linear features in the city along which movement occurred. These paths were generally major highways, and they served to connect the different elements on the map. Second, *edges* represented linear features that were not necessarily used for movement. Primarily, these edges were the boundaries between different parts of the landscape, closing one region off from another. Third, *districts* were distinctive areas of urban space that had certain unifying characteristics. Subjects could conceptualize being either inside or outside such districts. Fourth, *nodes* were strategic points or intensive foci within the city, usually associated with the intersection of major paths. Fifth, *landmarks* were another type of point reference, which often served to orient the observer when traveling through the city. Most maps contained these five elements, although different examples of each, and they provided Lynch with a vocabulary for examining urban form.

Using as an example a person's cognitive map of Madison, Wisconsin, we can see how the elements are combined to create an overall image (Figure 8.2). The major highways are denoted as paths, including the beltline that circles much of the city. The shores of Lakes Mendota and Monona represent the major edges of the map, and Capitol Square is the major node or focus of the city. Two districts are represented by the villages of Shorewood Hills and Maple Bluff, and the Capitol Building and Camp Randall, where the University of Wisconsin football team plays, are major landmarks.

For anybody familiar with Madison, it is obvious that this sketch map is not a completely accurate representation of the city. In particular, as we discussed when considering the relationship between the phenomenal and behavioral environments, it reflects a certain amount of selectivity and distortion. In terms of selectivity, for example, only the major highways and lakes have been portrayed. Also, only those districts, or neighborhoods, that are particularly distinctive have been shown. In terms of distortion, there are four major categories: first, there is distance distortion, as the elements on the map are not the correct distance apart; second, there is directional distortion, as the elements are not correctly oriented with respect to each other; third, there is size, or areal distortion; and fourth, the shapes of the elements

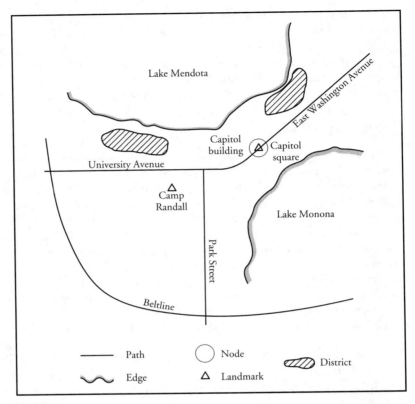

Figure 8.2 Cognitive map of Madison, Wisconsin.

have been changed. Later in this chapter we consider in some detail the characteristics of distance and directional distortion.

In research of this nature there is always some question concerning the degree to which the results are independent of methodology. Does, for example, a person's sketch map tell us more about cartographic ability than about urban images? In this respect, it is significant that Lynch used a variety of methodologies in order to compare different results. He found that the relationship between the cognitive map derived from a person's sketch map, on the one hand, and his or her interview, on the other, was in some cases rather weak. However, when the data from all the interviews were combined to form a composite map and this was compared with a composite map derived from individual sketches, the two maps were remarkably similar.

Some significant differences did remain, however (Lynch, 1960:144–5). The sketch maps tended to emphasize paths and excluded those elements that were especially difficult to draw. The sketch maps also tended to have a higher threshold, meaning that those elements that were least frequently mentioned in the interviews did not appear at all in the sketches. Finally, the sketch maps were unduly fragmented in terms of connections and overall organization, reflecting the difficulty of fitting everything together simultaneously.

In general, the methodologies employed by Lynch have all the advantages and disadvantages of the clinical approach in psychology. The information is gently teased out of the subject's consciousness and is not unduly constrained by a rigid framework of imposed instructions. The resulting maps reflect the schematic nature of imagery, including the selectivity and distortion that take place during the process of perception and memory. As with the clinical approach in psychology, however, the open-ended nature of the research design makes it difficult to quantify and generalize the kinds of selectivity and distortion that are represented.

In a practical context, one of Lynch's major concerns was to address the issue of urban design. To this end, a central concept of his work was the idea of *imageability*, or legibility. Cities were imageable if they could be comprehended as coherent patterns with interconnected parts, and imageable cities were desirable for a number of reasons. First, clarity is generally regarded as being esthetically attractive, whereas cluttered images are regarded as ugly. Second, an imageable city is easy to use and find one's way around in. Third, an imageable city provides an excellent framework for helping a person to incorporate new items of information. Fourth, clarity of image enhances the ability of politicians, planners, and residents to communicate ideas about their city.

Socioeconomic Status

Much of the work on cognitive maps, since Lynch's original pioneering effort, has focused on variations related to differences in socioeconomic status. As different socioeconomic groups live in different parts of the city, they tend to form different images of that city. In particular, sketch maps drawn by higher-status socioeconomic groups are far more extensive, detailed, and interconnected than those drawn by lower-status groups. Also, performance in terms of accuracy tends to improve with increasing socioeconomic status.

One reason for these differences is that high socioeconomic status is often associated with more developed cartographic and verbal skills. A second and probably more important difference, however, is that the lower income groups have more restricted activity spaces. They generally travel shorter distances to work or recreation and are more likely to use public transportation than private automobiles. The latter point is significant, as the flexibility of private transportation encourages the formulation of more detailed and extensive images.

One study, in particular, has empirically verified these socioeconomic differences. In an investigation carried out by the Los Angeles City Planning Commission and reported by Orleans (1973), a series of five composite maps of Los Angeles were constructed from sketch maps drawn by distinctive samples of residents in five different locations. Subjects in the primarily black, lower-socioeconomic-status sample, located in a southeastern neighborhood of Los Angeles known as Avalon near Watts, had a rather restricted spatial image of the city. Most of the north–south roads linking Avalon and the city center were noted, but surprisingly few subjects mentioned such major edges as the Pacific Ocean and the Santa Monica Mountains. Much of the information was represented in the form of districts, but these were not well connected, and there was remarkably little interstitial material.

By contrast, the composite map produced by the upper-class sample from Westwood, near the UCLA campus in West Los Angeles, was much more detailed and extensive. In fact, most subjects had a fairly detailed image of the entire Los Angeles basin. A wide variety of paths, edges, nodes, districts, and landmarks were included, and these were well integrated into an overall conceptual schema that linked the different elements of the spatial image.

Similar variations in cognitive maps are also related to gender differences, as a result of the different roles played by husbands and wives. These variations due to role differences have been demonstrated using the concept of *home area analysis* (Everitt and Cadwallader, 1981), which involves asking subjects to mark on a map "the boundaries of your home area," where home area is described as being "the area in which you feel at home." The study was conducted in the Mar Vista area of West Los Angeles using a sample of 65 couples, and it was assumed that people think in terms of contiguous home areas rather than a system of individual pathways (Aitken and Prosser, 1990). This assumption proved reasonable, as all but one subject was able to define such an area. The one exception did not drive, but took buses, so her home area map consisted of a set of bus routes.

The results of the analysis supported the hypothesis that the wives would demonstrate greater attachment to the local neighborhood by designating a larger home area. First, the average home area size was 3.4 square kilometers for wives and only 1.8 square kilometers for husbands. Second, 28 percent of the wives, as against 20 percent of the husbands, drew at least one of their boundaries outside the base map. Third, 56 percent of the wives' maps were larger than their respective husbands' maps, as against 41 percent of the cases where the reverse was true.

The difference in home area size between husbands and wives was related to role differences, as 94 percent of the husbands went out to work, whereas only one-third of the wives did so. This relationship was reinforced by the fact that the workplace significantly influenced other major behavior patterns. For example, the development of friendships and club memberships often directly resulted from the workplace, and this was also true for some forms of recreation. By contrast, the wives tended to have greater social interaction within the local neighborhood.

Spatial Learning

Obviously, the structure of a person's cognitive map of a city develops over time (Humphreys, 1990), and Golledge (1978; 1990) has argued that this development is associated with the relative connectivity of points, lines, and areas within the cognitive map. When a person first moves into a new city, he or she initially identifies a home, a place of employment, and the location of various food sources (Figure 8.3a). The home and workplace are very quickly established as dominant nodes in the emerging cognitive map, whereas shopping involves a trial-and-error process and is more subject to change.

Once the dominant nodes have been established, together with the major links connecting them, there is a spread effect on the areas surrounding these nodes. Minor nodes are established within the vicinity of the major nodes, joined by a set of minor links. Eventually, the patterns of shopping and recreational behavior lead to consolidated areas, or districts, of detailed information (Figure 8.3b) involving an articulated network of primary, secondary, and minor nodes. This hierarchical development of nodes and links is based on the reasonable supposition that people learn about the urban environment primarily by interacting with it.

The learning process is not necessarily a continuous one, although the cognitive map tends to emerge gradually and smoothly as more information is gathered about the local area.

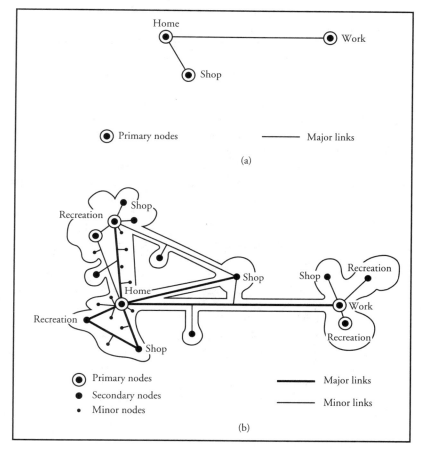

Figure 8.3 Length of residence and the changing cognitive map: (a) skeletal node–path relations; (b) nodes, paths, and neighborhoods. (From R. G. Golledge, "Learning about urban environments," in T. Carlstein, D. Parkes, and N. Thrift, eds., *Making Sense of Time*, Edward Arnold, London, 1978, Fig. 1, p. 78, and Fig. 3, p. 80.)

Sudden increases in information can occur when a trip is made to a previously unfamiliar part of town, and a whole new area of the cognitive map begins to be sketched in. The locations of the minor nodes and links, and even some of the major ones, are relatively flexible in the early stages of cognitive map development, but as time passes the cognitive representation tends to become increasingly stable, and new information results in only minor adjustments to the basic structure of the map.

Golledge's suggestion that initial knowledge about nodes and links eventually leads to more detailed information about districts or neighborhoods implies partly that people develop from a route-mapping stage, involving a network of major nodes and links, to a survey-mapping stage, involving a knowledge of both areas and the interconnections between them. This conceptualization is somewhat similar to Appleyard's (1969) attempt to derive a taxonomy for sketch maps. In particular, he distinguished between maps that emphasize sequential elements, such as paths, and maps that emphasize spatial elements, such as districts (Figure 8.4). In sequential maps the constituent parts are relatively connected, whereas in spatial maps the constituent parts are more disconnected and scattered.

According to Appleyard, the most primitive kind of *sequential map* is a fragmented map, which consists of fragmented paths. A chain map is equally simple, but tends to be more schematic, with impressionistic bends. Branch and loop maps contain branches and loops as outcrops from the basically linear system, while a netted map is the most complete in terms of its representation of the major paths, or links.

The simplest form of *spatial map* is a scattered map, containing such elements as individual buildings, without any drawn connections. A mosaic map is characterized by the enclosure of districts using schematic boundaries and looks something like a neighborhood map. A linked map consists of places, or districts, connected by schematic linkages, and a patterned map represents the most complete and accurate form of spatial map.

In a study of Ciudad Guayana in Venezuela, Appleyard (1970) found that sequential maps were the dominant form of representation, although this result might be due partly to two factors peculiar to his particular study. First, Ciudad Guayana is described as being a generally linear city with no dominant center, thus increasing the potential for sequential or path-oriented sketch maps. Second, the subjects were instructed to draw a map of the city between the steel mill and San Felix. The wording of this instruction tends to emphasize the linearity of the city, and if the subjects had simply been asked to draw a map of Ciudad Guayana, a rather different set of sketches might have emerged. Finally, in this context, recent work suggests that it is only in those situations where the task is relatively complex that people move from a sequential map to a more developed spatial map; otherwise, there appear to be strong and consistent individual preferences for mapping style that persist throughout the learning process (Spencer and Weetman, 1981; Lee and Schmidt, 1988).

8.3   Designative Perceptions

The information contained within cognitive maps can be categorized as being either designative or appraisive (Smith, 1992). Purely *designative perceptions* of places involve such attributes as distance, direction, size, and shape and are devoid of evaluative content. *Appraisive perceptions*, on the other hand, involve the value judgments that we have of different places. This section focuses on the designative information contained within cognitive maps, especially as it pertains to the cognition of distance and direction; in the next section we consider the appraisive information contained within cognitive maps.

Cognitive Distance

At this point in the discussion it is important to note the difference between cognitive distance and perceived distance. The notion of perceived distance has attracted the attention of psychologists for a number of years, but what we are concerned with here are those physical distances that are too large to be perceived in a single glance. When working at the urban scale, it is assumed that people can think about distances in the abstract, without actually "seeing" them. The term cognitive distance is more appropriate for these unseen distances. Our

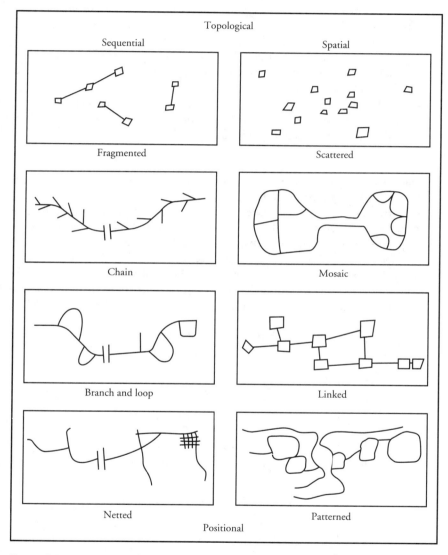

Figure 8.4 Types of sketch maps. (From D. Appleyard, "City designers and the pluralistic society," in L. Rodwin, ed., *Planning Urban Growth and Regional Development: The Experience of the Guayana Program of Venezuela*, MIT Press, Cambridge, Mass., 1969, Fig. 23.2, p. 437. Copyright © 1969 by The Massachusetts Institute of Technology.)

cognitions include not only the information gathered from direct experience, as in perceived distance, but also information gathered from various other sources, such as road maps (Thorndyke, 1981). In this way people are able to estimate the distances to places they may never have actually visited.

Cognitive distance is likely to be a factor in at least three major types of decisions concerning spatial interaction: (1) it affects the decision to stay or go, (2) it affects the decision of where to go, and (3) it affects the decision of which route to take. Given the influential role of cognitive distance in spatial decision making, it is important to understand the nature of the relationship between physical distance and cognitive distance, as this will help us to transform physical space into the relevant cognitive space.

Most research on cognitive distance indicates that the relationship between physical distance and cognitive distance can be satisfactorily approximated by either a linear function or the following power function (Cadwallader, 1976):

$$CD = a\,(PD)^{\,b} \tag{8.1}$$

where CD is cognitive distance, PD is physical distance, and a and b are constants. The dashed line in Figure 8.5 represents the situation where there is no difference between physical and cognitive distance, so the power function indicates that, whereas short distances are overestimated, longer distances tend to be underestimated. The point of changeover in this relationship varies, with a distance of between 6 and 7 miles being suggested as the breakpoint for London and only about 3 miles for Dundee (Pocock and Hudson, 1978:53).

A variety of reasons have been advanced to explain the discrepancies between cognitive and physical distances, although the empirical evidence for them has often been less than impressive. The explanatory variables can be categorized as representing the attributes of the

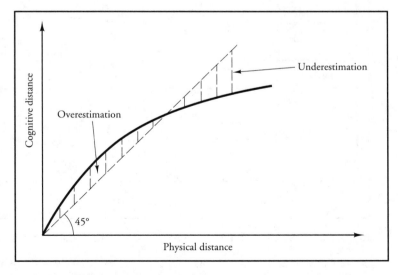

Figure 8.5 Relationship between cognitive distance and physical distance.

stimuli or locations, the attributes of the subjects, and the attributes of the routes. In terms of the *characteristics of the stimuli*, it has been suggested that cognitive distance is influenced by the direction of the stimulus from the subject, although there is some confusion as to the precise nature of this relationship. Lee (1970) compared the cognitive distances involved in journeys toward downtown and journeys away from downtown and concluded that there was a much greater overestimation of outward journeys. Experiments conducted by Golledge et al. (1969), however, yielded results that directly contradict this finding. Although noting that cognitive distance does vary with direction, they found that distances away from the central business district tend to be underestimated, whereas those toward the central business district tend to be overestimated. Similarly, the relative attractiveness of the stimuli, or destination, has also been found to account for some of the deviations between physical and cognitive distance.

Characteristics of the subjects are also significant in terms of explaining distance judgments. A person's familiarity with a particular city will obviously influence his or her distance estimates within that city. As one might expect, it has been shown that an effective learning process enables long-term residents to estimate distances more accurately than do recent arrivals.

Finally, the *characteristics of the route* and the mode of transportation tend to influence distance judgments. Generally, the greater the number of turns or intersections in a journey, the greater the imagined length of that journey. In terms of mode of transportation, Canter (1977:99) reports that bus travelers in London tend to relatively underestimate distances. In a wider context, Canter and Tagg (1975) suggest that the general layout or imageability of a city is related to the pattern of distance distortion. A city with a confusing image, like Tokyo, often leads to the overestimation of distances, whereas a more legible city, like Glasgow, may lead to underestimation.

Any overall summary of the work on cognitive distance is exceedingly difficult, however, as a number of methodological differences preclude the direct comparison of results (Cadwallader, 1973; Ewing, 1981). First, there have been differences of opinion as to whether subjects should be asked to estimate route distances or straight-line distances. For example, Lee (1970) used estimates of the shortest walking distance, whereas Canter and Tagg (1975) used straight-line or "crow-flight" estimates. It seems that straight-line distances are more appropriate, however, because, if it is maintained that subjects possess some kind of spatial representation of the physical environment, straight-line distances can be drawn fairly easily from this abstract representation, whereas a knowledge of route distances might require direct experience with those routes. This argument follows from the theoretical assumption that people tend to work from some kind of cognitive map, rather than retracing in their mind specific trips that they might have taken in the past. Although it is unclear what actually transpires inside the black box of the human brain, most researchers assume that some form of externalized map can be drawn from the spatial relations held in the mind.

Second, there has been considerable variation with respect to the spatial scale involved in cognitive distance studies. For example, Lundberg (1973) asked his subjects to estimate the distances between 13 cities located in different parts of the world, whereas others have focused on distances within individual cities. These differences make it difficult to compare results, because the threshold distance at which overestimation changes to underestimation is partly a function of spatial scale.

Third, most studies have aggregated the individual estimates, consequently masking the possibility that a variety of relationships exists at the individual level. For example, some subjects might exhibit linear relationships between physical and cognitive distance, whereas for others the relationship might be curvilinear. In addition, it has been shown that aggregated data significantly exaggerate the strength of the relationship between physical and cognitive distance (Cadwallader, 1973).

Finally, the method of obtaining distance estimates has varied from study to study, as there are a variety of ways in which subjects can be asked to scale physical distances subjectively. For example, they can be asked to draw maps from which the experimenter can extract the appropriate cognitive distance information, or they can be asked to arrange scale models into their real-world pattern, again allowing the experimenter to measure cognitive distance. The most popular methodology, however, has involved a number of scaling techniques, such as ratio estimation and direct magnitude estimation (Phipps, 1979).

Research has indicated that the precise form of the relationship between cognitive distance and physical distance is not independent of methodology (Baird et al., 1982). For example, Day (1976) compared distance estimates derived in four different ways. The subjects were asked to draw maps, provide verbal estimates, mark scales of a specified length, and use a ratio scaling method. Although the relationship between physical and cognitive distance was approximately linear when using all four estimation techniques, the slope of the least-squares line varied.

The present author (Cadwallader, 1979a) addressed this problem of methodological inconsistency by comparing results obtained at the individual level to see whether subjects give different distance estimates according to the form in which they are asked to provide those estimates. The subjects were asked to estimate the distances between pairs of cities using both the method of direct magnitude estimation and category rating scales. *Direct magnitude estimation* is a technique used by psychologists to measure such sensory magnitudes as loudness (Gescheider, 1988), and it has already been mentioned in the context of the consumer behavior study reported in Section 8.1. The experimenter usually selects a particular single stimulus on a given physical continuum, such as distance, and assigns a number to its subjective magnitude. The subject is then presented with a set of variable stimuli and is instructed to assign to each stimulus a number that is proportional to its subjective magnitude as compared to the standard stimulus. In the present experiment the stimuli were the pairs of cities which the subjects were asked to estimate the distance between, while the standard stimulus was the particular pair of cities chosen by the experimenter.

A brief example will clarify the use of this technique. Assume that there are 10 physical distances (A, ..., J) for which estimates are needed. The distance E can be assigned the standard value of 100, and then the subjects are asked to rate the other distances accordingly. It is advisable to select the standard stimulus so that it is in the middle range of the variable stimuli and to assign it a number like 100, as this is easy to divide into simple ratios and allows the subjects to think in terms of percentages. If a person thinks that distance I is twice as long as distance E, he or she will assign I the value 200; if he or she thinks that I is only half as long as E, then I will be assigned the value 50, and so on, until all the distances have been estimated.

A *category rating scale*, on the other hand, is produced when a subject judges a set of stimuli, in this case physical distances, in terms of a set of categories distinguished by either numbers or adjectives. The subject is requested to assign the smallest stimulus to the first category, the largest to the last category, and to use the other categories in such a way that they are equally spaced subjectively. An odd number of categories are normally used, and the stimuli are chosen so that the subjects tend to choose the various categories equally often.

The consistency of the estimates across these two methodologies was assessed by correlating the two sets of responses for each person, with a strong positive correlation indicating a high degree of similarity between the estimates. It should be noted, however, that for most variables the relationship between a magnitude scale and a category scale lies somewhere between a linear and a logarithmic function. The category widths seem to increase as one moves up the category scale, meaning that two stimuli close together on the low end of the scale are more easily discriminated between than those close together on the high end of the scale. In other words, it becomes increasingly difficult to distinguish between two stimuli as their magnitudes increase.

Following this reasoning, the two sets of distance estimates for each person were correlated in two different ways. First, the magnitude scale was correlated with the category scale (Table 8.2). In both cases the majority of correlations fell within the range from 0.61 to 0.80, indicating that the individual estimates were not completely consistent across the two measurement techniques. That is, the distance estimates provided by an individual subject were

Table 8.2 Comparison of Different Methodologies for Measuring Cognitive Distances

Pearson correlation coefficients for the relationship between magnitude scales and category scales:

Coefficients	Frequency
0.40 or less	0
0.41 to 0.60	7
0.61 to 0.80	32
0.81 to 1.00	26

Pearson correlation coefficients for the relationship between the logarithm of magnitude scales and category scales:

Coefficients	Frequency
0.40 or less	0
0.41 to 0.60	7
0.61 to 0.80	35
0.81 to 1.00	23

Source: M. T. Cadwallader, "Problems in cognitive distance: Implications for cognitive mapping," *Environment and Behavior,* 11 (1979), Tables 1 and 2, p. 568.

not always independent of the way in which he or she was asked to provide those estimates, thus reflecting one of the major difficulties involved in attempting to calibrate the relationship between physical and cognitive distance.

Intransitivity and Noncommutativity in Cognitive Distance

Efforts to calibrate the relationship between physical and cognitive distance have also been thwarted by the problems of intransitivity and noncommutativity. *Intransitivity* occurs if the following conditions hold: A is estimated to be greater than B, B is estimated to be greater than C, but C is estimated to be greater than A, where A, B, and C are interpoint distances. The degree of intransitivity in different kinds of judgmental behavior is an empirical question, and it is important to know to what extent distance estimates are intransitive, as any inherent intransitivity means that cognitive distance, and therefore cognitive maps, cannot be represented in terms of Euclidean geometry.

In an experiment designed to address this problem of intransitivity (Cadwallader, 1979a), 56 subjects made a series of paired comparison judgments in which they were required to estimate which one of two intercity distances was longer. In all, there were six intercity distances, so each subject made 15 paired comparison judgments. The data from the experiment were first analyzed at the aggregate level to determine the relative amounts of strong, moderate, and weak stochastic transitivity. Strong stochastic transitivity asserts that if the probabilities of estimating A greater than B and B greater than C are both equal to or greater than 0.5, then the probability of estimating A greater than C is equal to or greater than the larger of the other two probabilities. Moderate stochastic transitivity asserts that, if the probabilities of estimating A greater than B and B greater than C are both equal to or greater than 0.5, then the probability of estimating A greater than C is equal to or greater than the smaller of the other two probabilities. While weak stochastic transitivity simply asserts that if the probabilities of estimating A greater than B and B greater than C are both equal to or greater than 0.5, then the probability of estimating A greater than C is also equal to or greater than 0.5. The six intercity distances used in the experiment created 20 different distance triads, or combinations of three distances, of which 17 met the conditions of strong stochastic transitivity, while the other 3 met the conditions of moderate stochastic transitivity.

A different picture emerged, however, when intransitivity was investigated at the level of individual subjects. As there were 56 subjects and 20 triads per subject, there were 1120 triads in the complete experiment. Of these, 92 triads, or 8 percent, were intransitive. This represents a large percentage, as within each triad of distances there are six possible transitive orderings and only two possible intransitive orderings. Table 8.3 is a frequency table indicating the number of intransitive triads for each subject. Of the 56 subjects, 29 had at least one intransitive triad, and of the 27 subjects who had no intransitive triads, not one of them had all six stimuli, or intercity distances, in the correct order. Indeed, of 1028 triads that were transitive, only 339 or 33 percent contained the correct ordering of the three stimuli. An extreme example of this phenomenon was the subject who exhibited complete transitivity but had the stimuli within every triad in an incorrect order.

In summary, then, research on cognitive distance and ultimately cognitive maps is faced with the problem that distance estimates are sometimes intransitive. This intransitivity means

Table 8.3 Cognitive Distance and Intransitivity

Number of Intransitive Triads	Frequency
0	27
1	6
2	6
3	6
4	6
5	0
6	3
7	2

Source: M. T. Cadwallader, "Problems in cognitive distance: Implications for cognitive mapping," *Environment and Behavior*, 11 (1979), Table 4, p. 572.

that the subjects are unable to order the stimuli on a single continuum. As a result, it appears that efforts to uncover a simple linear or curvilinear relationship between cognitive distance and physical distance are oversimplifying a complex situation.

The second major problem in analyzing cognitive distance is the issue of noncommutativity (Burroughs and Sadalla, 1979). Cognitive distance is noncommutative if the following condition holds:

$$CD\,(A, B) \neq CD\,(B, A) \tag{8.2}$$

where A and B are point locations, and CD is cognitive distance. Thus noncommutativity means that the estimated distance from A to B does not equal the estimated distance from B to A. An example of such a situation would be the often-felt experience that the journey to work seems shorter than the journey home from work.

To investigate the degree to which distance estimates are noncommutative, an experiment was conducted in which subjects were asked to estimate the distance between 30 pairs of cities in the United States (Cadwallader, 1979a). Later, as part of the same experiment, they were asked to repeat this task, but this time the order of each pair of cities had been reversed. The difference between each pair of estimates was calculated as a percentage of the largest estimate. In 259 instances, both estimates were exactly the same (Table 8.4). Most pairs of estimates, however, revealed a difference of somewhere between 10 and 40 percent.

These results indicate that distance estimates are often noncommutative, suggesting that people do not possess internal spatial representations of the environment that can be portrayed in terms of Euclidean geometry (Golledge and Hubert, 1982). The only other conclusion is to speculate that noncommutativity merely represents some form of measurement error. On the basis of present evidence, however, it seems dangerous not to at least entertain the idea that noncommutativity is a genuine property of cognitive maps.

Table 8.4 Cognitive Distance and Noncommutativity

Percentage Difference	Frequency
0	259
0.1 to 9.9	255
10.1 to 19.9	358
20.0 to 29.9	301
30.0 to 39.9	158
40.0 to 49.9	97
50.0 or more	72

Source: M. T. Cadwallader, "Problems in cognitive distance:
Implications for cognitive mapping," *Environment and
Behavior,* 11 (1979), Table 5, p. 574.

Frame Dependency in Cognitive Maps

Less attention has been given to the directional component of cognitive maps, although the locational information contained in such maps is a composite of both distance and directional distortions (Lloyd, 1989b). It has been argued that the directional distortions can be analyzed in terms of the frame-of-reference concept employed in psychological investigations of the perception of verticality (Cadwallader, 1977). Psychologists have long been concerned with the distortion of the vertical within visible fames of reference, but far less consideration has been given to frame dependency in the horizontal plane. Also, when discussing cognitive maps, it should be remembered that we are involved with the influence of unseen frames of reference, that is, frames of reference that exist in physical space but whose size prevents them from being encompassed in a single glance. These frames of reference are represented in the cognitive map, and any displacement of their orientation from the true orientation can result in a corresponding displacement of all the directional information contained within that map.

Various typologies for *frames of reference* have been developed, and what we are concerned with here are those reference systems that are based in the environment. In such systems, different elements of the social or physical environment may act as the frame of reference. For example, Trowbridge (1913), in a pioneering investigation of directional orientation in Manhattan, suggested that the differences between the true directions and the estimated directions were due to the belief among his subjects that the longitudinal streets of Manhattan are isomorphic with a north–south axis. He argued that the subjects were using the street system as their frame of reference, and, as they believed that the street system corresponded to a north–south and east–west grid system, their directional estimates were displaced from the true directions in direct association with this inaccurate frame of reference.

A similar study was undertaken by the present author (Cadwallader, 1977) to determine the extent to which a group of residents located in West Los Angeles orient themselves according to major features of the physical landscape. More specifically, it was hypothesized that the edge of the Pacific Ocean would provide the major frame of reference for their cognitive maps. The experimental design was similar to that used by Trowbridge (1913) and required each subject to estimate the direction to 30 cities in the Los Angeles basin. The resulting data were analyzed by means of directional statistics, as we are dealing with circular rather than linear distributions.

The mean estimated direction to each city was determined, and these directions were compared to the real directions. On the average, the estimated directions were 18.57 degrees less than the real directions. This difference suggests that the coastline was being used as a frame of reference, as its orientation is northwest to southeast. In other words, the subjects appeared to think that the coastline runs north-south, although it is actually displaced from this orientation, with the result that the estimated directions were similarly displaced. An analysis of the difference between the estimated direction and the real direction for each of the 30 cities indicated that 27 of those cities had a mean estimated direction that was less than their real direction, thus reflecting a displacement toward the frame of reference. Moreover, the difference between the estimated direction and the real direction appeared to be greatest for those cities closest to the coastline. That is, the degree of frame dependency varied according to the distance of the stimuli from the frame of reference.

Although this experiment provided some evidence that cognitive maps are frame dependent, there are still a number of unanswered questions. For example, it is unclear whether one frame of reference is usually applied to the whole map or whether different frames of reference are utilized for different parts of the map. Also, we as yet know very little about why particular elements in the environment are chosen as a frame of reference or whether the chosen frame of reference changes over time.

8.4   Appraisive Perceptions

Having discussed the designative component of cognitive maps, we now turn our attention to the appraisive information contained within such maps. In particular, we will focus on the amount of information and the perceived attractiveness associated with different parts of the cognitive map (Gale et al., 1990; Thill and Sui, 1993). Much of the pioneering work in this area was carried out by Gould (Gould and White, 1986), who investigated the perceived residential desirability of different regions within the United States.

Residential Preference Patterns

If the residential desirability of each state, as rated by a group of subjects, is measured on a scale from zero to 100, these values can be used to construct a residential preference surface. Isolines connect the points of equal value, thus creating a surface that reflects the hills and valleys of desirability for a particular group of people. For example, Figure 8.6 shows the residential desirability surface for a group of subjects located in California. Where the isolines are far apart, it means that the perceived residential desirability changes slowly, while closely spaced isolines indicate more dramatic changes in the preference surface.

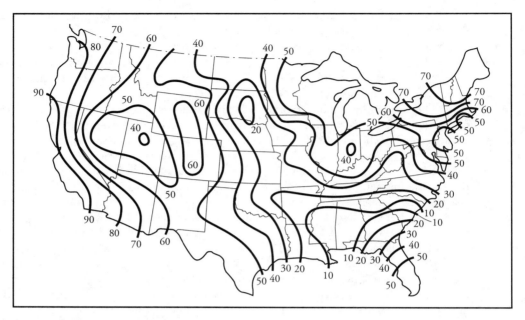

Figure 8.6 Residential desirability surface from California. (From R. Abler, J. S. Adams, and P. R. Gould, *Spatial Organization: The Geographer's View of the World*, Prentice Hall, Englewood Cliffs, N.J., 1971, Fig. 13.31, p. 520.)

Similar surfaces have also been constructed for individual cities, and they have implications for the issue of neighborhood revitalization (Day and Walmsley, 1981). Clark and Cadwallader (1973a) analyzed a series of residential preference maps of Los Angeles. The sample consisted of 1024 persons spread throughout the metropolitan area. Each subject was shown a map of Los Angeles that included the freeway system and the Santa Monica mountains. In addition, approximately 180 neighborhoods were indicated by name on the map, although their boundaries were not given. After looking at the map, each subject was asked to do the following: "Taking your family income into consideration, please show me on this map the three neighborhoods where you would most like to live, starting with the one you would like to live in most." The income constraint was used so that people would not automatically choose such obviously attractive, but realistically unattainable communities as Beverly Hills, Bel Air, or Palos Verdes.

The first-choice preferences of the subjects are shown in Figure 8.7. The neighborhoods are mapped with an intensity according to the number of times they were chosen by the 1024 subjects; the most heavily shaded areas represent communities that were chosen by the greatest number of people. A minimum of six subjects was used as the cutoff point for the highest preference category, as this was the number of choices that every community would receive if they were all perceived as being equally attractive. The preference map contains four distinctive regions of highly preferred neighborhoods. First, there is a ridge extending from Santa Monica in the west to Hollywood in the east, reflecting the physically attractive nature of this

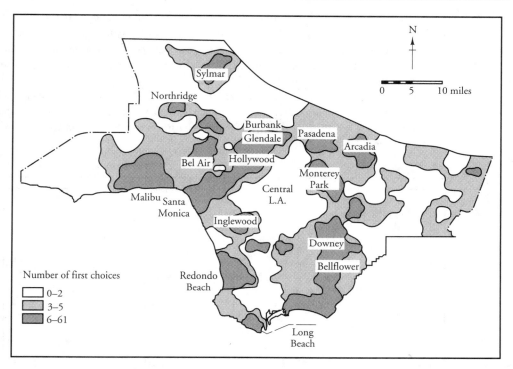

Figure 8.7 First-choice preferences of sample respondents in the Los Angeles metropolitan area. (From W. A. V. Clark and M. T. Cadwallader, "Residential preferences: An alternate view of intraurban space," *Environment and Planning*, 5, 1973, Fig. 3, p. 697.)

area and its relatively good access to the ocean and mountains. Second, the beach communities, such as Malibu and Redondo Beach, are also highly preferred. Third, there is a highly preferred region in the eastern portion of the Los Angeles basin, including such communities as Arcadia and Pasadena, where Arcadia represents newer housing opportunities and Pasadena represents a very established residential community. Finally, the belt of highly preferred communities from Long Beach to Downey reflects the residential preferences of lower-income whites, blacks, and Mexican Americans.

Urban Information and Preference Surfaces

A more detailed investigation of urban information and preference surfaces in the Los Angeles basin was undertaken by the present author (Cadwallader, 1978b) in order to look backward from these "invisible landscapes" to some of the causal influences that form them. Whereas a *preference surface* reflects the varying attractiveness that people assign to different locations or places, an *information surface* reflects the varying amount of information that people possess about those same locations or places. In this particular instance, the aim was to account for any regularities in these surfaces, in the hope of identifying variables involved in the evaluation process. To this end, trend surface analysis was used to generate the explanatory hypotheses, which were then tested by means of correlation and regression analysis.

The data for the research were collected from residents of West Los Angeles, and the subjects all lived within three blocks of each other. In this way, individual differences in the information and preference patterns could be compared while holding location constant. Both the information and preference data were obtained by the previously described method of direct magnitude estimation, with each subject being asked to rate his or her familiarity with and preference for 30 cities in the Los Angeles basin (Figure 8.8), using Culver City and Hollywood as the standard stimuli.

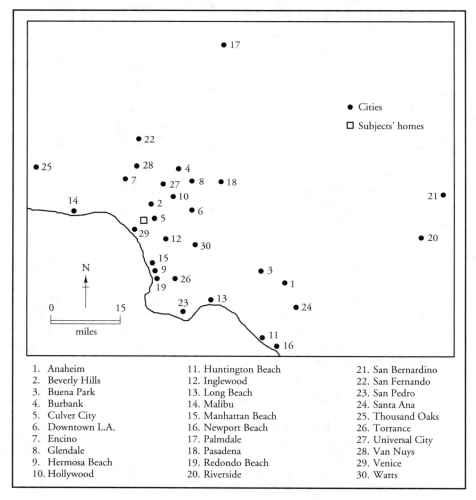

1. Anaheim	11. Huntington Beach	21. San Bernardino
2. Beverly Hills	12. Inglewood	22. San Fernando
3. Buena Park	13. Long Beach	23. San Pedro
4. Burbank	14. Malibu	24. Santa Ana
5. Culver City	15. Manhattan Beach	25. Thousand Oaks
6. Downtown L.A.	16. Newport Beach	26. Torrance
7. Encino	17. Palmdale	27. Universal City
8. Glendale	18. Pasadena	28. Van Nuys
9. Hermosa Beach	19. Redondo Beach	29. Venice
10. Hollywood	20. Riverside	30. Watts

Figure 8.8 Thirty cities involved in the information and preference study of Los Angeles. (From M. T. Cadwallader, "Urban information and preference surfaces: Their patterns, structures, and interrelationships," *Geografiska Annaler*, 60 B, 1978, Fig. 1, p. 99.)

Trend surface analysis (see Section 5.4) was used to fit linear, quadratic, and cubic surfaces to the information and preference data (Table 8.5). The best fit for both the information and preference surfaces was provided by a cubic surface, although even the cubic surfaces only accounted for just over half the total variation in the data. The comparatively small coefficients of determination indicate that the surfaces were extremely convoluted. Of greater interest, however, is the fact that the trend surface analysis successively identified a number of important explanatory variables.

Inspection of the preference surfaces and their residuals suggested the significance of distance from the Pacific Ocean as a determinant of residential desirability. Also, some measure of affluence appeared to be important, as cities such as Beverly Hills and Malibu were given consistently high ratings. In terms of the information surfaces, distance again appeared to be a prominent factor, although in this case it was distance from the subjects' homes that was important rather than distance from the ocean. City size emerged as a second variable influencing information levels, suggesting that the larger cities tended to be more familiar than the smaller ones.

The extent to which these four variables accounted for the variation in the information and preference surfaces was investigated by means of multiple correlation and regression analysis (see Section 4.1). For the preference surface, the multiple regression model was as follows:

$$Y_i = a - b_1 X_{i1} + b_2 X_{i2} \tag{8.3}$$

where Y_i is the attractiveness of city i; X_{i1} is the distance of city i from the Pacific Ocean; X_{i2} is the average income of city i; and a, b_1, and b_2 are constants. The multiple correlation coefficient for this equation was 0.85, and the partial correlation coefficients suggest that average income was more important than distance from the ocean (Table 8.6). Note that the coefficient for average income was positive, indicating that the affluent cities were more highly preferred, whereas the coefficient for distance from the ocean was negative, indicating that residential desirability decreased with increasing distance from the ocean. It should be remembered, however, that extreme caution is required when interpreting the coefficients in models of this type, as empirical estimates of the distance coefficient can be influenced by the particular configuration of points, or cities (Sheppard, 1984).

The variation in information levels was analyzed using the following multiple regression model:

Table 8.5 Trend Surface Analysis of the Information and Preference Surfaces for Los Angeles

	Information Surface (R^2)	Preference Surface (R^2)
Linear	0.13	0.15
Quadratic	0.24	0.34
Cubic	0.53	0.60

Source: M. T. Cadwallader, "Urban information and preference surfaces: Their patterns, structures, and interrelationships," *Geografiska Annaler,* 60 B (1978), Table 1, p. 101.

Table 8.6 Partial Correlation Coefficients Associated with the Information and Preference Surfaces for Los Angeles

Information Surface	
Distance from the subjects' homes	−0.54
City size	0.15
Preference Surface	
Distance from the ocean	−0.44
Average income for city	0.80

Source: M. T. Cadwallader, "Urban information and preference surfaces: Their patterns, structures, and interrelationships," *Geografiska Annaler*, 60 B (1978), Table 2, p. 101.

$$Y_i = a - b_1 X_{i1} + b_2 X_{i2} \tag{8.4}$$

where Y_i is the information level for city i; X_{i1} is the distance of city i from the subjects' homes; X_{i2} is the population of city i; and a, b_1, and b_2 are constants. The multiple correlation coefficient for this equation was 0.58, with distance from the subjects' homes being more important than population size (Table 8.6). Again, note that information levels decreased with increasing distance, but increased with increasing population size.

The latter results correspond closely with those obtained by Gould (1975) when investigating information surfaces at a national scale. Using data from Swedish schoolchildren, Gould found that the configuration of the information surfaces could be satisfactorily accounted for by population and distance variables. Similar results have also been reported when measurements are made concerning the accuracy of information, rather than just the quantity.

Cognitive Structures

A more thorough investigation of the information and preference surfaces can be undertaken by identifying the underlying evaluative dimensions. Principal components analysis (see Section 6.4) provides a suitable methodology for this purpose. With the subjects as cases and the cities as variables, an orthogonal rotation was performed in order to emphasize most clearly the groupings of cities. Those cities with highly correlated familiarity, or preference ratings, would be associated with the same components, and it was hoped that the interpretation of these components would suggest the underlying dimensions that the subjects had used to evaluate the cities.

The underlying structure of the preference surface is represented in Table 8.7, and the highest loading for each city, on any of the four components, was used to interpret those components. The four components describing the preference pattern accounted for approximately 87 percent of the total variance, while the first two components alone accounted for over 65 percent. The first component was associated with interior cities such as San Bernardino,

Table 8.7 Preference Structure of Cities in the Los Angeles Basin

City	Components			
	I	II	III	IV
Anaheim	<u>0.808</u>	0.383	0.289	−0.066
Beverly Hills	0.656	<u>0.715</u>	−0.090	−0.093
Buena Park	<u>0.778</u>	0.562	0.153	−0.120
Burbank	<u>0.890</u>	0.369	0.140	−0.034
Downtown L.A.	−0.031	−0.005	<u>0.816</u>	0.260
Encino	<u>0.724</u>	0.603	0.070	−0.153
Glendale	<u>0.855</u>	0.317	0.215	0.223
Hermosa Beach	0.481	<u>0.817</u>	0.094	0.174
Huntington Beach	0.415	<u>0.766</u>	−0.001	0.269
Inglewood	0.660	<u>0.681</u>	0.062	0.018
Long Beach	0.442	<u>0.780</u>	0.323	−0.227
Malibu	<u>0.786</u>	0.549	−0.100	0.023
Manhattan Beach	0.442	<u>0.834</u>	0.061	0.193
Newport Beach	0.370	<u>0.770</u>	−0.053	0.006
Palmdale	0.253	−0.120	<u>0.768</u>	0.115
Pasadena	<u>0.844</u>	0.293	0.213	0.217
Redondo Beach	0.491	<u>0.786</u>	0.096	0.203
Riverside	<u>0.830</u>	0.380	0.326	0.008
San Bernardino	<u>0.723</u>	0.570	0.290	−0.041
San Fernando	0.502	0.075	<u>0.736</u>	−0.072
San Pedro	0.014	<u>0.624</u>	0.632	0.071
Santa Ana	<u>0.857</u>	0.350	0.328	−0.023
Thousand Oaks	<u>0.864</u>	0.449	−0.029	−0.060
Torrance	0.345	<u>0.780</u>	0.357	−0.208
Universal City	0.519	0.249	<u>0.623</u>	−0.149
Van Nuys	0.328	0.185	<u>0.719</u>	−0.347
Venice	0.064	0.153	0.334	<u>0.817</u>
Watts	−0.122	0.096	<u>0.810</u>	0.278
Percent of total variance:	36.4	29.0	16.7	5.0

Source: M. T. Cadwallader, "Urban information and preference surfaces: Their patterns, structures, and interrelationships," *Geografiska Annaler,* 60 B (1978), Table 3, p. 102. *Note:* Underscores indicate highest factor loading for each variable.

Thousand Oaks, and Riverside, whereas the second component was associated with cities on or close to the coastline (Figure 8.8). The latter cities included Hermosa Beach, Redondo Beach, San Pedro, and Long Beach. The only exception to this general rule was Malibu, which loaded more heavily on the first component. The overwhelming importance of the first two components in terms of variance accounted for and their easy interpretation is significant. It suggests that the subjects, when asked to evaluate the cities according to their relative attractiveness, simply thought in terms of two dimensions: coastal cities and interior cities. As a result, the coastal cities all received similar ratings, as did the interior cities. However, the relative attractiveness of coastal cities in general, as opposed to interior cities in general, varied across individuals.

The results of the principal components analysis of the information surface were less easily interpreted. It required six components to account for 76 percent of the total variance, and no single component accounted for more than approximately 17 percent (Table 8.8). It is noteworthy, however, that cities grouped together in information space also tended to be grouped together in physical space (Figure 8.8). For example, San Bernardino and Riverside were associated with the same component, as were Long Beach and San Pedro, and Hermosa Beach, Manhattan Beach, and Redondo Beach. This finding was not unexpected, as it means that, if a person possesses a great deal of information about a city in one particular part of the Los Angeles basin, he or she is also likely to be very familiar with other cities in the same area. As with the preference structure, however, the relative amount of information associated with each group of cities varied from person to person.

The final step in the analysis involves an investigation of the interrelationships between the two underlying structures. Canonical correlation analysis was deemed appropriate for this task, as it allows one to measure the relationships between two sets of variables or, as in this case, two sets of components (Kachigan, 1986). Canonical analysis identifies linear composites of each set of variables such that the relationships between the composites are maximized. By tradition, the two groups of variables are labeled the criterion, or dependent set, and the predictor, or independent set. Canonical analysis is most often used in those instances where a single observed variable is inadequate to represent the criterion dimension of interest. In all cases, however, the overriding concern when employing canonical analysis is to identify the structural relationships between the two groups of variables as a whole, rather than the relationships between individual variables.

The procedure involves deriving successive linear composites of both sets of variables in such a way that the correlation between each successive pair of composites is maximized. The first pair of composites identifies the most important pattern that is common to both groups of variables, and succeeding pairs of composites are uncorrelated with each other. In other words, the two sets of variables are transformed into orthogonal canonical vectors, where the canonical vectors are linear functions of the original variables, such that the correlations between certain variables of the two sets are maximized.

The canonical correlation expresses the degree of association between the two sets of variables for each canonical vector. The square of the largest canonical correlation is the proportion of the variance in the first composite of the criterion set that is accounted for by the first composite of the predictor set. In this sense, it is important to note that the canonical correlation provides a measure of the association between a pair of linear composites derived

Table 8.8 Information Structure of Cities in the Los Angeles Basin

Cities	Components					
	I	II	III	IV	V	VI
Anaheim	0.333	0.611	0.261	0.141	0.504	0.234
Beverly Hills	0.624	0.268	0.153	0.425	−0.145	0.000
Buena Park	0.792	0.103	0.253	0.030	0.363	0.214
Burbank	0.232	0.566	−0.024	0.031	0.493	−0.008
Downtown L.A.	0.324	−0.017	0.105	0.749	0.206	0.188
Encino	0.644	0.041	0.566	0.115	−0.141	0.154
Glendale	0.071	0.131	0.183	0.193	0.789	0.119
Hermosa Beach	0.358	−0.069	0.116	0.015	0.285	0.779
Huntington Beach	0.750	0.036	0.006	0.390	0.246	0.272
Inglewood	0.605	0.113	0.037	−0.017	0.139	0.040
Long Beach	0.281	0.124	0.571	0.215	0.507	0.297
Malibu	0.672	0.423	0.046	0.376	−0.114	0.236
Manhattan Beach	0.132	0.234	0.123	0.038	0.026	0.905
Newport Beach	0.112	0.205	0.068	0.365	0.067	0.657
Palmdale	0.143	0.259	0.568	0.073	0.427	0.152
Pasadena	0.196	0.044	0.425	0.642	0.132	0.237
Redondo Beach	0.132	−0.001	0.303	0.502	0.150	0.581
Riverside	0.438	0.588	0.263	0.022	0.436	0.288
San Bernardino	0.064	0.936	0.094	0.177	0.077	0.058
San Fernando	0.155	0.879	0.183	−0.082	0.009	0.055
San Pedro	0.296	0.140	0.512	0.258	0.506	0.175
Santa Ana	0.616	0.155	0.280	0.176	0.522	0.206
Thousand Oaks	0.801	0.063	0.356	0.053	0.205	0.163
Torrance	−0.048	−0.036	−0.285	0.862	0.040	0.007
Universal City	0.029	0.897	0.168	−0.027	0.127	0.118
Van Nuys	0.088	0.399	0.777	−0.025	0.001	0.125
Venice	0.203	0.204	0.324	0.473	0.326	0.125
Watts	0.254	0.073	0.638	−0.036	0.322	0.034
Percent of total variance:	17.3	15.1	11.7	11.0	10.7	10.4

Source: M. T. Cadwallader, "Urban information and preference surfaces: Their patterns, structures, and interrelationships," *Geografiska Annaler,* 60 B (1978), Table 4, p. 103.
Note: Underscores indicate highest factor loading for each variable.

from the criterion and predictor variables, rather than a measure of association between the two sets of variables themselves. Finally, the canonical weights express the degree of association between the original variables and the canonical vectors. These weights are like the loadings in a factor matrix and thus represent standardized measures that enable one to determine the relative contribution of each variable to each canonical vector.

In the present instance, the component scores, rather than measurements on the original variables, were used as input to the canonical analysis. The major problem with using measurements on the individual variables as input data is that the importance of the canonical vectors is not determined by how well they account for the variation within the two sets of variables; they simply optimize the relationships between the two sets. It seems logical, therefore, to first determine the underlying structures and then to relate those structures.

The first four components in the canonical analysis represent the preference structure, while the remaining six represent the information structure (Table 8.9). The canonical correlations associated with three of the four canonical vectors are very low, suggesting that the two structures are only loosely related. Consequently, no attempt is made to interpret the canonical vectors. The results of the canonical analysis are thus somewhat disappointing, as some evidence of an interrelationship between the two structures was expected. To explore this line of reasoning more fully, the correlations between the information and preference patterns for each individual were computed (Table 8.10). The preponderance of correlation coefficients below 0.51, however, provides further evidence that no simple relationship exists between information and preferences.

In summary, then, having distinguished between the behavioral and phenomenal environments, we explored in some detail the concept of cognitive maps. More specifically, we dis-

Table 8.9   Results of the Canonical Analysis of Cities in the Los Angeles Basin

Components	Canonical Vectors			
	I	II	III	IV
I	0.048	0.350	0.116	0.929
II	0.248	0.494	−0.830	−0.080
III	−0.218	0.803	0.455	−0.319
IV	0.930	0.056	0.354	−0.097
V	0.005	−0.066	−0.046	−0.081
VI	−0.104	−0.095	0.373	−0.788
VII	−0.541	−0.143	0.400	−0.145
VIII	0.876	−0.117	0.283	−0.213
IX	−0.059	0.006	0.759	0.574
X	−0.063	−0.973	−0.136	0.159
Canonical Correlations:	0.714	0.450	0.270	0.100

Source: M. Cadwallader, "Urban information and preference surfaces: Their patterns, structures, and interrelationships," *Geografiska Annaler,* 60 B (1978b), Table 5.

Table 8.10   Individual Correlations between Information and
Preferences for Cities in the Los Angeles Basin

Correlation Coefficient	Frequency (no. of individuals)
0 to 0.25	23
0.26 to 0.50	18
0.51 to 0.75	9
0.76 to 1.00	0

Source: M. Cadwallader, "Urban information and preference surfaces:
Their patterns, structures, and interrelationships," *Geografiska Annaler*,
60 B (1978b), Table 6.

cussed how such maps might be elicited from a set of subjects, how they vary according to
socioeconomic status, and how they change over time. We concluded by distinguishing
between the designative information contained in cognitive maps, especially with respect to
distance and direction, and the appraisive information, which involves the value judgments
that we have of different places.

Chapter 9

Movement Patterns within Cities

9.1   Traffic Forecasting Models

Movement patterns within cities can be categorized as being of either short- or long-term duration. Short-term or daily movement consists of trips involving such activities as work, shopping, and recreation, whereas long-term or more permanent movement involves changing residence. In this chapter we focus on daily movement patterns, especially the journey to shop; residential mobility is considered in Chapters 10 and 11. After first considering the nature of traffic forecasting models in general, we then consider the trip distribution component of such models in more detail. Within the context of the journey to shop, we discuss the traditional gravity model approach, the revealed preference approach, the behavioral approach, and a dynamic, or learning-based, approach. The exploration of movement patterns is vital for a true understanding of urban areas. Movement both creates and reflects the spatial structure of cities described in the earlier chapters. For example, the hierarchy of shopping centers in any city is partly a response to consumer preferences with respect to travel behavior, but also a determinant of that travel behavior. Similarly, social neighborhoods within cities are simultaneously created by movement patterns, while also helping to mold those patterns.

Attempts to forecast daily movement patterns involve the construction of four interrelated submodels: trip generation and attraction, trip distribution, modal choice, and trip assignment (Black and Blunden, 1984; Pas, 1986). These submodels are usually used to forecast traffic flows between regions known as traffic zones, and they parallel the kinds of decisions facing the intended traveler. First, there is the initial decision to make a trip, thus contributing to the overall trip generation associated with a particular traffic zone. Second, a destination must be selected, thus contributing to the overall distribution of trips between various pairs of traffic zones. Third, a particular form of transportation must be chosen. Fourth, a particular route must be selected.

Trip Generation and Attraction

A variety of variables influence the amount of *traffic that is generated* by individual traffic zones (Sheppard, 1986). In particular, the number of automobile trips per family is inversely related to population density. This relationship is partly due to the negative correlation between population density and income, but it also reflects the fact that high-density zones are characterized by a large proportion of walking trips, which are usually excluded from trip generation equations. In addition, the number of automobile trips per family is related to car ownership rates. All other things being equal, as the number of cars owned by a household increases, the daily trips per household also tend to increase. The relationship between car ownership rates and family income also implies that high-income regions within a city tend to generate more than their share of automobile traffic.

Each trip generation variable is usually combined within a multiple regression model in order to predict the overall amount of traffic that will be generated by any individual traffic zone. Some of these regression models are very satisfactory, in terms of the associated coefficients of multiple determination, but certain important methodological problems detract from their reliability. For example, the explanatory or independent variables are often highly intercorrelated. As suggested previously, there is usually a positive relationship between car ownership and income, whereas population density and income tend to be negatively correlated. Where possible, it is desirable to purge the multiple regression model of highly correlated independent variables.

Trip attraction rates for individual traffic zones are related primarily to type of land use. Residential, manufacturing, and public lands appear to generate about the same number of person trip destinations per square mile, but the number associated with commercial land use is often considerably higher. As one would expect, the higher the land use density, the greater the trip attraction rate.

Trip Distribution

Having forecast the number of trips that will be generated by and attracted to each individual traffic zone, the next step is to forecast the distribution of trips between pairs of traffic zones. This problem can be conceptualized in terms of an interaction matrix (Figure 9.1). There are N traffic zones in this hypothetical example. The rows are the origins, representing the traffic generated by each zone, and the columns are the destinations, representing the traffic attracted to each zone. The purpose of any trip distribution model is to forecast the number of trips in each cell of the interaction matrix, such as t_{32}, which indicates the number of trips beginning in zone 3 and ending in zone 2.

Two types of models, or equations, are generally used to predict the distribution of trips within a city. The simplest of these is an equation designed to capture the *friction-of-distance* effect, which reflects the fact that trip numbers are inversely related to increasing distance. More specifically, a curvilinear relationship is usually observed, with the number of trips

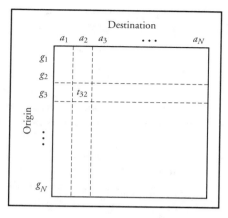

Figure 9.1 Interaction matrix.

decreasing rapidly at first and then more gradually (Figure 9.2). Such a relationship is conveniently expressed by a power function, discussed in Chapter 3:

$$T_{ij} = aD_{ij}^{-b} \qquad (9.1)$$

where T_{ij} is the number of trips between traffic zones i and j; D_{ij} is the distance between traffic zones i and j; and a and b are constants. Recall that the constants can be estimated by least-squares analysis if the equation is transformed into the following linear function:

$$\log T_{ij} = \log a - b\,(\log D_{ij}) \qquad (9.2)$$

where log refers to common logarithms to the base 10 and the remaining notation is the same as in equation (9.1). The value of b represents the slope of the line and provides a numerical value for the frictional effect of distance. In those situations where b has a relatively high value, indicating a steep slope, it means that the number of trips declines very rapidly with increasing distance.

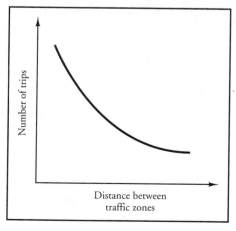

Figure 9.2 Friction-of-distance effect.

The *gravity model*, however, is the most popular trip distribution forecasting tool (Haynes and Fotheringham, 1984; Fotheringham and O'Kelly, 1989). In its simplest form it is represented as follows:

$$T_{ij} \propto \frac{P_i P_j}{D_{ij}} \tag{9.3}$$

where P_i is the population of zone i, P_j is the population of zone j, $\propto$ means "proportional to," and the remaining notation is the same as in equation (9.1). A simple example will help clarify how it is used for traffic forecasting.

Imagine that we have just three traffic zones, labeled A, B, and C, with populations and distances as represented in Figure 9.3. Substituting in equation (9.3), we can predict the proportion of traffic that will flow between each pair of traffic zones. Starting with zones A and B, we obtain:

$$T_{AB} \propto \frac{P_A P_B}{D_{AB}} = \frac{(50)\,(20)}{10} = \frac{1000}{10} = 100 \tag{9.4}$$

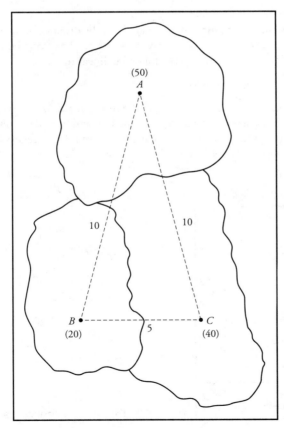

Figure 9.3 Hypothetical gravity model example.

Then, for zones A and C,

$$T_{AC} \propto \frac{P_A P_C}{D_{AC}} = \frac{(50)\ (40)}{10} = \frac{2000}{10} = 200 \tag{9.5}$$

Finally, for zones B and C,

$$T_{BC} \propto \frac{P_B P_C}{D_{BC}} = \frac{(20)\ (40)}{5} = \frac{800}{5} = 160 \tag{9.6}$$

Therefore, if for any given day there are 1000 trips within this hypothetical city, then, using equations (9.4), (9.5), and (9.6), we would predict that

$$T_{AB} = \frac{100}{460}(1000) = 217 \tag{9.7}$$

$$T_{AC} = \frac{200}{460}(1000) = 435 \tag{9.8}$$

and

$$T_{BC} = \frac{160}{460}(1000) = 348 \tag{9.9}$$

This general model can be adapted to predict specific types of trips. For example, when considering the journey to work, the gravity model is often rewritten as follows:

$$T_{ij} = a\frac{P_i E_j}{D_{ij}^b} \tag{9.10}$$

where E_j is the number of employment opportunities in zone j, a and b are constants, and the rest of the notation is the same as in equations (9.1) and (9.3). Note that trip generation is measured by population, as it refers to the home end of the work trip, trip attraction is measured by employment opportunities, and the friction of distance effect is included as the denominator.

Modal Choice

Modal choice models attempt to predict which mode of transportation will be chosen for particular journeys. For example, Daniels and Warnes (1980:199) identified five sets of variables that influence the choice of travel mode for the journey to work (Figure 9.4). First, the *location* of the person with respect to his or her workplace will obviously make a difference. If the person lives only a few blocks away from work, he or she might well walk or ride a bicycle. Second, and related to the location factor, are a set of *trip factors*. These factors include the trip length, the travel cost, and the travel-time ratio. The travel-time ratio is often particularly significant, as it compares the travel times associated with different modes of transportation.

Third, there are a number of *private transportation factors*. Parking facilities near the workplace and the cost of parking are of primary concern. Also, the availability of suitable car

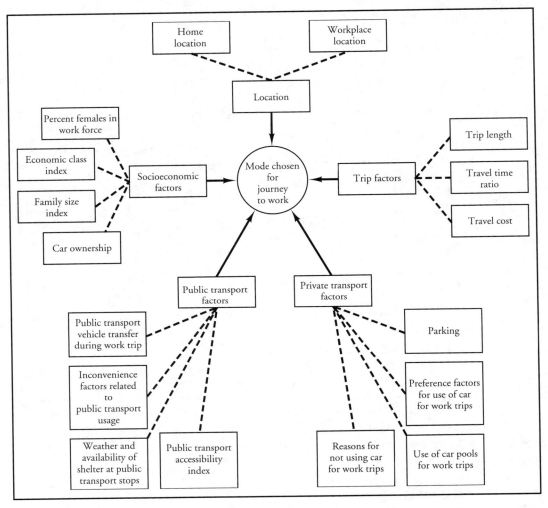

Figure 9.4   Some factors affecting the choice of transportation mode for the journey to work. (From P. W. Daniels and A. M. Warnes, *Movement in Cities: Spatial Perspectives on Urban Transport and Travel,* Methuen, New York, 1980, Fig. 7.4, p. 198.)

pools will influence the decision of whether to use private transportation. Fourth, a set of *public transportation factors* is involved. These factors include the accessibility of the person to public transportation facilities, such as buses and trains, and the inconveniences associated with using public transportation. For example, during cold weather the decision to use mass transit systems is often dependent on the availability of bus shelters and the like. Fifth, and finally, modal choice is influenced by a set of *socioeconomic factors*. In particular, automobile ownership rates and economic status are important. Areas characterized by high economic status tend to be less dependent on public transportation than are areas characterized by low economic status.

Trip Assignment

The final task, in terms of forecasting traffic flows, is to assign trips to particular roads or routes within the city. In other words, trip assignment involves models of route choice (Bovy and Stern, 1990). The simplest such model, when predicting flows between traffic zones, postulates that everybody will choose the route with the shortest travel time. This approach is termed an *all-or-nothing assignment*, as one route is assumed to carry all the traffic. A variety of problems occurs when using all-or-nothing assignments. First, as with modal choice, people not only consider travel time when selecting a route, but also travel cost. Second, individual perceptions of travel time will vary. Third, traffic conditions and therefore travel time will also vary according to the time of day. Fourth, and most important, when two or more routes are close together, one route has to be only marginally quicker than the others for it to be assigned all the trips.

In an effort to address this last problem, many transportation models use *proportional assignment* rather than all-or-nothing assignment. Basically, proportional assignment involves assigning proportions of traffic among a number of alternative routes, as a function of the time differences between those routes. Thus, although this approach is more realistic in terms of producing a multipath solution, it remains rather simplistic, as travel time is still the only determinant of route choice. A more sophisticated, but somewhat cumbersome model of route choice might include variables that attempt to capture travel time in terms of the number of traffic lights or expected congestion. Similarly, the perceived attractiveness of different routes might be included in the model. In the latter context, the desire to avoid dangerous areas of the city would need to be calibrated in some way.

9.2   The Gravity Model Approach

In this section and the following three, we focus our attention on the trip-distribution component of the traffic forecasting model. In particular, we explore the traditional gravity model approach, the revealed preference approach, the behavioral approach, and a dynamic or learning-based approach. Each of these approaches will be examined within the context of a particular kind of spatial interaction, that of shopping behavior, to facilitate comparisons. In this sense, the problem becomes one of describing and explaining how consumers choose between various kinds of spatial alternatives, as represented by individual supermarkets or shopping centers (Pipkin, 1986; Halperin, 1988). At the same time, however, it should be remembered that these same approaches can be applied to a variety of other kinds of movement patterns.

Reilly's Law of Retail Gravitation

Reilly (1931) presented one of the first mathematical formulations of the gravity concept of human interaction in his study of the retail trade areas associated with cities in Texas.

He postulated that two cities would attract consumers from some smaller, intermediate city in direct proportion to their population sizes and in inverse proportion to the square of their distances from the intermediate city. More formally, he expressed these relationships as follows:

$$\frac{B_a}{B_b} = \frac{P_a}{P_b}\left(\frac{D_b}{D_a}\right)^2 \tag{9.11}$$

where B represents the proportion of retail business from the intermediate city attracted to city a and city b, P is population, and D is distance from the intermediate city.

This same model can be applied just as easily to shopping behavior within a particular city by imagining that a and b represent supermarkets, or shopping centers, and that there is some neighborhood, or census tract, located between them. In this context, the model predicts the proportion of households from the intermediate neighborhood that patronize each of the two supermarkets. The population size variable can be replaced by retail floor space in the case of supermarkets and by number of individual establishments in the case of shopping centers.

A simple, hypothetical example will serve to clarify the use of equation (9.11). Figure 9.5 depicts a situation in which there are two supermarkets, a and b, that draw consumers from some intermediate neighborhood, X. The sizes and distances are represented, and the symbol S, for supermarket size, has replaced the population term in Reilly's original formulation. By substituting the appropriate values in the equation, it can be calculated that supermarket a should receive 50 times as much trade from the intermediate neighborhood as does supermarket b.

Reilly's law of retail gravitation is representative of what is known as the *social physics approach* to understanding human behavior, as an effort is made to construct a theory of interaction based on an analogy with Newtonian physics (Brown, 1992). In this respect, Reilly suggested that interaction can be understood in terms of a mass-divided-by-distance formulation, and to maintain the analogy with Newton's laws of motion, the distance term was squared. The use of the distance variable is comparable to the friction-of-distance concept described earlier, except that in the latter context the exponent associated with distance is not determined a priori, but is empirically estimated for a particular set of data.

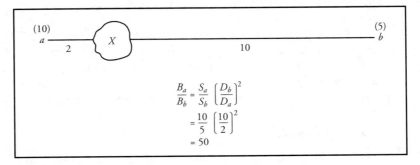

Figure 9.5 Reilly's law of retail gravitation.

Reilly's law of retail gravitation was later modified by Converse (1949) to create the *breaking-point formula*. This modification makes it possible to predict the point between two competing supermarkets of shopping centers where the trading influence of each is equal. The breaking-point formula is expressed as follows:

$$D_a = \frac{D_{ab}}{1 + \sqrt{S_b / S_a}} \qquad (9.12)$$

where D_a is the breaking point between supermarkets a and b measured in miles from supermarket a, S is supermarket size, and D_{ab} is the distance separating the two supermarkets. Using the values in Figure 9.5 and substituting them in equation (9.12), we obtain the following:

$$D_a = \frac{12}{1 + \sqrt{5/10}} = 7.03 \qquad (9.13)$$

Thus the breaking point is approximately 7 miles from supermarket a. Note that if we wish to calculate the distance of the breaking point from supermarket b, we simply divide the size of a by the size of b in the denominator of equation (9.12).

In the presence of a set of competing supermarkets, or shopping centers, the breaking-point formula can be used to predict the retail trade area associated with any individual supermarket or shopping center. Figure 9.6 illustrates the situation where there are five competing supermarkets of various sizes and depicts supermarket c's trade area by lines drawn through the breaking points associated with supermarket c and each of the other four supermarkets. As can be judged from the regular geometrical shape of this trade area, however, the delimitation of trade areas using the breaking-point formula can be regarded only as a highly generalized representation of reality. Actual trade areas will be far more irregularly shaped, as discussed in Section 5.2.

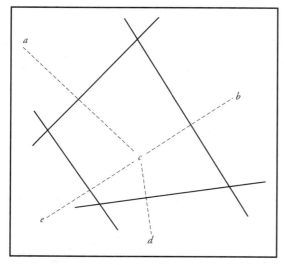

Figure 9.6 Estimating a retail trade area using the breaking-point formula.

A Probabilistic Model

Huff (1964) has suggested an alternative model for delimiting retail trade areas in which the trade area is represented in terms of probability contours (Figure 9.7a). A person located at X has a probability between 0.9 and 1.0 of patronizing supermarket a, while a person located at Y has a probability of 0.8. The probability contours for competing supermarkets within a given area can be superimposed on each other, so that a person located at X in Figure 9.7b has a 0.75 probability of going to supermarket a and a 0.25 probability of going to supermarket b. Note that the probabilities associated with the alternative opportunities, or supermarkets, sum to 1.

Huff suggested that the necessary probabilities to create the probability contours can be estimated by calculating the ratio of the utility, or attractiveness, of a particular supermarket to the total utility of all the supermarkets. Symbolically, this ratio can be expressed as follows:

$$P_{ij} = \frac{U_j}{\sum\limits_{j=1}^{M} U_j} \tag{9.14}$$

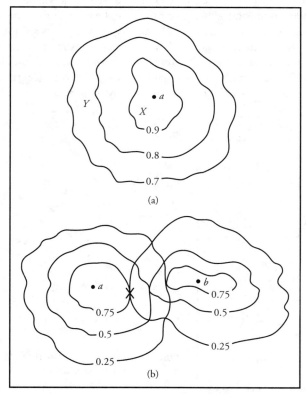

Figure 9.7 Retail trade areas represented by probability contours.

where P_{ij} is the probability of a consumer at location i traveling to supermarket j, U is the utility value associated with each supermarket, and M is the number of supermarkets. Huff postulated that the utility value of any supermarket for any particular consumer is directly proportional to the size of that supermarket and inversely proportional to its distance from the consumer. Thus the utility associated with supermarket j can be expressed as follows:

$$U_{ij} = \frac{S_j}{T_{ij}^b} \tag{9.15}$$

where U_{ij} is the utility value of supermarket j for consumer i; S_j is the size of supermarket j as measured by retail floor space; T_{ij} is the travel time involved in getting from i to j; and b is an exponent reflecting the frictional effect of distance as discussed in Section 9.1. By substituting the size and distance terms from equation (9.15) in equation (9.14), we get the following expression:

$$P_{ij} = \frac{S_j / T_{ij}^b}{\displaystyle\sum_{j=1}^{M} S_j / T_{ij}^b} \tag{9.16}$$

where the notation is the same as in equations (9.14) and (9.15).

A hypothetical example will help to clarify how equation (9.16) is used to obtain the required probabilities. Imagine that we have an individual consumer located at i who has to choose between three supermarkets, labeled A, B, and C (Figure 9.8). For computational ease we will assume that b, representing the frictional effect of distance, is 1. Then, substituting in equation (9.16), we can calculate the probability associated with supermarket A as follows:

$$P_{iA} = \frac{20/5}{20/5 + 10/2 + 5/5} = 0.4 \tag{9.17}$$

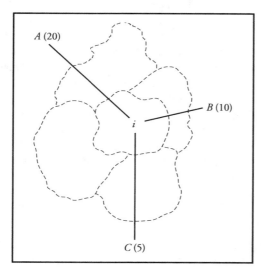

Figure 9.8 Hypothetical example to illustrate Huff's probability model.

Similarly, the probabilities associated with supermarkets B and C turn out to be 0.5 and 0.1, respectively.

This model can also be used at an aggregate as well as an individual level of analysis. Suppose that we are dealing with census tracts rather than individual consumers, as represented by the dashed lines in Figure 9.8. Then the expected number of consumers within a given census tract i that will patronize a particular supermarket j is equal to the number of consumers within i multiplied by the probability that any one of them will select supermarket j. That is,

$$E_{ij} = P_{ij}C_i \tag{9.18}$$

where E_{ij} is the expected number of consumers within census tract i who will patronize supermarket j, P_{ij} is the probability that any given consumer within census tract i will patronize supermarket j, and C_i is the total number of consumers within census tract i. Similarly, we can estimate the total demand for any particular supermarket by simply adding up the number of expected consumers from each census tract, as follows:

$$TD_j = \sum_{i=1}^{N} P_{ij}C_i \tag{9.19}$$

where TD_j is the total number of consumers expected to patronize supermarket j, N is the number of census tracts, and the remaining notation is the same as in equation (9.18).

Despite its probabilistic nature, Huff's model is very similar to Reilly's law of retail gravitation, as the utilities associated with the different supermarkets are simply a function of size and distance. Huff's formulation does have certain distinct advantages, however. First, it allows the retail trade areas to be graduated in terms of probabilities. Second, as reflected by the denominator in equation (9.16), the attractiveness of any individual supermarket takes into account the attractiveness of all other supermarkets (Vandell and Carter, 1993). Third, the exponent reflecting the frictional effect of distance is free to vary, rather than being arbitrarily set at 2.

However, although it has been demonstrated that the gravity model is a useful predictor of aggregate consumer behavior, it has often been argued that such an approach is incapable of articulating an explanation of that behavior in terms of the underlying decision-making process. Although efforts have been made to erect a theoretical foundation for the gravity model, especially within the contexts of utility theory and statistical mechanics, in many ways it is still little more than a crude physical analogy of the individual choice strategy. For example, it is probably naive to presume that people choose stores to patronize by simply making a psychological trade-off between size and distance. In other words, the gravity model can be regarded as a kind of black box that provides useful information about aggregate consumer behavior for reasons that are not altogether clear at the level of individual consumers, although it is worthwhile remembering that the original gravity model was not intended to be a model of micro-level behavior. Attempts to probe inside this black box in order to understand something about the decision-making process itself have led to an interest in the revealed preference approach, and it is to this particular approach to understanding consumer spatial behavior that we now turn our attention.

9.3   The Revealed Preference Approach

The revealed preference approach involves the examination of observed behavior in order to uncover the underlying preference structure and so establish rules of spatial behavior. In this respect, the use of revealed preferences can be viewed as a reaction to the gravity type of models, where the exponents are dependent on the spatial structure of the area within which the model is calibrated and so provide only descriptions of behavior in space, rather than fundamental parameters of some spatial choice theory. The concept of revealed preferences has been extensively utilized within the theory of consumer demand, where it has been argued that it is possible to obtain a unique ranking of objects from any consistent statement of preferences derived from the paired comparisons of those objects. Such a ranking can be graphically portrayed in terms of an *indifference surface*, which shows different combinations of goods between which the consumer is indifferent. In the present context, the objects to be ranked are supermarkets or shopping centers, and we have already come across the notion of an indifference curve when discussing the concept of bid-rents (see Section 3.1).

The gravity model can be explored via the revealed preference approach, as it implies that consumers make a trade-off between the size of a supermarket or shopping center and the distance to that supermarket or shopping center. Figure 9.9a illustrates such a trade-off, where the indifference lines join equal values of size divided by distance and thus represent various combinations of size and distance that consumers rank equally attractive. Consumers would prefer to be on the highest indifference line, as this line indicates those supermarkets or shopping centers that are among the largest and closest. Reilly's particular formulation of the gravity model, in which distance is squared, would imply the series of indifference curves shown in Figure 9.9b. The use of an exponent greater than 1 indicates that longer distances play a proportionately more important role than do shorter distances in terms of establishing the underlying preference structure.

Similarly, central place theory (see Section 5.2) implies the set of indifference lines shown in Figure 9.9c. Recall that one of the fundamental assumptions of central place theory is that consumers will patronize the nearest store offering the desired good. This assumption results in a set of vertical indifference lines, as it means that consumers care only about distance and are completely indifferent to variations in size. All consumers are assumed to prefer the left-hand indifference lines, because these are associated with the shortest distances. The greater the degree of substitution between the variables, in this case size and distance, the more the indifference curves will depart from the vertical or the horizontal.

The aim of the revealed preference approach is to derive the form of the indifference curves empirically from an analysis of overt or observed behavior. In this way, the actual trade-offs between various combinations of variables, in any given situation, can be identified (Rushton, 1981). A hypothetical example will help to clarify the underlying methodology. Assume that we ask a person to rank order the five available supermarkets in a particular city, according to how often he or she uses them. Such a task might generate the data shown in Figure 9.10a, where the distance and size associated with each supermarket are also represented. From these data the indifference curves can be mapped (Figure 9.10b), noting that the tied ranks for supermarkets *A* and *D* indicate that they both lie on the same curve. The consumer prefers to be on the uppermost curve, as this represents the most attractive combination of size and dis-

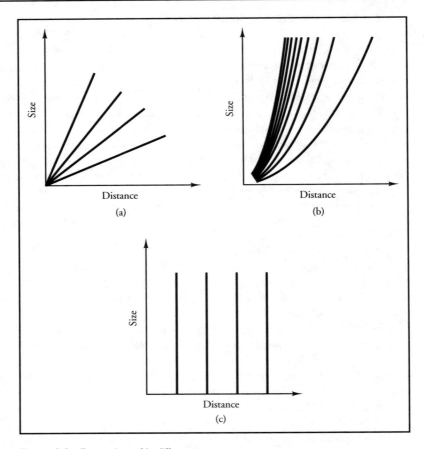

Figure 9.9 Examples of indifference curves.

tance. In general, however, we wish to generate the indifference curves for a group of consumers, rather than a single consumer, in order to filter out the effect of individual idiosyncracies. The next section describes some of the results obtained from one such aggregate study.

An Empirical Example

The data used in the analysis were obtained via questionnaires from 521 households in the city of Christchurch, New Zealand (Clark and Rushton, 1970). Each household was requested to identify its major supply center for six commodities, including groceries, and data were collected regarding the size of these supply centers and their distance from the household. It was argued that if a particular spatial opportunity was chosen whenever it was present that opportunity is the most preferred one in the consumer's mental ranking of alternatives. On the other hand, a spatial opportunity that was never chosen, even when available, must occupy the lowest rank in the consumer's evaluation of alternatives.

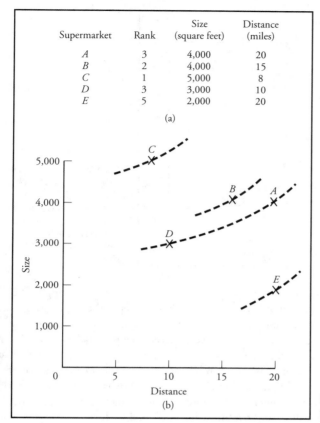

Supermarket	Rank	Size (square feet)	Distance (miles)
A	3	4,000	20
B	2	4,000	15
C	1	5,000	8
D	3	3,000	10
E	5	2,000	20

(a)

(b)

Figure 9.10 Deriving a person's indifference curves.

With this argument in mind, the following index was constructed to reflect the mental ordering of spatial alternatives:

$$I_{ij} = \frac{A_{ij}}{P_{ij}} \qquad (9.20)$$

where I_{ij} is the index value for center size class i and distance class j; A_{ij} is the number of households that chose to patronize a center in the ith center size class and the jth distance class; and P_{ij} is the total number of households that could have chosen a center in the ith center size class and the jth distance class. So the index represents the ratio of the number of times a spatial opportunity was chosen to the number of times it was available to be chosen. A pictorial representation of the resulting index values is provided in Figure 9.11. The lines are indifference curves, implying that a household is indifferent among all the spatial opportunities represented on any one indifference curve. Also, each household should prefer any alternative on a higher curve to one on a lower curve.

These indifference curves, then, represent the actual trade-off between size and distance in the context of grocery shopping. The behavior with respect to a number of different kinds

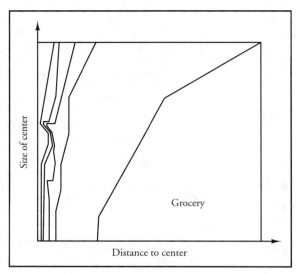

Figure 9.11 Indifference curves for grocery purchases. (From W. A. V. Clark and G. Rushton, "Models of intra-urban consumer behavior and their implications for central place theory," *Economic Geography*, 46, 1970, Fig. 6, p. 495.)

of commodities can be compared, using these indifference curves, as long as the size and distance scales are kept the same. Relatively speaking, the steeper the slope, the greater the influence of distance, and we might expect the slope to vary according to the type of good. In general, convenience goods are characterized by steeper slopes than are luxury items, as consumers are prepared to travel much farther to purchase a car than to purchase a loaf of bread.

The revealed preference approach has been subjected to a certain amount of criticism because it has been argued that only purely discretionary behavior can be analyzed meaningfully in terms of revealed preferences. In those situations where choice is constrained in some way, such as when the shopping trip must be combined with the journey to work, it is the study of the constraints themselves that may prove to be the most rewarding. Furthermore, it has also been argued that, as space itself is a constraint on choice, because of imperfect information and awareness spaces, as well as being a dimension of choice, as represented by distance, we cannot assume that preferences are independent of the environment in which they are studied (Maclennan and Williams, 1980). Until the extent of various information constraints on decision making can be identified, we cannot truly know the degree to which observed behavior reflects underlying preferences.

Since the revealed preference approach enables one only to deduce a preferential ordering for the range of spatial opportunities that are available in the study area, some researchers have attempted to develop experimental designs that allow attribute values to be manipulated such that a variety of abstract combinations can be produced, thus creating a set of hypothetical alternatives that is independent of any particular spatial structure. Perhaps the most popular approaches in this context have been information integration theory and the conjoint measurement model. *Information integration theory* and its associated method of functional measurement have been applied to a wide variety of behaviors (Anderson, 1981; 1982). The procedure essentially involves constructing an algebraic model of human information processing and then testing that model by means of analysis of variance (Timmermans, 1982). One major disadvantage of the technique, however, is that the experimental levels associated with each attribute have to be predetermined by the experimenter, and these levels will not necessarily coincide with the internalized thresholds for these attributes held by the subjects themselves.

The *conjoint measurement model*, like information integration theory, is also used in an experimental context and provides a method for defining a utility value for each alternative as a joint effect of its constituent attributes (Timmermans et al., 1984). The coefficients associated with these attributes indicate their individual contributions to the overall utility value. Studies of supermarket choice have successfully utilized the conjoint measurement technique, and its main advantage over the revealed preference approach is that subjects are directly evaluating the attributes assumed to underlie supermarket or shopping center preferences, rather than the alternatives themselves. It is in this respect that the derived rules of spatial choice are considered to be independent of the particular opportunity set being considered. As in information integration theory, however, the subject is presented with combinations of hypothetical levels of different attributes, and it is the necessary prior selection of these levels that constitutes one of the weaknesses of this approach.

The issue of external validity has frequently been raised in the context of decompositional models, such as those involving information integration theory and conjoint measurement, as decision-making under experimental conditions might well differ from actual behavior in real-world situations (Horowitz and Louviere, 1990; Timmermans et al., 1992). There is empirical evidence to suggest, however, that stated preferences in hypothetical experimental contexts are systematically related to the actual choice of supermarkets and shopping centers (Timmermans et al., 1983). In addition, although the reliability and validity of stated preference judgments are obviously compromised in those situations where the experimental task is excessively complex and the subjects are only weakly motivated, discrete choice models based on revealed preference run the considerable risk of conflating real utilities and spatial constraints. That is, the estimated parameters of a discrete choice model will be influenced by the spatial pattern of relevant alternatives. Fotheringham (1983) and Eagle (1988a), for example, have noted that spatial choices are influenced by the arrangement of competing destinations due to the role of competition and agglomeration effects.

Decompositional multiattribute preference models also have the advantage that they can refer to choice alternatives beyond the realm of direct experience, and they can therefore increase the variation in key explanatory variables (Louviere and Timmermans, 1990a). Thus, in experimental situations the choice alternatives can be designed in such a way as to maximize the number of marginal decisions that have to be made by the potential consumer (Moore, 1989). When analyzing revealed preference data, one alternative might be clearly chosen over all others, thus limiting the amount of information that is gained concerning the underlying decision-making process. It is necessary to find a sufficiently large sample of people who actually have a genuine choice and who are therefore in a position to contemplate a trade-off among the constituent attributes (Bates, 1988).

In the next section we consider an approach that is rather different from the revealed preference approach, although the overall aims are very similar. In the behavioral approach, the underlying rules or preferences associated with spatial behavior are not deduced from the analysis of observed behavior. Rather, they are hypothesized in an a priori fashion by constructing a decision-making model and then are tested by comparing the predictions derived from the model with actual behavior.

9.4   The Behavioral Approach

Like the revealed preference approach, the behavioral approach is based on the premise that there are regularities in spatial behavior and that these regularities are invariant across different spatial structures. That is, it is assumed that there are some innate rules of spatial behavior, regardless of structure. Consequently, it is argued, we need to identify these basic behavioral postulates, in contrast to the strategy of merely explaining behavior in terms of the structure in which it is operating.

A Behavioral Model of Consumer Spatial Decision Making

The present author (Cadwallader, 1975) has constructed and tested a decision-making model that conveniently illustrates the behavioral approach to understanding consumer spatial behavior. For our purposes, *decision making* is defined as the cognitive process of selecting from among alternatives, and in this particular example the alternatives are represented by five different supermarkets. The general form of the model has already been discussed within the context of the model-building process (see Section 2.3).

Around every household or group of households, there exists an objectively defined *opportunity set* (Thill, 1992). That is, within a certain radius around every household, there is a finite number of supermarkets. The problem then becomes one of analyzing how the household chooses among the alternatives presented by the opportunity set. In the present instance the aim is to construct a model that is capable of predicting the proportion of consumers who will choose each of the five different supermarkets. If this is successfully accomplished, it suggests that some of the major variables involved in the decision-making process have been identified.

The *conceptual form* of the model postulates that a prospective consumer sorts his or her information about each store in order to form judgments about their relative attractiveness and accessibility. The interplay between these two factors, attractiveness and distance, is then translated into an overt response. After a store has been patronized, there will be a feedback effect as regards the amount and nature of the information that the consumer possesses concerning his or her opportunities. This new information will cause the consumer to reassess his or her judgments regarding the relative merits of each store. In this way the decision-making process keeps repeating itself, and the stores are continually reevaluated in the light of new information. Initially, in the search stage there will be major fluctuations in the response pattern as consumers try to acquaint themselves with the system of opportunities. Eventually, however, a more consistent response pattern will develop. Any changes in the system of opportunities, such as the appearance of a new store, are likely to result in renewed search behavior.

The model suggests that the proportion of consumers patronizing a particular store will depend on the interplay among attractiveness, distance, and information. Symbolically, these relationships can be expressed as follows:

$$P_i = f(A_i, D_i, I_i) \tag{9.21}$$

where P_i is the proportion of consumers patronizing supermarket i; A_i is some measure of the attractiveness of supermarket i; D_i is some measure of the distance to supermarket i from the consumers' homes; and I_i is some measure of the amount of information generated by supermarket i. More specifically, it is postulated that the proportion of consumers patronizing any particular store increases with increasing attractiveness and decreases with increasing distance. This trade-off between distance and attractiveness operates in conjunction with an information variable. The stores that are known to exist by more consumers will tend to attract more trade; a consumer cannot patronize a store that is outside his or her field of information. These particular relationships can be expressed as follows:

$$P_i = \frac{A_i}{D_i} I_i \qquad (9.22)$$

where the notation is the same as in equation (9.21).

It is noteworthy that this kind of formulation is very different from that contained in central place theory, where it is assumed that consumer behavior is solely distance minimizing and that the attributes of the stores are identical. It is, however, similar to a gravity model formulation, with the addition of the information variable. The major difference between this model and the conventional gravity model lies in the specification of the mass and distance variables. In the present model these variables are measured subjectively by the consumers, whereas in the more conventional forms of the gravity model the mass is measured according to some objective criterion, such as retail floor space, and distance is measured in terms of physical distance. Also, the conventional models generally specify the distance variable in exponential form. This is not done in the present model, however, as the intention is to show how the distance variable may be most suitably defined, rather than to search for the best exponent. In this context it should be noted that the exponent attached to the distance variable may well vary according to the particular pattern of spatial alternatives under investigation. However, the most appropriate definition of distance, either cognitive time distance, cognitive distance, or physical distance, is likely to remain the same, irrespective of any particular configuration of spatial opportunities.

The following paragraphs describe how each part of the model can be *operationalized*, and then the output from the model is compared with observed behavior. This comparison provides an indication of the predictive capacity of the model. The data for this research were collected by questionnaires from 53 households located in West Los Angeles. The questions were answered by the member of the family who was normally responsible for the grocery shopping. Only the static form of the model is tested, as the data collected preclude analysis of the feedback effect.

In accordance with the behavioral approach, *supermarket attractiveness* is evaluated by the consumers themselves. In the present instance supermarket attractiveness is measured across four variables: (1) speed of checkout service, (2) range of goods sold, (3) quality of goods sold, and (4) prices. These variables were chosen on the basis of a pilot study in which consumers were asked to list those factors that they considered most important when selecting a supermarket. The items mentioned most often were included in the final questionnaire, and they correspond closely with those variables used in other studies of consumer spatial behavior (Fotheringham, 1988; Moore, 1988).

Each supermarket was rated by the consumers with respect to each of the four variables. This was accomplished by using a seven-point rating scale going from very unsatisfactory to very satisfactory. The resulting scale values for each store on each variable were then aggregated, using the median scale values, into a 4 by 5 attractiveness matrix (Table 9.1). This matrix indicates, for example, that the average level of satisfaction with supermarket B, with respect to prices, is 6.0. It is noteworthy that in some cases there are relatively small disparities between the stores with respect to a particular variable. For example, the highest value for the checkout service is 5.71 and the lowest value is 4.62. There are two possible reasons for this. First, it could be that the differences between the supermarkets as regards the checkout service are not sufficiently large to warrant significant differentiation. Second, it could be related to the methodology. As already noted, the subjects were provided with a seven-point scale, and the tendency is to choose the middle points on the scale. The effect of this is magnified when the median values are used, as is the case here.

The consumers were also asked to rank the variables in order of their importance in selecting a supermarket. This ranking produced a row weighting vector, the values of which have been standardized so that they add to 1 (Table 9.2). This vector shows, for example, that the prices and quality of goods sold are judged to be of equal importance. The attractiveness matrix was then premultiplied by this row weighting vector. In effect, this means that the scores on each variable and the resulting aggregate attractiveness of each store are adjusted by the importance consumers attach to each variable. The resulting row vector describes the rela-

Table 9.1   Attractiveness Matrix

	Supermarkets				
	A	B	C	D	E
Checkout service	4.62	5.39	5.13	5.71	4.78
Price	3.34	6.00	5.33	6.00	3.82
Quality of goods	6.57	5.24	5.33	5.86	5.33
Range of goods	6.74	5.16	5.40	5.29	4.77

Source: M. T. Cadwallader, "A behavioral model of consumer spatial decision making," *Economic Geography*, 51 (1975), Table 3, p. 343.

Table 9.2   Weighting Vector

Checkout service	0.13
Price	0.32
Quality of goods	0.32
Range of goods	0.23

Source: M. T. Cadwallader, "A behavioral model of consumer spatial decision making," *Economic Geography*, 51 (1975), Table 4, p. 343.

tive attractiveness of each supermarket (Table 9.3). This vector indicates that supermarket D is perceived to be the most attractive supermarket. It is significant that, in order for the variables to be additive, they are assumed to be independent. In general, this seems to be reasonable. For example, there is no reason to suppose that the range of goods sold is related to the quality of goods sold. The only exceptions to this assumption might be the quality of goods sold and prices, where one might expect some interdependence.

The next step in the model construction is to operationalize the concept of *distance*. This is accomplished by using three different measures of distance: cognitive distance, cognitive time distance, and physical distance. This means that the trade-off between store attractiveness and distance can be expressed in the following three ways:

$$A_i / CD_i \tag{9.23}$$

$$A_i / CTD_i \tag{9.24}$$

$$A_i / PD_i \tag{9.25}$$

where A_i is the attractiveness of store i, CD_i is the cognitive distance to store i as measured by the method of direct magnitude estimation (see Section 8.3), CTD_i is the cognitive time distance to store i, and PD_i is the physical distance to store i. The median value for each distance measure to each supermarket was computed (Table 9.4). It is evident that there is some varia-

Table 9.3 Attractiveness Vector

Supermarket A	5.322
Supermarket B	5.485
Supermarket C	5.321
Supermarket D	5.754
Supermarket E	4.646

Source: M. T. Cadwallader, "A behavioral model of consumer spatial decision making," *Economic Geography*, 51 (1975), Table 5, p. 344.

Table 9.4 Cognitive and Physical Distances to Each Supermarket

Supermarket	CD	CTD (minutes)	PD (miles)
A	100	5.18	0.55
B	100	5.26	0.43
C	258	10.48	1.60
D	289	10.22	1.75
E	192	7.42	0.80

Source: M. T. Cadwallader, "A behavioral model of consumer spatial decision making," *Economic Geography*, 51 (1975), Table 6, p. 345.

tion in the distance measures as regards the ordering of alternatives. For example, supermarket *D* is perceived to be closer than supermarket *C* in terms of cognitive time distance, but farther away in terms of cognitive distance. This difference is not surprising, as the two variables are measuring two different kinds of cognitive distances. In this situation the aim is to identify which of these distance measures should be inserted in models of consumer behavior.

Having operationalized the concepts of attractiveness and distance, all that remains is to specify how the level of *information* is measured. In the present study this is accomplished very simply. Information is regarded as a dichotomous variable; a consumer is either aware of a particular store or is unaware of that store. Following from this, the level of information associated with each store is measured by the proportion of consumers who are aware of that store (Table 9.5). This vector indicates, for example, that whereas all the consumers are aware of supermarket A, only 60 percent are aware of supermarket D. It is obvious that these disparities in levels of information must be incorporated within any worthwhile model, as without this constraint the attractiveness matrix is grossly misleading.

The final step is to combine the measure of attractiveness, distance, and information into the *complete model*. Because of the three different distance measures, we obtain the following three formulations:

$$P_i = \frac{A_i}{CD_i} I_i \tag{9.26}$$

$$P_i = \frac{A_i}{CTD_i} I_i \tag{9.27}$$

$$P_i = \frac{A_i}{PD_i} I_i \tag{9.28}$$

where the notation is the same as in equations (9.21) through (9.25). The results obtained from equations (9.26), (9.27), and (9.28) can be compared with the actual proportion of consumers patronizing each supermarket by using the index of dissimilarity, which is constructed in a fashion similar to the index of segregation described in Section 6.1. These results can also be compared with the predicted proportion of consumers patronizing each supermarket as

Table 9.5 Information Vector

Supermarket A	1.00
Supermarket B	0.97
Supermarket C	0.68
Supermarket D	0.60
Supermarket E	0.74

Source: M. T. Cadwallader, "A behavioral model of consumer spatial decision making," *Economic Geography*, 51 (1975), Table 7, p. 345.

calculated from a classical gravity model formulation, in which the variables are measured in objective terms. For this purpose the following equation was used:

$$P_i = \frac{OA_i}{PD_i} \qquad (9.29)$$

where OA_i is the objective attractiveness of supermarket i, as measured by retail floor space, and the remaining notation is the same as in equation (9.28).

From the results it is evident that the predictive capacity of the model is very high (Table 9.6). When using equation (9.26), three of the five predicted values are exactly equal to their observed counterparts. Also, the use of the cognitive distance measures gives better results than when using physical distance. When using cognitive distance, the index of dissimilarity is 3, whereas when using physical distance, it is 8. It can also be seen that the purely objective measures, equation (9.29), provide the least satisfactory predictions.

All the equations have comparatively small values for the index of dissimilarity, given that the index has a range of from 0 to 100. At first glance this would suggest that all four formulations work almost equally well. However, if we know nothing about the underlying decision-making process, our best prediction would be that an equal proportion of consumers will patronize each supermarket. That is, 20 percent of the consumers will patronize each of the five supermarkets. If this prediction is compared to the observed behavior, the index of dissimilarity is 27. Thus 27 is a more plausible extreme value for the index, in this particular context, than is 100. Viewed in this light, the results suggest that the predictive power of equations (9.26) and (9.27) is genuinely superior to that of equations (9.28) and (9.29).

In general, then, these results substantiate the claim that consumer spatial behavior can be better understood in terms of subjectively measured variables than in terms of their more objective counterparts. It is pertinent to bear in mind, however, that, although the model shows an encouraging predictive capacity, full explanation has yet to be reached. The model is based on the premise that the subjectively distorted environment is a better predictor of

Table 9.6   Observed versus Predicted Behavior (Percent)

Supermarket	Observed	Predicted by Equation			
		(9.26)	(9.27)	(9.28)	(9.29)
A	32	35	32	32	40
B	35	35	32	41	31
C	13	10	11	7	9
D	8	8	11	6	7
E	12	12	14	14	13
		$D = 3$	$D = 5$	$D = 8$	$D = 9$

Source: M. T. Cadwallader, "A behavioral model of consumer spatial decision making," *Economic Geography*, 51 (1975), Table 8, p. 346.
Note: D refers to the index of dissimilarity.

human behavior than is the objective environment. However, it does not go so far as to explain why the environment is subjectively distorted in the way that it is.

Further Modifications

The behavioral model described above is an example of what are generally known as multiattribute attitude models. Such models use a compositional approach, in that the utility of some object is postulated to be a weighted sum of that object's perceived attribute levels. In other words, a particular spatial alternative, such as a supermarket, can be viewed as a bundle of attributes, and the overall attitude toward that alternative is presumed to reflect the net resolution of a person's cognitions as to the degree to which it possesses those attributes, weighted by the importance of each attribute to the person. These weighted additive models are also compensatory in the sense that low ratings on one attribute can be compensated for by high ratings on another attribute.

The traditional weighted additive model can be modified in at least two major ways, however. First, the *differential weighting* of the items is usually represented by simple numbers, but the weighting function can also be calibrated by incorporating some kind of sliding scale to reflect the widespread phenomenon of decreasing marginal utilities associated with higher attainment levels. Second, multiplicative versions of the model can also be formulated to reflect the fact that subjective judgments often conform to a multiplying rule between factors. A major characteristic of the multiplicative model is that if any attribute is at a near-zero psychological value the overall attractiveness value will be very low irrespective of how high the other attributes might be rated. Recent research on consumer behavior suggests that the additive and multiplicative functions often perform equally well (Cadwallader, 1981b), and additive models appear to provide satisfactory predictions whenever the predictor variables are monotonically related to the dependent variable.

The same basic model presented above has also been tested for *different income groups* (Lloyd and Jennings, 1978). Predictions of the proportion of consumers patronizing each of five stores were computed for a high-income sample and a low-income sample. The model worked well for the low-income sample, but was less successful in terms of predicting the behavior of the high-income sample (Table 9.7). In the latter case, the proportion patronizing store A was seriously overestimated.

9.5   Dynamic Models of Consumer Behavior

The models of consumer behavior discussed thus far have been static in nature. That is, they have not involved a sequence of decisions and have therefore not considered the kind of learning process that is involved in consumer behavior. In this section we explore the use of Markov chain analysis in the context of consumer behavior and then discuss a simple learning model.

Markov Chain Models

Markov chain analysis is a useful technique for modeling sequences of decisions, as any individual choice is assumed to be dependent on some preceding choice. For example, imag-

Table 9.7 Observed and Predicted Percentages of Consumers Patronizing Five Common Grocery Stores

Store	High-income Sample		Low-income Sample	
	Observed	Predicted	Observed	Predicted
A	22	70	73	78
B	37	14	0	6
C	37	15	17	12
D	4	1	7	3
E	0	0	3	1

Source: R. Lloyd and D. Jennings, "Shopping behavior and income: Comparisons in an urban environment," *Economic Geography*, 54 (1978), Table 4, p. 163.
Note: Predicted values are based on Cadwallader's model.

ine that we have two supermarkets, A and B, in a particular area and that an individual consumer makes the following sequence of decisions regarding these two stores: *AABABAABAA*. In other words, A is chosen, then A again, then B, and so on. This sequence can be expressed in terms of probabilities: (1) the probability of B following A is 0.5, (2) the probability of A following A is 0.5, (3) the probability of A following B is 1.0, and (4) the probability of B following B is 0.0.

These same probabilities can also be expressed in terms of a *transition probability matrix* (Figure 9.12a), where stores A and B represent states that the system can be in at any particular point in time. The value 0.5 in the upper-right cell of this matrix is the probability of moving, or making the transition from state A to state B. Note that each event in the sequence of decisions depends on the event immediately prior to it. In general, the one-step transition probabilities for a set of states S_1, S_2, S_3 are expressed as in Figure 9.12b, where P_{13} represents the probability of making the transition from S_1 to S_3.

$$
\begin{array}{cc}
 & \begin{array}{cc} A & B \end{array} \\
\begin{array}{c} A \\ B \end{array} & \begin{bmatrix} 0.5 & 0.5 \\ 1.0 & 0 \end{bmatrix}
\end{array}
\qquad
\begin{array}{cccc}
 & & \begin{array}{ccc} & t+1 & \end{array} & \\
 & & \begin{array}{ccc} S_1 & S_2 & S_3 \end{array} & \\
t & \begin{array}{c} S_1 \\ S_2 \\ S_3 \end{array} & \begin{bmatrix} P_{11} & P_{12} & P_{13} \\ P_{21} & P_{22} & P_{23} \\ P_{31} & P_{32} & P_{33} \end{bmatrix} &
\end{array}
$$

(a) (b)

$$
\begin{array}{cccc}
 & \begin{array}{ccc} A & B & C \end{array} \\
\begin{array}{c} A \\ B \\ C \end{array} & \begin{bmatrix} 0.4 & 0.2 & 0.4 \\ 0 & 1 & 0 \\ 0.3 & 0.3 & 0.4 \end{bmatrix}
\end{array}
\qquad
\begin{array}{cccc}
 & \begin{array}{ccc} A & B & C \end{array} \\
\begin{array}{c} A \\ B \\ C \end{array} & \begin{bmatrix} 0.4 & 0 & 0.6 \\ 0.7 & 0 & 0.3 \\ 0.8 & 0 & 0.2 \end{bmatrix}
\end{array}
$$

(c) (d)

Figure 9.12 Transition probability matrices.

Certain rules are associated with such transition probability matrices. First, an item can only be in one of the states at any given time. Second, the sum of the elements in any row must equal unity, meaning that all possible alternatives have been specified. Third, an item moves successively within a state or from one state to another in constant intervals, and each of these moves is referred to as a *step*.

Two special situations can occur with respect to transition probability matrices. First, if an element on the diagonal of such a matrix has a value of unity, the state to which it pertains is called an *absorbing state*, as it would be impossible to leave that state. Figure 9.12c illustrates an absorbing state, as once the item is in state B, in other words store B has been chosen, it is impossible to get out of that state. Second, an *inaccessible state* is one in which all the elements of the associated column are zero. For example, in Figure 9.12d it is impossible to enter state B, or choose store B.

The transition probabilities can be obtained either empirically or theoretically. If they are to be empirically derived, one starts with a tally matrix, which is then converted into a transition probability matrix. We discussed this procedure when describing the filtering-of-housing concept in Section 6.2. In that example the transition probabilities represented the probability that a home will be occupied by a particular type of family (Table 6.5). The transition probabilities can also be derived from some theory, such as the gravity model, in which case the probabilities would reflect the sizes and distances to the various states, or supermarkets.

Besides the transition probabilities, we also need to know the values in the *starting vector* before we can analyze a Markov chain model. The starting vector summarizes the distribution of the item or items among the various states at the beginning of the time period under investigation. For example, the following starting vector illustrates the situation where we have just one consumer who is patronizing store number 2 at the beginning of the time period:

$$[S_1 \ S_2 \ S_3] = [0 \ 1 \ 0] \tag{9.30}$$

Similarly, the following starting vector illustrates the situation where we are monitoring the behavior of a number of consumers, so the distribution is depicted in terms of the proportion of consumers initially patronizing each of the three stores:

$$\left[S_1 \ S_2 \ S_3\right] = \left[0.12 \ 0.53 \ 0.35\right] \tag{9.31}$$

Abler et al. (1971:506–8) have developed a simple example to illustrate the use of Markov chain analysis in the context of consumer behavior. Imagine that we have an individual consumer who is faced with choosing between three supermarkets, A, B, and C. If the starting vector at time t_0 is multiplied by the transition probability matrix, we obtain the state vector for time t_1, as follows:

$$
\begin{array}{c}
A\ B\ C \\
[0\ 1\ 0]
\end{array}
\begin{array}{c}
 \\
A \\
B \\
C
\end{array}
\begin{bmatrix}
A & B & C \\
0.72 & 0.20 & 0.08 \\
0.49 & 0.35 & 0.16 \\
0.47 & 0.29 & 0.24
\end{bmatrix}
=
\begin{array}{c}
A \quad\ B \quad\ C \\
[0.49\ 0.35\ 0.16]
\end{array}
\tag{9.32}
$$

Note that, after initially choosing supermarket B at time t_0, there is an approximately 50 percent chance that the consumer will switch to supermarket A at time t_1, a 0.35 probability that he or she will again patronize supermarket B, and a 0.16 probability that he or she will switch to supermarket C.

If we continue to multiply each new state vector for the succeeding time periods by the transition probability matrix, the state vector eventually converges on the following values:

$$\begin{matrix} A & B & C \\ [0.633 & 0.248 & 0.119] \end{matrix}$$

In other words, our consumer eventually learns to choose supermarket A about two-thirds of the time, supermarket B about one-fourth of the time, and supermarket C for the remaining 12 percent. In this sense, he or she progresses from a search stage, in which all three supermarkets are used frequently, to a much more predictable pattern of behavior that is focused on supermarket A.

This model is very simplistic, as it assumes that the transition probabilities remain constant. In reality the transition probabilities will tend to change over time, as a result of the consumer's experiences of success or failure with each supermarket. In any event, the amount of time it takes the consumer to move from the search stage to a more stereotyped pattern of response, or habitual state, is a reflection of the rate of learning, which can be explored by a variety of learning models.

Learning Models

Golledge (1981b) has suggested a number of learning models that might be profitably used in the context of consumer behavior. Perhaps the simplest of these is a *single-operator linear model*, which can be described by a first-order difference equation of the following form:

$$P_{t+1} = P_t + (1-X)(1-P_t) \tag{9.33}$$

where P_{t+1} is the probability of selecting a particular alternative on trial $t+1$; P_t is the probability of selecting that alternative on the previous trial, t; and X is a fraction representing the probability of making an error on successive trials. In other words, the model assumes that the probability of choosing a particular alternative on trial $t+1$ is the sum of the probability of choosing that alternative on trial t and an increment that is a proportion of the maximum possible increase, $1 - P_t$.

A hypothetical example will help to clarify the use of this model. Suppose that, for a given person, P_t is 0.5 and X is 0.8; then, using equation (9.33), we have

$$P_{t+1} = 0.5 + (1-0.8)(1-0.5) = 0.6 \tag{9.34}$$

Using the calculated value of P_{t+1} to calculate P_{t+2}, we obtain

$$P_{t+2} = 0.6 + (1-0.8)(1-0.6) = 0.68 \tag{9.35}$$

Then, continuing to P_{t+3}, we obtain

$$P_{t+3} = 0.68 + (1-0.8)(1-0.68) = 0.744 \tag{9.36}$$

and so on. Thus, the probability of choosing a new alternative, or store, on the next trial decreases at a decreasing rate (Figure 9.13a). Note that the learning curve will not intersect the horizontal axis, as the learning parameter X generates an asymptotic relationship.

A second hypothetical example will provide a useful comparison with the first. Imagine that this time P_t is 0.5 and X is 0.2; then, using equation (9.33), we have

$$P_{t+1} = 0.5 + (1-0.2)(1-0.5) = 0.9 \tag{9.37}$$

and

$$P_{t+2} = 0.9 + (1-0.2)(1-0.9) = 0.98 \tag{9.38}$$

In this case, where the probability of making an error has been reduced from 0.8 to 0.2, habitual behavior is approached more rapidly (Figure 9.13b).

Social scientists are interested in how and why the learning parameter X varies across subgroups and different kinds of behavior. As we have noted, the magnitude of this learning parameter influences the rate of approach toward habitual behavior. A large value for the learning parameter indicates that the decision maker quickly terminates search activity and assumes a stable behavioral pattern in which he or she regularly patronizes one particular alternative. Such a situation may occur where previous experience, latent learning, or urgency of action are factors that influence the decisions made on each trial. A small value for the learning parameter indicates that the habitual stage is preceded by a long period of search behavior, such as might occur in those situations where there are an especially large number of alternatives to choose among.

In summary, we have explored the traditional gravity model approach, the revealed preference approach, the behavioral approach, and a learning-based approach for analyzing the trip-distribution component of the traffic forecasting model. The gravity model is gener-

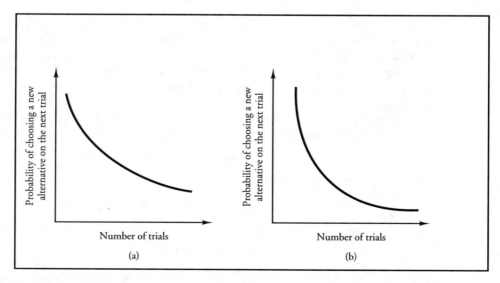

Figure 9.13 Learning curves.

ally a useful predictor of aggregate behavior, but is less successful in explaining that behavior in terms of the underlying decision-making process. As a reaction to the gravity model, revealed preferences have been used to analyze observed behavior in order to uncover the underlying preference structure. Although providing descriptive insights into the various trade-offs involved in the decision-making process, however, it is often difficult to distinguish true preferences from underlying constraints. As a result, in the behavioral approach the preferences associated with spatial behavior are not deduced from an analysis of observed behavior, but are hypothesized in an a priori fashion and then compared with actual behavior. Finally, the learning approach attempts to address the process involved in making a sequence of decisions.

A number of other issues with respect to consumer spatial choice models also require further elaboration. A desirable property of any choice model is that it be able to predict choice behavior in other geographical locations. That is, the parameter estimates of the model should be spatially transferable. The issue of transferability, however, is exceedingly complex and not easily resolved by empirical examination, as much depends on the selection of appropriate study areas (van der Heijden and Timmermans, 1988). Recent research suggests that individual decision-making is influenced by the composition of the choice set, with individuals placing more importance on those attributes exhibiting a higher variance (Eagle, 1988b). Such weight-shifting effects imply that the parameters of spatial choice models are context dependent.

An additional problem of transferability is introduced by the fact that the multinomial logit model, for example, predicts that the probability of choosing a particular spatial alternative is independent of the overall spatial pattern of alternatives. There is evidence, however, that competition effects involving the relative accessibility of competing alternatives need to be incorporated (Borgers and Timmermans, 1988). In this respect, current consumer choice models can also be criticized for failing to take into account complex multipurpose and multi-stop behavior (Timmermans and van der Waerden, 1992). Timmermans (1988) has attempted to address this problem by requesting individuals to state their preferences for a set of choice alternatives that consists of either a single or a combination of alternatives. These preferences can then be represented by a utility function that incorporates the effect of accumulative attributes.

Many of these issues would be more appropriately explored through the use of longitudinal data. Consumer choice models have traditionally been estimated using cross-sectional survey data, but it is now possible to recognize a class of dynamic discrete choice models that are longitudinal analogs of their cross-sectional counterparts (Dunn et al., 1987). The examination of repeated choices and choice sequences, however, requires an understanding of several interrelated sources of variation (Wrigley et al., 1988). First, nonstationarity, involving the variation over time in individual choice probabilities, may result from temporal shifts in an individual's attitude structure or from changes in the choice alternatives themselves. Second, heterogeneity, involving variation across individuals, might be generated by both observed and unobserved exogenous variables, such as age (Smith, 1988). Third, intertemporal state-dependence will reflect the relationship between present and past behavior. Dynamic models provide the potential for elaborating the process by which consumers use new information to continuously update and reevaluate their attitudes toward choice alternatives (Meyer and Cooper, 1988).

Chapter 10

Residential Mobility

10.1 Patterns of Mobility

With approximately 20 percent of the U.S. population changing residence every year, residential mobility is a characteristic to varying degrees of all urban neighborhoods. Indeed, it is this mobility that is largely responsible for the changing socioeconomic structure of neighborhoods and is generally associated with the deterioration and decline of particular regions within cities. More recently, however, the annual mobility rate for the United States has exhibited a general downward trend (Rogerson, 1987). There are a number of potential reasons for declining mobility during the 1970s and 1980s. First, the slow economic growth during parts of this period may have reduced the employment opportunities for prospective migrants. Second, unfavorable housing market conditions, sometimes including rapid inflation and high interest rates, made moving more difficult. Third, and perhaps most important, the increased labor force participation by females tended to foster declining mobility rates, as it has been empirically demonstrated that two-worker households are less likely to move than their single-worker counterparts (Krumm, 1983; Linneman and Graves, 1983).

Finally, Rogerson (1987) suggests that an inverse relationship exists between national mobility levels and generation size. In particular, he argues that large generations are characterized by lower mobility rates because of the increased competition in the labor and housing markets. For the 28 years for which data were available to him, he showed a negative correlation between the fraction of the total population in the 20 to 24 age group and the overall mobility rate. Note that generation size was assumed to be a surrogate for housing and labor market conditions, so economic variables were not controlled for. A more explicit model of the underlying causal relations would postulate that the demographic variable of generation size has an indirect effect on mobility rates through the operation of labor and housing market variables. In any event, demographic impacts have often been neglected in investigations of residential mobility.

When local *mobility rates* are computed for different cities, substantial variation is observed. In general, data are consistent with arguments that higher turnover rates are associated with high immigration, employment, and rates of housing growth (Moore and Clark, 1986). More specifically, a strong positive correlation has been shown between in-migration rates for metropolitan areas and the local mobility rates for those same areas (Goodman, 1982). There are two main explanations for this relationship. First, the repeat migration theory suggests that a number of moves by the same individuals are responsible for the aggregate-level correlations. In particular, long-distance migrants are often uncertain concerning the nature of the local housing market when they first arrive in a city. As a result, their initial choice of housing is often temporary while they explore different neighborhoods and housing opportunities. Second, it has also been argued that in-migration to an area often induces the original residents to leave. That is, according to the place-effect theory, in-migration can increase the mobility of current residents by altering the demographic or social composition of the neighborhood and by changing housing market conditions. Goodman (1982) has provided empirical support for both the repeat mover and the place-effect theory.

Of course, the overall pattern of *residential mobility* in any given city is composed of individual decisions to move and their associated spatial flows. The most frequently reported correlate of movement propensity is stage in the life cycle (Doorn and Van Rietbergen, 1990; Gober et al., 1991). This relationship between stage in the life cycle and residential mobility is primarily a response to the changing dwelling-space needs of the family (Clark and Onaka, 1983). The highest probability of moving occurs between the ages of 20 and 30, with the beginning of married life and the arrival of children. There tends to be greater stability while the children are at school and the head of the household is consolidating his or her career, and then mobility often increases again when the children leave home and less living space is required.

Most descriptions of individual mobility have focused on the biases associated with *distance and direction* (Clark, 1982a). There is strong evidence to suggest that the distribution of distances can be adequately represented by a family of negative exponential functions, reflecting the greater frequency of short- as opposed to long-distance moves (Knox, 1987:175–6). In terms of directional biases associated with mobility patterns, it has been suggested that although central-area moves appear to be random in direction, the moves within the suburban areas of the city tend to be biased in a sectoral fashion. This sectoral bias has been related to the idea that urban residents might possess mental images of the city that are predominantly sectoral rather than zonal (Smith and Ford, 1985) and also to the fact that the underlying socioeconomic structure of the city often has a strong sectoral component. Efforts to generalize the nature of directional biases across different cities have failed, however, largely due to the fact that the direction of moves will be as sensitive to the idiosyncratic location of new housing opportunities in a particular city as it is to the overall spatial pattern of cities in general.

Perhaps the most significant regularity in terms of mobility patterns is that households seem to move between areas of similar socioeconomic status. Most moves take place within or between census tracts of similar economic characteristics. This phenomenon emphasizes the considerable economic constraints provided by income and housing costs and suggests that, at the aggregate level at least, intraurban migration flows are remarkably predictable. Similar conclusions are implied by Alperovich (1983), who constructed a multiple regression model to assess the influence of several variables on migration within the Israeli city of Tel Aviv–Yafo. The estimated coefficients suggested the importance of such origin and destination characteristics as housing quality and resident age and educational level.

This chapter begins by focusing on the spatial distribution of mobility rates within cities and then goes on to explore the interrelationships between mobility rates and other features of the urban environment, such as socioeconomic, demographic, and housing characteristics. A simultaneous equations approach is then introduced to address the possibility of feedback effects, or reciprocal causation, between residential mobility and these other variables. Finally, the notion of structural equation models with unobserved variables is introduced, and a LISREL model of residential mobility is briefly described.

The Spatial Distribution of Mobility Rates

The amount of mobility or population turnover within urban areas is closely associated with different types of neighborhoods and thus varies substantially from one part of the city to another (Morrill, 1988). The present author (Cadwallader, 1982) has analyzed the spatial pattern of mobility rates for Portland, Oregon, for three different time periods. Data for the study were derived from census tract material for Portland, using Multnomah County to identify the spatial extent of the city. The amount of residential mobility, or rather lack of it, was defined to be the number of people residing in the same house in 1970 as in 1965, as a percentage of persons 5 years old and over in 1970, for each census tract. Values for 1950 and 1960 were derived in exactly the same fashion as for 1970, except that the data for 1950 involved the number of people residing in the same house in 1950 as in 1949. Although this modification obviously changes the absolute values, the spatial patterns associated with 1950, 1960, and 1970 can still be meaningfully compared in a relative, or distributional, sense.

The initial analysis involved investigating the relationship between residential mobility and distance from the central business district by calibrating the following three equations:

$$RM = a + bD \tag{10.1}$$

$$RM = ae^{bD} \tag{10.2}$$

$$RM = aD^b \tag{10.3}$$

where RM is residential mobility, as defined previously, and D is straight-line distance from the central business district. These three equations represent linear, exponential, and power functions, respectively, and the latter two can be fitted using least-squares analysis by making the following logarithmic transformations (as discussed in Section 3.4):

$$\ln RM = \ln a - bD \tag{10.4}$$

$$\log RM = \log a - b(\log D) \tag{10.5}$$

where the notation is the same as in equations (10.1), (10.2), and (10.3); ln refers to natural, or naperian, logarithms, and log refers to common logarithms to the base 10.

The results of this curve-fitting exercise (Table 10.1) show that for all three time periods the amount of residential stability increases with increasing distance. The coefficients of determination indicate that the power function is the most appropriate form for specifying this relationship, although clearly, even in this case, there is still a comparatively large proportion of unexplained variation, suggesting that more complex functional relationships might be profitably explored. Trend surface analysis (see Section 5.4) was used to facilitate this exploration, as it allows the pattern of residential mobility to be conceptualized as a surface, rather than simply averaging the mobility rates across different directions from the city center.

The goodness of fit between the computed surface and the mapped variable, in this case residential mobility, is usually stated in terms of the percentage reduction in the total sum of squares attributable to the fitted surface. In this context it should be remembered that the distribution of data points in a trend surface analysis is of some importance, as a clustering of data points will tend to inflate the R^2 values. The distribution of data points is not a problem in the present study, however, as they represent the centroids of census tracts and are thus fairly evenly distributed throughout the city. What little clustering there is is associated with the smaller tracts toward the center of the city and is not considered critical, as there is experimental evidence to suggest that trend surface analysis is fairly robust to small departures from a completely uniform distribution of data points.

The results of fitting linear, quadratic, and cubic surfaces are given in Table 10.2, although the significance levels associated with each surface are not reported, as these are notoriously difficult to interpret in the case of trend surface analysis. More complex surfaces

Table 10.1 Coefficients of Determination for Residential Mobility and Distance from the CBD

	1950	1960	1970
Linear function	0.09	0.17	0.17
Power function	0.26	0.33	0.34
Exponential function	0.10	0.19	0.18

Source: M. T. Cadwallader, "Urban residential mobility: A simultaneous equations approach," *Transactions of the Institute of British Geographers*, New Series, 7 (1982), Table I, p. 461.

Table 10.2 Coefficients of Determination for the Trend Surface Analyses of the Residential Mobility Surfaces

	1950	1960	1970
Linear surface	0.24	0.08	0.02
Quadratic surface	0.45	0.40	0.10
Cubic surface	0.57	0.52	0.21

Source: M. T. Cadwallader, "Urban residential mobility: A simultaneous equations approach," *Transactions of the Institute of British Geographers*, New Series, 7 (1982), Table II, p. 462.

were not explored, as it is often extremely difficult to determine the empirical meaning of anything beyond a third-order surface. Also, degrees of freedom begin to become a problem with higher-order surfaces, as a perfect fit will result whenever the number of terms in the trend equation equals the number of data points.

For all three years the best fit is provided by the cubic surface, with between 21 and 57 percent of the variation in mobility rates being accounted for, although the comparatively small coefficients of determination indicate that the actual surfaces are extremely convoluted. Of greater interest, however, is the change over time. For all three surfaces, the linear, quadratic, and cubic, the amount of explained variation is greatest for 1950 and least for 1970. This situation, which remains the same even after calculating the corrected coefficients of determination, in order to take into account the different number of census tracts for each time period, indicates that the mobility surface has become increasingly complex over time. Significantly, in previous research, similar kinds of results have been obtained when investigating the configuration of population density surfaces over time (see Section 5.4).

10.2   The Relationships among Housing Patterns, Social Patterns, and Residential Mobility

As suggested earlier, in addition to identifying the spatial pattern of mobility rates, it is also of interest to establish the interrelationships among mobility rates and other features of the urban environment, such as socioeconomic, demographic, and housing characteristics. Although the spatial distribution of housing characteristics and the nature of housing markets within cities have become one of the major research foci of urban geography (Chapter 4), there has been little effort to relate this research on housing markets to the ecological literature concerning social areas in cities (Chapter 6). In this section the intention is to explore explicitly the interaction between the demand for housing as expressed by different social groups, and the supply of housing, as represented by different types and quality of housing, and to identify the interrelationships among housing patterns and social patterns, on the one hand, and rates of residential mobility, on the other. These interrelationships are expressed in the form of a causal model (Figure 10.1), based on research conducted by the present author (Cadwallader, 1981a), which is analyzed by means of path analysis.

The general framework of the causal model is based on the essential interplay between *households and housing stock*. As the housing stock of an area changes, we might expect a simultaneous adjustment of the population characteristics of that area. For example, the aging and extensive subdivision of a particular neighborhood will lead to a higher proportion of multifamily dwelling units and thus a higher proportion of small and often low-income families. In many ways, this association between housing stock and household characteristics is dynamically articulated by the filtering process, whereby housing that is occupied by one income group deteriorates over time and thus becomes available to the next lower income group (see Section 6.2). Intermediaries, such as financial institutions and real estate agents, also play an important role in this matching of demand and supply in the housing market (see Section 4.2), although these social institutions are not explicitly considered in the present model.

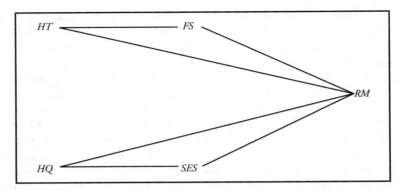

Figure 10.1 Causal model representing the interrelationships among housing patterns, social patterns, and residential mobility. *HT*, housing type; *HQ*, housing quality; *FS*, family status; *SES*, socioeconomic status; *RM*, residential mobility.

It seems plausible to suggest that housing characteristics should be placed causally prior to population characteristics. In other words, the housing stock is considered to be a constraint on the pattern of housing opportunities and thus is a major mechanism responsible for the evolution of residential differentiation in general and social areas in particular. Changes in the supply, or evolution, of the housing stock arise largely because of the aging process. Differences in the timing of housing development, due to building cycles and prevailing economic conditions, thus produce different residential patterns within cities.

Housing stock can be divided according to type, ownership, and costs. For the purpose of the present model, type and ownership, or tenure status, are collapsed into one dimension and costs are reinterpreted more broadly as housing quality. Population characteristics are also divided into two major dimensions, socioeconomic status and family status, based on the classic social area analysis studies (see Section 6.3). Thus it is postulated that when choosing a home the prospective buyer must select in terms of the type and quality of housing and that this decision will be influenced primarily by the buyer's stage in the life cycle and socioeconomic status. For example, those areas of the city containing spacious homes and yards will tend to be particularly attractive to large families with young children. Hence, in general, a neighborhood occupying a particular location in social space, defined in terms of socioeconomic status and family status, is likely to occupy an analogous, or equivalent, location in housing space, where the major axes represent the type and quality of housing. Also, we might expect socioeconomic status to be especially associated with quality of housing and family status to be especially associated with type of housing. In other words, it is assumed that socioeconomic status primarily determines the quality of housing purchased, while stage in the life cycle primarily determines the type of housing purchased.

Finally, residential mobility is interpreted as a phenomenon of the housing market, with families changing their housing stock as they experience changes in terms of both family status and socioeconomic status. More specifically, the patterns of residential mobility, as measured by turnover rates, are expected to be related primarily to housing type and family status, rather than to housing quality and socioeconomic status, as previous research has shown stage in the life cycle to be one of the dominant forces behind the decision to move (Webber, 1983). As Rossi (1980:61) stresses, residential mobility is the process by which families adjust their housing to meet the demands for space generated by changing family composition. In an aggregate sense, low-family-status areas are disproportionately inhabited by young people, renters, and apartment dwellers, all of which are characteristics normally related to high mobility rates. High-family-status areas, on the other hand, are comprised mainly of families with children, homeowners, and single-family dwelling units, characteristics usually associated with high rates of stability.

This overall conceptual schema can be formally expressed as follows:

$$H_i = f(HT_i, HQ_i) \tag{10.6}$$

$$S_i = f(SES_i, FS_i) \tag{10.7}$$

where H_i is the location of area i in housing space; HT_i is the type of housing associated with area i; HQ_i is the quality of housing associated with area i; S_i is the location of area i in social space; SES_i is the socioeconomic status associated with area i; and FS_i is the family status associated with area i. In addition,

$$S_i = f(H_i) \tag{10.8}$$

and, more specifically,

$$FS_i = f(HT_i) \tag{10.9}$$

$$SES_i = f(HQ_i) \tag{10.10}$$

where the notation is the same as in equations (10.6) and (10.7). Finally,

$$RM_i = f(FS_i, HT_i) \tag{10.11}$$

where RM_i is the amount of residential mobility associated with area i. The corresponding hypothesized causal model (Figure 10.1) also shows links between housing quality and residential mobility and between socioeconomic status and residential mobility, but these were not expected to be statistically significant.

In accordance with the proposed theoretical framework, the empirical analyses are divided into four parts: (1) the postulated major dimensions, or axes, of social space are verified via principal components analysis; (2) the major dimensions of housing space are examined in a fashion similar to the social space dimensions; (3) the hypothesized interrelationships between the social and housing patterns are tested using correlation analysis; and (4) their relationships with residential mobility, as expressed in the causal model, are explored by means of path analysis. As in the factorial ecological studies, census-tract-level information was used, with appropriate data being collected from the 1970 U.S. Census of Population and Housing for four U.S. cities: Canton (Ohio), Des Moines (Iowa), Knoxville (Tennessee), and Portland (Oregon). Four cities were chosen in order to test the robustness of the conceptual schema in different contexts, and these particular cities were chosen as they provide variation in terms of both size, with one large and three smaller, and spatial location. The data for each city included split tracts, but excluded those tracts containing zero for any of the selected variables.

Social Space and Housing Space

The major dimensions of *social space* for these four cities, as identified by principal components analysis, have already been described (Table 6.8), and the coefficients of congruence, for both socioeconomic status and family status, are all extremely high, indicating that the components are almost identical for all four cities (Table 6.9). Similar analyses to those undertaken for social space were used to identify and describe the major dimensions of *housing space*. In this case, however, the previous literature was far less helpful as a guide to the selection of variables, because there has been no consistent approach to the definition of housing submarkets, especially in a spatial context (Adams, 1991). As a result, two major criteria were used when choosing the housing variables. First, they should be representative of the mix of attributes that together make up the housing "package" or "bundle." Second, they should have theoretical implications with respect to patterns of residential mobility.

With these criteria in mind, the following six variables were chosen to represent housing space: percentage of all-year-round housing units owner-occupied; number of single-family dwelling units as a percentage of all-year-round units; median number of rooms; median housing value for owner-occupied dwelling units; percentage of all-year-round housing units with more than one bathroom; and percentage of all-year-round housing units built in 1939 or earlier. The tenure status and single-family dwelling unit variables were chosen because they are among the most powerful predictors of residential mobility. Part of the reason for the stability of owner–occupiers lies in their higher moving costs, combined with a greater flexibility in terms of being able to adjust in situ by remodeling and their greater social commitment to the local neighborhood. Dwelling unit size, especially when related to household size, is also an important determinant of mobility (Rossi, 1980). The remaining three variables — housing value, number of bathrooms, and housing age — are generally indicative of housing quality. Housing value in particular is obviously a very direct indicator of housing quality, while housing values are also associated with a variety of other housing attributes, such as age and number of bathrooms. The age of the dwelling unit can also be used as a surrogate for housing quality.

The principal components analyses of the housing variables, again using a varimax rotation, revealed two distinct dimensions, labeled housing type (HT) and housing quality (HQ) (Table 10.3). Housing type consists of owner-occupancy rates, number of single-family dwelling units, and number of rooms, and housing quality is made up of housing value, number of bathrooms, and housing age. It is interesting to note that the number of rooms loads consistently high on housing type rather than on housing quality, while housing age is associated predominantly with housing quality, although in the case of Des Moines it loads more highly on housing type. Overall, the variation in housing age is the least satisfactorily accounted for of the housing measures, with communalities ranging from 0.77 to 0.30.

The congruency coefficients between the housing dimensions reveal that the cities are remarkably similar in terms of housing structure, with the coefficients for both housing type and housing quality being as high as those previously reported for the social space dimensions (Table 10.4). For example, in the case of Knoxville and Canton the coefficient of congruence

Table 10.3   Housing Space

	HT	HQ	Comm.	HT	HQ	Comm.
		Canton			Des Moines	
Housing value	0.28	0.94	0.96	0.14	0.97	0.96
Number of bathrooms	0.24	0.94	0.94	0.23	0.96	0.97
Age of housing	−0.46	−0.75	0.77	−0.65	−0.40	0.58
Percent owner occupied	0.90	0.39	0.96	0.97	0.20	0.98
Percent single-family dwelling units	0.96	0.19	0.96	0.99	0.03	0.98
Number of rooms	0.75	0.35	0.69	0.70	0.55	0.79
Percent total variance	44.0	44.0		48.5	39.5	
		Knoxville			Portland	
Housing value	0.03	0.97	0.94	−0.02	0.95	0.90
Number of bathrooms	0.42	0.87	0.93	0.31	0.86	0.84
Age of housing	−0.53	−0.63	0.67	−0.30	−0.46	0.30
Percent owner occupied	0.94	0.29	0.96	0.96	0.23	0.97
Percent single-family dwelling units	0.99	0.06	0.97	0.98	0.05	0.96
Number of rooms	0.82	0.52	0.94	0.84	0.43	0.89
Percent total variance	49.8	40.8		46.2	34.8	

Source: M. T. Cadwallader, "A unified model of urban housing patterns, social patterns, and residential mobility," *Urban Geography*, 2 (1981), Table 3, p. 124.

Table 10.4   Coefficients of Congruence for Housing Space

	(1)	(2)	(3)	(4)
Canton (1)	—	0.96	0.99	0.98
Des Moines (2)	0.99	—	0.99	0.99
Knoxville (3)	0.98	0.99	—	0.99
Portland (4)	0.98	0.97	0.99	—

Source: M. T. Cadwallader, "A unified model of urban housing patterns, social patterns, and residential mobility," *Urban Geography*, 2 (1981), Table 4, p. 124.
Note: Coefficients for housing quality are above the diagonal; coefficients for housing type are below the diagonal.

is 0.99 for housing quality and 0.98 for housing type. The interpretability of the components, plus their obvious similarity across the four cities, suggests that an orthogonal rotation is capable of identifying simple structure, thus obviating the necessity for employing more complex and sophisticated oblique rotations (see Section 6.4).

Having identified two distinctive and recurring sets of dimensions associated with social space and housing space, the next step in the analysis was to investigate the interrelationships between these dimensions. These interrelationships were analyzed simply by calculating the zero-order correlation coefficients between the social and housing dimensions. The results of the zero-order correlation analysis indicate clearly that socioeconomic status is closely related to housing quality, and family status is equally closely related to housing type (Table 10.5). In the case of Portland, for example, socioeconomic status and housing quality have a correlation of 0.885, while family status and housing type have a correlation of 0.827. By contrast, family status and housing quality, and socioeconomic status and housing type, have correlations of only 0.018 and 0.044, respectively. The unambiguous nature of these results is especially encouraging, as they provide extremely strong evidence for the relationships postulated in equations (10.9) and (10.10).

Social Space, Housing Space, and Residential Mobility

The interrelationships among the social space and housing space dimensions and residential mobility were analyzed by means of path analysis, with the amount of residential mobility, or rather lack of it, being defined as in Section 10.1. Path analysis (see Section 4.4) allows one to estimate the magnitude of the linkages between variables, which then provide information about the underlying causal processes. In the present instance, the causal model being postulated is recursive, as the dependent variables have an unambiguous causal ordering. In this situation, ordinary least-squares regression techniques can be used to obtain the path coefficients, provided that the normal regression assumptions are met. The path coefficient associated with the disturbance terms, incorporating the combined effect of all unspecified variables, is simply the square root of the unexplained variation in the dependent variable under consideration.

The *causal model* to be calibrated is represented by the following set of structural equations (see also Figure 10.1):

$$FS = b_1 HT \tag{10.12}$$

$$SES = b_1 HQ \tag{10.13}$$

$$RM = b_1 HT + b_2 HQ + b_3 SES + b_4 FS \tag{10.14}$$

Table 10.5 Correlation Coefficients between the Social Space and Housing Space Dimensions

	Canton	Des Moines	Knoxville	Portland
Socioeconomic status and housing quality	0.855	0.928	0.910	0.885
Socioeconomic status and housing type	0.296	0.023	0.114	0.044
Family status and housing quality	0.015	0.115	0.057	0.018
Family status and housing type	0.685	0.857	0.723	0.827

Source: M. T. Cadwallader, "A unified model of urban housing patterns, social patterns, and residential mobility," *Urban Geography*, 2 (1981), Table 5, p. 125.

where the notation is the same as in equations (10.6) through (10.11). There are no values for the intercepts, as the variables are all in standardized form, and the standardized regression coefficients for these equations provide the path coefficients. Note that there are no links between housing type and housing quality and family status and socioeconomic status, as these variables represent independent components derived from the principal components analyses. Also, there are no paths between housing type and socioeconomic status and housing quality and family status, as these interrelationships were found to be insignificant in the correlation analysis (Table 10.5).

The results of the *path analysis* (Figures 10.2 to 10.5) are strikingly similar for all four cities. In every case, the path coefficients for housing type and family status and housing quality and socioeconomic status are very large and in the expected positive direction. Similarly, in all cases the path coefficient between housing type and residential mobility is significant and positive, indicating that it is those census tracts with high proportions of owner-occupied and single-family dwelling units that experience the least amount of residential mobility. For three of the four cities, there are no other significant links between any of the causally prior variables and residential mobility; but in the case of Portland, housing type also has an indirect effect via family status. These results indicate that the same comparatively simple causal structure can be postulated for all four cities. Moreover, the path coefficients associated with this causal structure are strikingly similar for each city. Nevertheless, as the path coefficients con-

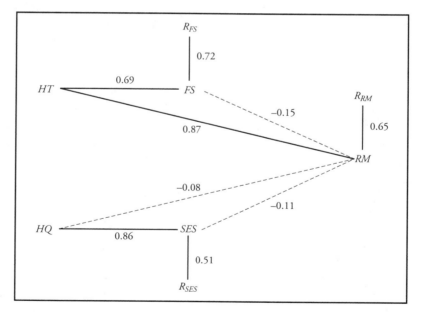

Figure 10.2 Path coefficients for Canton. *R*, residual; dashed lines represent paths that are not statistically significant at 0.05. (From M. T. Cadwallader, "A unified model of urban housing patterns, social patterns, and residential mobility," *Urban Geography*, 2, 1981, Fig. 1, p. 116.)

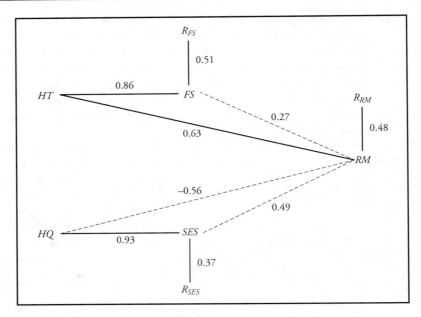

Figure 10.3 Path coefficients for Des Moines. (From M. T. Cadwallader, "A unified model of urban housing patterns, social patterns, and residential mobility," *Urban Geography*, 2, 1981, Fig. 2, p. 116.)

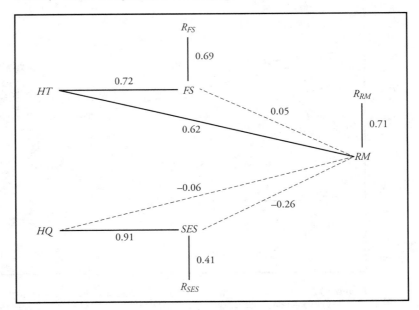

Figure 10.4 Path coefficients for Knoxville. (From M. T. Cadwallader, "A unified model of urban housing patterns, social patterns, and residential mobility," *Urban Geography*, 2, 1981, Fig. 3, p. 117.)

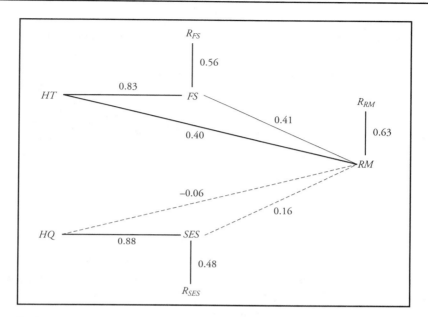

Figure 10.5 Path coefficients for Portland. (From M. T. Cadwallader, "A unified model of urban housing patterns, social patterns, and residential mobility," *Urban Geography,* 2, 1981, Fig. 4, p. 118.)

nected with the disturbance terms testify, there remain other, as yet unidentified, variables that should be included in the analysis.

In summary, there are three major results to this analysis. First, it has been demonstrated that the housing attributes of urban subareas can be decomposed into housing quality and housing type dimensions. Second, evidence has been presented to support the hypothesis that the social space and housing space dimensions are significantly interrelated and, more specifically, that socioeconomic status is a direct function of variation in housing quality, and family status is a direct function of variation in housing type. Third, a causal model, constructed to elucidate the interrelationships between the social space and housing space dimensions and residential mobility generated remarkably consistent results across four different cities, with the housing type dimension proving to be the single most important determinant of residential mobility.

10.3 A Simultaneous Equations Approach

The path models of residential mobility described in the preceding section are recursive in nature and thus do not consider the possibility of two-way or reciprocal causation. The exclusion of two-way causality presents a major theoretical problem, as it can be reasonably argued that, although the socioeconomic characteristics of urban subareas undoubtedly influence the magnitude of residential mobility rates, the reverse is also equally true. With this problem in mind, the purpose of the present section is to introduce the idea of simultaneous equations and to describe a simultaneous equation model of residential mobility.

Simultaneous Equations

An introduction to causal models in general was provided in Section 4.4, but the issue of nonrecursive models, involving two-way causation, was not examined explicitly. The causal structure depicted in Figure 10.6a represents a nonrecursive model, as there is a feedback relationship between variables X_1 and X_3. This causal structure can be expressed by the following three equations:

$$X_1 = a + b_3 X_3 + e_1 \qquad (10.15)$$

$$X_2 = a + b_1 X_1 + e_2 \qquad (10.16)$$

$$X_3 = a + b_1 X_1 + b_2 X_2 + e_3 \qquad (10.17)$$

Note that there is one equation for each variable, as the values for each variable are at least partly determined by the other variables within the causal system (see Section 4.4 for a discussion of the distinction between exogenous and endogenous variables). Each equation also contains an error term, summarizing the effects of those variables that are not explicitly included within the causal system.

Three major issues are involved when dealing with a simultaneous equation model: specification, identification, and estimation. Misspecification, or *specification error*, involves what might be more simply called using the wrong model (Duncan, 1975:101). Specification error includes the omission of relevant variables in the model and the misspecification of the correct form of the equations. In the latter context, equations (10.15), (10.16), and (10.17) assume that the relationships between the variables are both linear and additive. A linear rela-

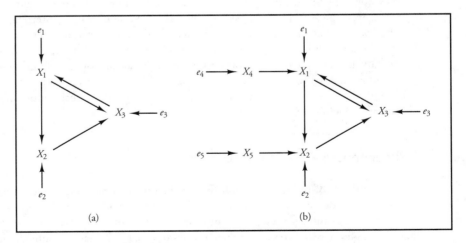

Figure 10.6 Causal models with reciprocal relationships.

tionship implies that a unit change in X_2 has the same effect on X_1, whatever the value of X_2, while an additive relationship implies that a unit change in X_2 has the same effect on X_1, whatever the values of the other variables in the equation.

Second, in order for the unknown coefficients in equations (10.15), (10.16), and (10.17) to be estimated, the model should be appropriately *identified*. The issue of identification is complex, involving the relationship between the number of equations and the number of unknowns. A necessary, although not sufficient condition for the identifiability of an individual equation within a given linear model is that the number of variables excluded from that equation should be at least equal to the number of equations, or, in other words, endogenous variables minus 1. Thus, if there are k equations, at least k - 1 of the coefficients must be set equal to zero. If precisely k - 1 are set equal to zero, the equation is exactly identified, whereas if more than k - 1 are set equal to zero, the equation is overidentified (Namboodiri et al., 1975:503). In those situations where the parameters cannot be identified in any given equation, and so the equation is underidentified, exogenous variables must be added to the system. With respect to the causal model represented in Figure 10.6a, all three equations, (10.15), (10.16), and (10.17), are underidentified, as none of them omit as many as two variables. However, by adding two exogenous variables, X_4 and X_5, to the causal system (Figure 10.6b), we obtain the following three equations:

$$X_1 = a + b_3X_3 + b_4X_4 + e_1 \tag{10.18}$$

$$X_2 = a + b_1X_1 + b_5X_5 + e_2 \tag{10.19}$$

$$X_3 = a + b_1X_1 + b_2X_2 + e_3 \tag{10.20}$$

Note that two of the five variables are omitted from each of the equations, leaving them all exactly identified.

The coefficients in these three equations must now be *estimated*. Ordinary least-squares analysis should not be used as an estimation procedure when dealing with nonrecursive systems, as the disturbance or error terms in each equation will ordinarily be correlated with the independent variables in that equation, thus leading to biased estimates. For example, in equations (10.18) and (10.20), e_1 influences X_1, which in turn influences X_3. Therefore, e_1 is not independent of X_3. An appropriate way to estimate the unknown coefficients, however, is to use two-stage least-squares analysis (Intriligator, 1978:385). As the name implies, this procedure estimates the coefficients in two stages. First, each of the endogenous variables is regressed on the exogenous variables, thus producing the *reduced-form equations*. Second, the predicted values for the endogenous variables, obtained from the reduced-form equations, are substituted on the right side of the original simultaneous equations in order to estimate the coefficients. In essence, then, the general idea behind two-stage least-squares analysis is to purify the endogenous variables that appear in the equation to be estimated in such a way that they become uncorrelated with the disturbance term in that equation (Namboodiri et al., 1975:514).

In particular, in the context of equations (10.18), (10.19), and (10.20), we have two exogenous variables, X_4 and X_5, and three endogenous variables, X_1, X_2, and X_3, thus producing the following three reduced-form equations:

$$\hat{X}_1 = a + b_4 X_4 + b_5 X_5 \tag{10.21}$$

$$\hat{X}_2 = a + b_4 X_4 + b_5 X_5 \tag{10.22}$$

$$\hat{X}_3 = a + b_4 X_4 + b_5 X_5 \tag{10.23}$$

where $\hat{X}_1$, $\hat{X}_2$, and $\hat{X}_3$ are the predicted values of X_1, X_2, and X_3, respectively, based on the use of ordinary least-squares analysis. These predicted values are then substituted into the right side of the original simultaneous equations (10.18), (10.19), and (10.20), giving the following equations:

$$X_1 = a + b_3 \hat{X}_3 + b_4 X_4 + e_1 \tag{10.24}$$

$$X_2 = a + b_1 \hat{X}_1 + b_5 X_5 + e_2 \tag{10.25}$$

$$X_3 = a + b_1 \hat{X}_1 + b_2 \hat{X}_2 + e_3 \tag{10.26}$$

These three equations are then estimated using ordinary least-squares analysis. Thus two-stage least-squares analysis essentially involves two separate applications of ordinary least-squares analysis.

An Empirical Example

Using the data for Portland, Oregon, described previously, the present author (Cadwallader, 1982) estimated a simultaneous equation model of mobility rates (Figure 10.7) for 1950, 1960, and 1970. Housing and population dimensions, similar to those described in Tables 10.3 and 6.8, respectively, served as input for the following simultaneous equations:

$$RM = a + b_1 SES + b_2 FS \tag{10.27}$$

$$SES = a + b_1 RM + b_2 HQ \tag{10.28}$$

$$FS = a + b_1 RM + b_2 HT \tag{10.29}$$

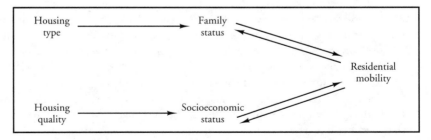

Figure 10.7 Causal structure represented by the simultaneous equation model. (From M. T. Cadwallader, "Urban residential mobility: A simultaneous equations approach," *Transactions of the Institute of British Geographers*, New Series, 7, 1982, Fig. 1, p. 466.)

where *RM* is residential mobility, *SES* is socioeconomic status, *FS* is family status, *HQ* is housing quality, and *HT* is housing type. Note that, unlike the path models described previously,
residential mobility is expected to exhibit reciprocal relationships with both socioeconomic
status and family status. Given the definition of residential mobility (see Section 10.1), all the
relationships were expected to be positive.

   Reduced-form equations were calibrated to derive estimates for the three endogenous
variables: residential mobility, socioeconomic status, and family status (Table 10.6). Housing
quality and housing type, the two exogenous or predetermined variables, were used to obtain
these estimates. In all cases the coefficients of determination are greater than 0.5, and most of
them are above 0.7. Housing type is a consistently more important predictor of residential

Table 10.6   Estimated Reduced-form Equations

1950			
RM = 76.05	+ 4.87*HT**	+ 1.02*HQ*	R^2 = 0.74
(0.41)	(0.42)	(0.42)	
SES = −0.01	+ 0.34*HT**	+ 0.82*HQ**	R^2 = 0.79
(0.06)	(0.06)	(0.06)	
FS = −0.01	+ 0.79*HT**	− 0.06*HQ*	R^2 = 0.62
(0.09)	(0.09)	(0.09)	
1960			
RM = 50.25	+ 6.07*HT**	− 1.19*HQ*	R^2 = 0.52
(0.64)	(0.64)	(0.64)	
SES = 0.02	+ 0.22*HT**	+ 0.88*HQ**	R^2 = 0.80
(0.05)	(0.05)	(0.05)	
FS = 0.00	+ 0.91*HT**	− 0.09*HQ*	R^2 = 0.84
(0.04)	(0.04)	(0.04)	
1970			
RM = 50.67	+ 6.99*HT**	+ 0.86*HQ*	R^2 = 0.55
(0.67)	(0.60)	(0.68)	
SES = −0.01	+ 0.03*HT*	+ 0.89*HQ**	R^2 = 0.78
(0.04)	(0.04)	(0.04)	
FS = 0.05	+ 0.74*HT**	+ 0.01 *HQ*	R^2 = 0.68
(0.05)	(0.05)	(0.05)	

Source: M. T. Cadwallader, "Urban residential mobility: A simultaneous
equations approach," *Transactions of the Institute of British Geographers*,
New Series, 7 (1982), Table VI, p 467.
Note: The values in parentheses are standard errors, and the asterisks indicate those coefficients that are significantly different from zero at 0.01.

mobility than is housing quality and, as anticipated, the direction of the relationship is positive. Also, as expected, housing type is by far the major determinant of family status, and housing quality is the major determinant of socioeconomic status, although in this latter case housing type also plays a minor role.

The structural equations, using the estimated values for the endogenous variables, exhibit reassuringly high coefficients of determination (Table 10.7). For all three time periods, family status is significantly related to residential mobility in the hypothesized positive direction. The coefficient for socioeconomic status, however, is not significantly different from zero at the 0.01 significance level. The equations for socioeconomic status are generally as hypothesized, with both housing quality and residential mobility exhibiting positive coefficients that are sig-

Table 10.7   Estimated Structural Equations

1950					
$RM =$	76.11	$+ 1.65\hat{SES}$	$+ 5.47\hat{FS}*$	$R^2 =$	0.40
	(0.63)	(0.75)	(0.85)		
$SES =$	−5.24	$+ 0.07\hat{RM}*$	$+ 0.75HQ*$	$R^2 =$	0.81
	(0.98)	(0.01)	(0.06)		
$FS =$	4.56	$− 0.06\hat{RM}$	$+ 1.08HT*$	$R^2 =$	0.65
	(6.33)	(0.08)	(0.42)		
1960					
$RM =$	50.27	$− 0.69\hat{SES}$	$+ 6.81\hat{FS}*$	$R^2 =$	0.29
	(0.77)	(0.86)	(0.85)		
$SES =$	−1.79	$+ 0.04\hat{RM}*$	$+ 0.92HQ*$	$R^2 =$	0.78
	(0.43)	(0.01)	(0.05)		
$FS =$	−3.61	$+ 0.07\hat{RM}$	$+ 0.48HT$	$R^2 =$	0.58
	(2.94)	(0.06)	(0.36)		
1970					
$RM =$	50.23	$+ 0.86\hat{SES}$	$+ 9.47\hat{FS}*$	$R^2 =$	0.52
	(0.69)	(0.79)	(0.84)		
$SES =$	−0.24	$+ 0.01\hat{RM}$	$+ 0.89HQ*$	$R^2 =$	0.78
	(0.28)	(0.01)	(0.04)		
$FS =$	−0.57	$+ 0.01\hat{RM}$	$+ 0.65HT$	$R^2 =$	0.71
	(3.01)	(0.06)	(0.42)		

Source: M. T. Cadwallader, "Urban residential mobility: A simultaneous equations approach," *Transactions of the Institute of British Geographers,* New Series, 7 (1982), Table VII, p. 468.
Note: $\hat{SES}$, $\hat{FS}$, and $\hat{RM}$ refer to the estimated values derived from the reduced-form equations.

nificantly different from zero. The only exception to this pattern is the coefficient associated with the residential mobility variable for 1970, which is positive, but not significantly different from zero. In contrast, family status is somewhat less satisfactorily accounted for in all three years. Housing type has the expected positive relationship with family status in all cases, but is only significantly different from zero in 1950. Residential mobility, on the other hand, has an unexpected negative relationship for 1950 and is not significantly different from zero for any of the three time periods. In other words, the amount of residential mobility appears to primarily influence the socioeconomic status of an area rather than the family status. In terms of change over time in the structural coefficients, the overall impression is one of remarkable stability. The coefficients have generally similar signs for all three years, and the pattern of statistically significant relationships is also noticeably similar. This temporal stability parallels that found for the housing and population dimensions themselves.

10.4 Models with Unobserved Variables

Models with unobserved, or latent, variables are usually specified in either of two contexts (Cadwallader, 1986a): where measured variables, like income and occupation, are subject to measurement error due to factors such as faulty recall or inaccurate coding and where accurately measured observed variables are considered to be merely indicators, or reflections, of underlying theoretical constructs that are inherently unobservable. In practice, it is difficult to distinguish between these two interpretations, as from a statistical point of view they are both treated in the same way.

Different disciplines tend to use different terminology in referring to observed and unobserved variables. In economics, the existence of either measurement error or hypothetical constructs is often recognized by referring to observed variables as proxy variables, while in sociology the term indicator has become popular for such proxy variables. In psychology, the distinction tends to be made between measured or manifest variables and latent variables, where latent variables represent hypothetical constructs, or factors, that cannot be directly measured.

Like models with observed variables, models containing unobserved variables can be either recursive or nonrecursive. It has been argued, however, that models using only measured variables are more appropriate for description and prediction than explanation and causal understanding, as measured variables will seldom correspond in a one-to-one fashion with the fundamental concepts of interest to the researcher (Bentler, 1980). Consequently, the parameters associated with such models might be expected to vary across populations, as they are only an inaccurate reflection of the true underlying causal mechanism. For this reason, causal models that are used to represent the basic theory of interest are often augmented by an *auxiliary or measurement model* that identifies the linkages between true scores and measured variables.

These auxiliary models might try deal with measurement error by using either single or multiple indicators. The traditional regression approach assumes perfect measurement of all but the dependent variable, but in recent years this restriction has been remedied by the development of confirmatory factor analysis and the more general model for the analysis of covariance structures. In particular, the development of *structural equation models* has integrated a number of originally disparate research traditions in econometrics, psychometrics, and sociometrics. While econometricians have traditionally favored simultaneous equation models, involving nonrecursive relationships among a set of variables that contain negligible measurement error, psychometricians have tended to emphasize the problems of measurement error and have thus pursued the areas of inquiry known as factor analysis and reliability theory. At the same time the work on path analysis in sociology has encouraged the realization that identification can be attained in the presence of both measurement error and simultaneous relationships.

The introduction of unmeasured variables into the causal system, however, tends to generate a large number of unknowns, and this in turn leads to problems of identification. In general, the more complex the causal model is the greater must be the ratio of measured to unmeasured variables. Thus we are often faced with a disturbing trade-off if we want to construct highly overidentified models that are easily rejectable. If our general theories are very complex, containing many unknown coefficients, then the measurement model must be simple. But if the measurement model is very complex, then the basic theory must be comparatively simple.

It is perhaps easiest to start by considering a model that contains measurement error in just one of the variables (Figure 10.8). The causal chain of interest is represented by the variables X_1, X_2^*, and X_3, but X_2^* is not observed directly. Rather, its observed counterpart X_2 is contaminated with an error, denoted as e. The term e refers to measurement errors, or errors-in-variables, as distinct from u and v, which refer to the unobserved disturbance terms, or error-in-equations. To simplify the example, it can be assumed that e is uncorrelated with X_2^* or with X_1 and X_3. In other words, the error is regarded as being random rather than systematic. This simple model can be represented by the following three equations:

$$X_2 = X_2^* + e \tag{10.30}$$

$$X_2^* = b_{21}X_1 + u \tag{10.31}$$

$$X_3 = b_{32}X_2^* + v \tag{10.32}$$

where equation (10.30) describes the measurement model and equations (10.31) and (10.32) constitute the causal model.

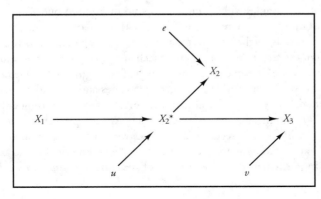

Figure 10.8 A model with measurement error in one of the variables. (From M. Cadwallader, "Structural equation models in human geography," *Progress in Human Geography,* 10, 1986, Fig. 7.)

More often, however, multiple rather than single indicators are utilized in the context of measurement error. Typically, no observed variable is an indicator of more than one latent variable, and the structural disturbances of the measurement equations are mutually independent (Bielby and Hauser, 1977). In some circumstances these specifications can be relaxed, thus allowing an observed variable to be an indicator of more than one latent variable and also allowing certain correlations to exist among the disturbances in the measurement equations.

The *confirmatory factor model* represents a measurement model containing multiple indicators that is not attached to a causal or structural equation model, as the factors, or latent variables, are merely correlated, rather than being linked by explicit causal connections. Figure 10.9 shows a confirmatory factor model that contains two factors, X_1 and X_2, that each have two indicators, with the arrows pointing from the latent variables to the measured variables. This diagram can be represented by the following four equations:

$$Y_1 = b_1 X_1 + e_1 \tag{10.33}$$

$$Y_2 = b_2 X_1 + e_2 \tag{10.34}$$

$$Y_3 = b_3 X_2 + e_3 \tag{10.35}$$

$$Y_4 = b_4 X_2 + e_4 \tag{10.36}$$

in which the predictors X and e are independent in each of the four equations and the e's are mutually independent, thus generating the classical factor analysis model. Note that the diagram contains more explicit information about the behavior of the disturbance terms than do the equations.

Traditionally, three different *types of factors* are distinguished (Kenny, 1979:110). A common factor appears in the equation of two or more variables, a specific factor appears in a single equation, and an error factor is simply the measurement error associated with a particular variable. In practical terms it is difficult to distinguish between specific and error factors, so they are usually combined to form the unique factor, represented by the e's in Figure 10.9. In our example, the two common factors X_1 and X_2 are expected to be correlated, implying an oblique, rather than an orthogonal, factor model.

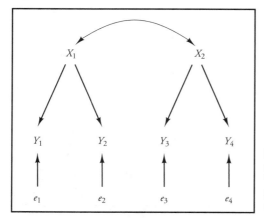

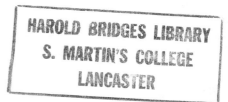

Figure 10.9 A confirmatory factor model. (From M. Cadwallader, "Structural equation models in human geography," *Progress in Human Geography*, 10, 1986, Fig. 8.)

In exploratory factor analysis the researcher is forced to make a number of restrictive assumptions (Carmines, 1986). First, all common factors, or latent variables, must be either correlated or uncorrelated. Second, all observed variables must be directly affected by all common factors. Third, the unique factors, or errors-in-variables, must be uncorrelated with one another. In contrast, the confirmatory factor model represents a more flexible approach, allowing the researcher to impose a series of substantively motivated, rather than statistically dictated, series of constraints. These constraints determine which pairs of common factors are correlated, which observed variables are affected by which common factors, and which pairs of unique factors are correlated. The model is confirmatory in the sense that statistical tests can be performed to confirm that sample data are consistent with the imposed constraints. That is, the investigator is required to hypothesize the relationships between the latent variables and observed indicators, and the model can always be rejected by the data.

Most often, *parameters* are given a fixed value, usually either zero or one, or two parameter estimates are forced to be equal. For example, a factor loading or the correlation between two factors can be set equal to zero, or a factor loading, or factor variance, can be set equal to one. In the context of Figure 10.9, one common factor regression weight must be fixed to identify each factor or else the factor variance must be fixed. In other words, b_1 and b_3 can be set equal to one, thus allowing the two factor variances to be free parameters, or the two factor variances can be set equal to one, thus allowing the four regression weights, b_1, b_2, b_3, and b_4, to be free parameters. The present model contains nine parameters (Bentler, 1980). If the common factor variances are fixed, then we have the four variances associated with the unique factors, one correlation, and four regression weights. Alternatively, if two of the regression weights are fixed, then we have the two common factor variances, the four unique factor variances, one correlation, and two regression weights.

By contrast, Figure 10.10 represents a simple two-factor, two-indicator model that indicates an explicit causal connection, rather than a mere correlation, between the two latent variables, X_1 and X_2. In factor analysis there is no structural model describing the relationships among the latent variables. In other words, they are simply correlated (oblique) or independent (orthogonal) factors. Figure 10.10, however, indicates a recursive structural model in which X_1 affects X_2. A disturbance term u has been added to X_2 in order to denote

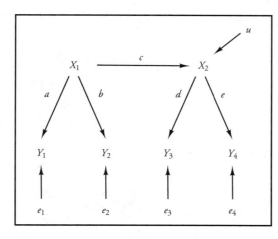

Figure 10.10 A causal model with latent variables. (From M. Cadwallader, "Structural equation models in human geography," *Progress in Human Geography,* 10, 1986, Fig. 9.)

the errors-in-equation term associated with the hypothesized structural relationship between X_1 and X_2.

Following Costner (1985), we can label the path coefficients a, b, c, d, and e and note that six correlations can be obtained from the data in order to estimate these five coefficients. Obviously, since X_1 and X_2 are both unmeasured and all the path coefficients connect some variable to either X_1 or X_2, none of the path coefficients can be directly computed from the data. The excess equation can provide a test of this overidentified system, as there is one degree of freedom. If additional indicators of each variable are introduced, then the system becomes highly overidentified, thus allowing multiple tests to be made. Also, additional indicators enable one to permit certain relatively simple kinds of nonrandom measurement errors.

Returning to Figure 10.10, we have two correlations for pairs of indicators of the same variable (Y_1, Y_2) and (Y_3, Y_4), plus four correlations using the pairs (Y_1, Y_3), $(Y_1\ Y_4)$, $(Y_2\ Y_3)$, and $(Y_2\ Y_4)$. The following equations can then be obtained for these six correlations:

$$r_{Y_1 Y_2} = ab \tag{10.37}$$

$$r_{Y_3 Y_4} = de \tag{10.38}$$

$$r_{Y_1 Y_3} = acd \tag{10.39}$$

$$r_{Y_1 Y_4} = ace \tag{10.40}$$

$$r_{Y_2 Y_3} = bcd \tag{10.41}$$

$$r_{Y_2 Y_4} = bce \tag{10.42}$$

If we have three indicators of both X_1 and X_2, and strictly random errors can be assumed for all indicators, then we have increased the number of empirical correlations to fifteen, while only increasing the number of unknown path coefficients to seven.

For such overidentified systems containing latent variables, Jöreskog (1982) has developed a procedure that provides *maximum-likelihood estimators*. An example of this LISREL model will be provided in the next section. In its most general form, the model allows one to estimate identified structural equation models that include both errors-in-variables and errors-in-equations. Because parameters within the model can be chosen to be fixed, free, or constrained, the general model contains a wide range of specific models, including the classical econometric models, where there are errors-in-equations but no errors-in-variables, and the factor analysis models in psychometrics, where there are errors-in-variables but no errors-in-equations.

10.5   A LISREL Model of Residential Mobility

The LISREL model (linear structural relations model) consists of two parts: the measurement model and the structural equation model (Jöreskog, 1982). The measurement model specifies the relationships between the latent variables and the observed variables, while the structural equation model specifies the causal relationships among the latent variables themselves. Using either a variance–covariance matrix or, less often, a correlation matrix, the approach allows one to estimate the parameters of both the causal model and the measurement model using a full information, maximum-likelihood method. The estimation technique generates parameter estimates such that the implied variances and covariances resemble, as closely as possible, their observed counterparts. That is, the aim is to generate estimates of the parameters that most closely reproduce the sample variance–covariance matrix of the observed variables.

Obviously, *estimation* is a complex task, as the measurement model represents a factor analytic approach, while the structural model involves multiple regression. It turns out, however, that the full-information methods of unweighted least squares, generalized least squares, and maximum likelihood can all be used to estimate covariance structure models, although maximum likelihood is generally preferred. Maximum likelihood is approximately normally distributed and has a comparatively small sampling variance (Long, 1983). Because of maximum likelihood's asymptotic properties, the sample size should be as large as possible, and Boomsma (1982) suggests that it is dangerous to use sample sizes of less than 100. As estimation is an iterative procedure, the necessary computer time can be very costly. The rate of convergence toward the final estimates, however, can be greatly facilitated by the selection of appropriate starting values.

The software to estimate such models is well documented in the widely available series of LISREL programs (Jöreskog and Sorbom, 1981), which includes a number of informative empirical examples. Input involves specifying eight parameter matrices, in which each element represents a fixed, constrained, or free parameter. The eight parameter matrices include four coefficient matrices and four covariance matrices. The coefficient matrices contain the following information: the loadings for the indicators of the latent exogenous variables, the loadings for the indicators of the latent endogenous variables, the paths from the latent exogenous variables to the latent endogenous variables, and the paths from the latent endogenous variables to other latent endogenous variables. Similarly, the four covariance matrices specify the covariances between the latent exogenous variables, the errors in equations, the errors in exogenous indicators, and the errors in endogenous indicators.

For determining *goodness of fit*, the difference between the implied and observed variance–covariance matrices can be statistically assessed using a chi-square statistic (Bentler and Weeks, 1980). The degrees of freedom for this likelihood ratio test statistic are equal to the number of overidentifying restrictions in the model, and a comparison is made between the constraints imposed by the model, or the null hypothesis, and the unrestricted moments matrix. Jöreskog's approach to the measurement model allows the researcher to generalize the traditional factor model by enabling the relaxation of the assumption of no correlations between the errors, as well as between factors (Zeller and Carmines, 1980:174). If one retains the assumption that the error terms are uncorrelated, then the common factors are considered to account for all the covariance in the observed variables, while the residuals only contribute to the variance of each variable (Long, 1981).

If the value of the likelihood ratio is large compared with the degrees of freedom, then it is concluded that the overall model does not represent the causal mechanisms that generated the data. It should be noted, however, that in very large samples even a minor discrepancy between model and data can lead one to reject the model, as the chi-square statistic is a function of sample size as well as the closeness of the estimated and observed covariance matrices. Consequently, the probability of rejecting a model increases with increasing sample size, even if the residual matrix contains only trivial discrepancies between observed and predicted values. In situations involving extremely large samples, virtually all models would be rejected on purely statistical grounds. This underlines the need for researchers to develop a goodness-of-fit test that is less sensitive to sample size.

The *standard error* associated with each parameter estimate provides an indication of the importance of that parameter to the model. Parameters whose estimates are small compared with their standard errors, as determined using the standard normal curve, can be eliminated from the model (Bentler, 1980). In this way certain paths can be removed from the causal structure, while an examination of the residuals might suggest the addition of other links, including those representing the correlations between errors. After such modifications the resulting model must be reestimated.

In those situations where an alternative model is a subset of the initial model, the difference in chi-square values between the two models is itself a chi-square statistic, which can thus be used to indicate the importance of the parameters that differentiate between the two models. In other words, in a series of *hierarchical models*, where restrictions are successively added or deleted in a systematic fashion, the likelihood-ratio statistics can be compared in order to test the significance of the restrictions imposed at each level of the hierarchy (Bielby and Hauser, 1977). The most powerful test of competing theories can be made when one theory is embedded within another such that their path diagrams are identical, except that certain paths are set equal to zero.

Higher-order measurement models can also be conceived. Here there are several levels, or orders, of latent variables, unlike the traditional factors model, which simply contains a first-order factor analytic measurement structure, where the measured variables are directly expressed via latent variables. In higher-order models, the influence of the higher-order latent variables on the measured variables is only indirect, thus blurring the distinction between the measurement model and the causal model, as the various levels of latent variables affect each other through regression structures (Bentler, 1982). Irrespective of whether one is using a first- or higher-order measurement structure, however, it is difficult to settle on the appropriate number of indicators for each latent variable. In theory, the more the better, but in practice a large number of indicators can make it difficult to fit a model to the data. One possible strategy is to estimate the measurement model in association with a just-identified structural model. Any significant lack of fit can then be attributed to the structural model (Kenny, 1979:182).

If a hypothesized model does not provide a good fit to the data, it can be modified in a number of ways. For example, parameters can be either eliminated or added to the model. A *modification index* is used to assess the likely result of relaxing a particular constraint (Jöreskog and Sorbom, 1981). The greatest improvement in goodness of fit is achieved by freeing the parameter with the largest modification index, although care should be taken to ensure that such inductively generated changes make theoretical sense. Latent exogenous variables can be left free to correlate, as in an oblique factor analysis. Similarly, an observed variable can become an indicator for more than one latent variable, although Browne (1982) criticizes the tendency to specify correlated error terms simply to obtain an improved fit;

there should always be sound theoretical reasons why particular errors are related. Finally, nonrecursive relationships can be specified, and selected parameters can be constrained to be equal.

Because of the presence of unobserved variables, *identification* can be a complicated matter in LISREL models, and unlike simultaneous equation and exploratory factor models, the LISREL model offers few general guidelines. Identification involves ensuring that the parameters of the model are uniquely determined; it addresses the question of whether or not there is a unique set of parameter values consistent with the data (Everitt, 1984:7). Attempts to estimate underidentified models can result in arbitrary estimates of the parameters. In all situations, therefore, it is important to remember that a LISREL model cannot be identified until the metric, or scale, of the latent variables has been established. In other words, one cannot estimate both the loadings and the variances associated with the latent variables. Thus a latent variable should be scaled either by using standard deviation units or by fixing one loading to a nonzero value. When a loading is fixed to one, the latent variable is given the same scale as the observed variable.

In general, identification is achieved by the imposition of constraints on the parameters. For example, in Figure 10.10 only two observed variables have nonzero loadings on each of the factors, thus restricting certain parameters to zero. Similarly, the error terms are unrelated, thus fixing another set of parameters. Equality constraints can also be imposed, whereby the values of the parameters are constrained to be equal. The only way to guarantee that parameters are uniquely identified, however, is to obtain algebraic solutions for them that incorporate the variances and covariances of the observed variables (Long, 1981). In some instances an overidentified structural model can help secure the identification of an underidentified measurement model (Kenny, 1979:182).

The general framework of the residential mobility model shown in Figure 10.11 is based on the previous discussion concerning the essential interplay between households and housing stock (Cadwallader, 1987). To briefly summarize, as the housing stock of an area changes, we might expect a simultaneous adjustment of the population characteristics of that area. Housing stock is divided according to housing type and housing quality, while population characteristics are classified into family status and socioeconomic status. It is postulated, therefore, that a prospective buyer choosing a home must select based on the type and quality of housing and that this decision will be primarily influenced by the buyer's stage in the life cycle and socioeconomic status. In particular, we might expect socioeconomic status to be associated with housing quality and family status to be associated with housing type. Finally, residential mobility is interpreted as a phenomenon of the housing market, with families changing their housing stock as they experience changes in both family status and socioeconomic status.

Each of the four major latent variables — housing type, housing quality, family status, and socioeconomic status — is measured by three indicator variables: percentage owner occupied (X_1), percentage single-family dwelling units (X_2), number of rooms (X_3); housing value (X_4), number of bathrooms (X_5), age of housing (X_6); persons per household (X_7), percentage under 18 (X_8), women in labor force (X_9); and education (X_{10}), income (X_{11}), and occupation (X_{12}). The remaining latent variable, residential mobility, is simply measured by the proportion of people residing in the same house as they did 5 years previously (X_{13}). Thus the *mea-*

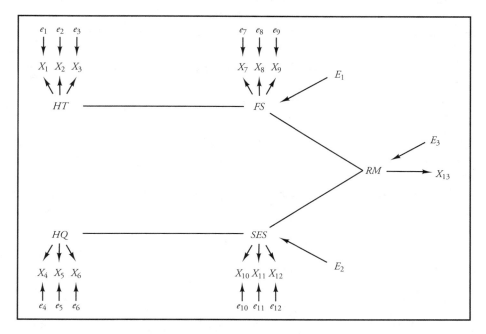

Figure 10.11   Residential mobility: Model 1. (From M. Cadwallader, "Linear structural relationships with latent variables: The LISREL model," *Professional Geographer*, 39, 1987, Fig. 2.)
Note: HT, HQ, FS, SES, and *RM* are latent variables representing housing type, housing quality, family status, socioeconomic status, and residential mobility, respectively; X_1 to X_{13} are measured indicators of the latent variables; e_1 to e_{12} represent errors in variables; and E_1 to E_3 represent errors in equations.

surement model, specifying the relationships between the observed and latent variables, can be formally expressed by the following set of equations:

$$X_1 = b_1 HT + e_1 \tag{10.43}$$

$$X_2 = b_2 HT + e_2 \tag{10.44}$$

$$X_3 = b_3 HT + e_3 \tag{10.45}$$

$$X_4 = b_4 HQ + e_4 \tag{10.46}$$

$$X_5 = b_5 HQ + e_5 \tag{10.47}$$

$$X_6 = b_6 HQ + e_6 \tag{10.48}$$

$$X_7 = b_7 FS + e_7 \tag{10.49}$$

$$X_8 = b_8 FS + e_8 \tag{10.50}$$

$$X_9 = b_9 FS + e_9 \tag{10.51}$$

$$X_{10} = b_{10} SES + e_{10} \tag{10.52}$$

$$X_{11} = b_{11} SES + e_{11} \tag{10.53}$$

$$X_{12} = b_{12} SES + e_{12} \tag{10.54}$$

$$X_{13} = b_{13} RM \tag{10.55}$$

where the notation is the same as in Figure 10.11.

Housing type and housing quality are treated as latent exogenous variables, while family status, socioeconomic status, and residential mobility are all latent endogenous variables. There are consequently three structural equations, one for each latent endogenous variable:

$$FS = b_{14} HT + E_1 \tag{10.56}$$

$$SES = b_{15} HQ + E_2 \tag{10.57}$$

$$RM = b_{16} FS + b_{17} SES + E_3 \tag{10.58}$$

The hypothesized structural model is recursive, as there are no reciprocal relationships between the endogenous variables.

Using Version 5 of the LISREL computer program (Jöreskog and Sorbom, 1981), this model was tested with 1970 census tract data for the city of Portland, Oregon. Examination of the residuals and modification indexes suggested that certain respecifications might improve the fit of the model. In general, the residuals indicate discrepancies between the covariances implied by the estimated model and the corresponding covariances that are observed in the data. These residuals can be examined one by one, and large residual covariances imply a specification error in that part of the model involving those two variables. The modification index, on the other hand, focuses on the fixed parameters of the model and indicates by how much the overall chi-square would decrease if any of these fixed parameters were made free. A relatively large modification index, therefore, suggests that the fit of the model can be substantially improved by freeing that particular parameter.

In this instance, the modification indexes indicated that the model should be changed in order to include direct effects between the two exogenous latent variables, housing type and housing quality, and residential mobility (Figure 10.12). The original structural model had thus been misspecified; housing type and housing quality not only influence residential mobility via family status and socioeconomic status, respectively, but also have a direct influence on residential mobility. Freeing a previously fixed parameter means losing one degree of freedom for chi-square, so in this case the degrees of freedom were reduced by two (Table 10.8).

Fortunately, the difference between the chi-square values of two nested models is also distributed as a chi-square distribution, with degrees of freedom equal to the difference in degrees of freedom between the two alternative models. This property is thus extremely useful when the researcher is involved in the process of model building (Herting, 1985). Adding parameters creates a less restrictive model, while removing paths would create a more restrictive model. The chi-square test thus addresses the issue of whether the improvement of fit in

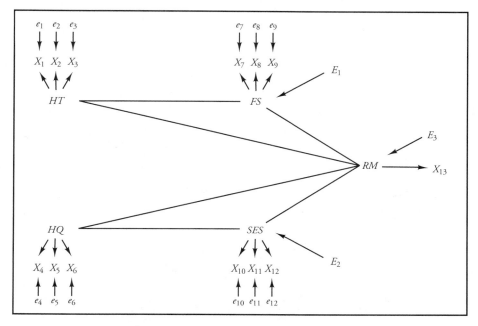

Figure 10.12 Residential mobility: Model 2. (From M. Cadwallader, "Linear structural relationships with latent variables: The LISREL model," *Professional Geographer,* 39, 1987, Fig. 3.)

Table 10.8 Successive Improvements in the Residential Mobility Model

	Reduction in Chi-Square	Reduction in Degrees of Freedom
Model 1 to model 2	136	2
Model 2 to model 3	10	1
Model 3 to model 4	222	2

Source: M. Cadwallader, "Linear structural relationships with latent variables: The LISREL model", *Professional Geographer*, 39 (1987), Table 1.

the less restrictive model is statistically significant. In this particular instance, the reestimated model showed a reduction in chi-square of 136, which is a statistically significant improvement (Table 10.8).

Examination of the modification indexes for model 2 suggested that further improvement might be achieved by allowing the two exogenous latent variables, housing type and housing quality, to be correlated, as in model 3 (Figure 10.13). This change provides a further reduction in chi-square of 10 for the loss of one degree of freedom. Finally, the modification indexes and residuals associated with model 3 suggest that correlated errors might be profit-

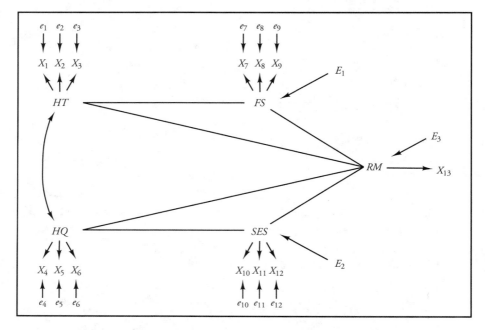

Figure 10.13   Residential mobility: Model 3. (From M. Cadwallader, "Linear structural relationships with latent variables: The LISREL model," *Professional Geographer,* 39, 1987, Fig. 4.)

ably introduced into the measurement model (Figure 10.14). In particular, variables X_1 and X_4 and X_2 and X_3 appear to have correlated errors, thus illustrating the fact that some of the indicators have spurious sources of covariation with other indicators. That is, pairs of indicators may have common sources of variation that have been omitted from the estimated models. Such appears to be the case in the present example, as the introduction of correlated errors reduces the chi-square value by 222 for the loss of only two more degrees of freedom.

In general, many different types of *specification error* can occur in causal models involving unmeasured variables (Herting and Costner, 1985). First, additivity specification error occurs if one mistakenly assumes that the effects of two or more variables are additive. Similarly, linearity specification error is present if one mistakenly assumes that all the relationships are linear. Second, direction-of-effect specification error is encountered if a stated dependency between two variables is in the opposite direction to the true dependence. Third, omitted path specification error entails omitting a causal link between two variables. Fourth, recursive specification error involves assuming a unidirectional effect when the true effect is reciprocal. Finally, spurious association specification error involves omitting a common source of variation between two variables.

It can be seen, therefore, that the LISREL model represents an extremely general approach that allows one to estimate structural equation models that contain both errors-in-variables and errors-in-equations. By judiciously fixing or freeing selected parameters, the gen-

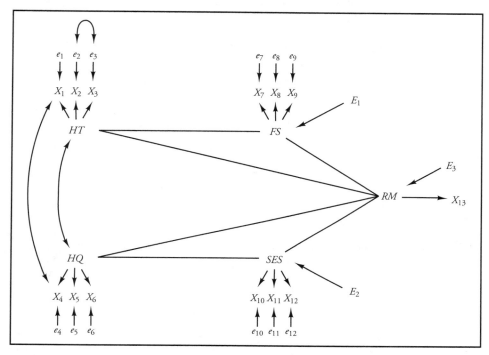

Figure 10.14   Residential mobility: Model 4. (From M. Cadwallader, "Linear structural relationships with latent variables: The LISREL model," *Professional Geographer*, 39, 1987, Fig. 5.)

eral model contains a wide range of more specific models, such as simultaneous equation systems and confirmatory factor analysis. Higher-order factor models, in which the various levels of latent variables affect each other through regression structures, can also be constructed (Bentler, 1982). In addition, certain nonlinearities can be introduced, as long as they can be transformed to achieve linearity. For example, psychophysicists often express the relationships between estimates of magnitudes and true magnitudes as power functions. Such power functions can be incorporated into measurement models by logarithmically transforming the observed scores prior to calculating the covariance matrix (Dwyer, 1983:273).

Chapter 11

The Decision-making Process in Residential Mobility

11.1 The Decision to Move

In contrast to the aggregated models described in the previous chapter, behavioral models of residential mobility focus on the individual decision maker. Traditionally, the decision-making process has been conceptually partitioned into three stages: the decision to move, the search for available alternatives, and the evaluation of those alternatives. Although this compartmentalization of the decision-making process obviously represents an oversimplification, it has had the advantage of allowing researchers to focus their attention on different parts of the whole. The present chapter parallels this framework by discussing, in sequence, the decision to move, residential search, neighborhood evaluation, and neighborhood choice.

Investigations of the initial decision to seek a new residence have emphasized the importance of prior mobility history as a determinant of subsequent behavior. In particular, the duration-of-residence effect, whereby the longer a household remains in a particular location the less likely it is to move, has been termed the *principle of cumulative inertia*. As always, there are exceptions to the rule, and for certain subpopulations the probability of a move appears to actually increase with increasing duration of stay. In general, however, several variables can be used to distinguish between movers and stayers (Poot, 1987). For example, besides the duration-of-residence effect, renters are more likely to move than owners, and there is some evidence that females are less mobile than males (Stapleton-Concord, 1982).

In the context of voluntary mobility, classifications of the reasons for moving often make a distinction between adjustment and induced moves. *Adjustment moves* are related to dissatisfaction with various attributes of the housing unit, such as its size, quality, neighborhood location, or accessibility. Space considerations are a particularly central stimulus in the mobility process, with the amount of space per person being a consistent predictor of the propensity to move (Clark et al., 1984).

Induced moves, on the other hand, are related to specific household events, such as a change in marital status or retirement. Indeed, Sell (1983) has estimated that as much as 25 percent of residential mobility occurs under conditions of substantial constraint associated with family dynamics. In this context, the traditional concept of a linear progression through a conventional life cycle has become increasingly dated. Patterns of family structure and behavior have become progressively diversified, with an increase in the number of single-headed and single-person households. An expanded conceptualization of the family life cycle is required in order to encompass the wide variety of potential living arrangements and the transitions between them (Stapleton, 1980). In addition, the complexity and the relationship to mobility patterns of such transitions can only be fully appreciated through the analysis of various interaction effects. For example, Clark et al. (1986) effectively utilized logit analysis to demonstrate that the evaluation of interaction effects among such variables as age, income, and tenure status can significantly enhance our understanding of the reasons to move or stay. The importance of interaction effects has been similarly documented in the context of residential mobility of the aged (Stapleton-Concord, 1984).

Residential Stress

Attempts to formally model the propensity to move of individual households have often revolved around the concept of locational or residential stress (Brummell, 1981; Phipps, 1989). The decision to move can be viewed as a function both of the household's present level of satisfaction and of the level of satisfaction it believes may be attained elsewhere. The difference between these levels can then be viewed as a measure of stress created by the present residential location. The decision of the household to actually seek a new residence represents an adjustment to that stress. The term *stress*, of course, has many different meanings, and our techniques for measuring stress and its effects are as yet rather crude compared with our ability to measure intelligence, attitude, or perceptual skills. Selye (1956), however, has proposed a set of terms that have been accepted by many psychologists and physiologists. He uses *stress* to refer to the state of the organism, *stressors* to refer to stress-producing agents, and *stress reactions* to refer to those responses that characterize a stressed organism.

Using this terminology, we can turn our attention to identifying the stressors. The overall amount of stress experienced by an individual household can be measured across a fairly simple set of stressors. In one study (Clark and Cadwallader, 1973b), the level of stress was measured with regard to the size and facilities of the dwelling unit, the kind of people living in the neighborhood, the accessibility to friends and relatives, proximity to the workplace, and the amount of air pollution in the neighborhood. These stressors represent both the physical aspects of the dwelling and the social conditions of the neighborhood, but additional and alternative stressors are not precluded in extensions of the model.

The size and facilities of the dwelling unit were included because these, in conjunction with the household's position in the life cycle, have been found to be of great importance for residential satisfaction. For example, Rossi (1980:61) has indicated that the major function of mobility is to enable families to adjust their housing to the housing needs that are generated by the shifts in family composition that accompany changes in the life cycle. The kind of people living in the neighborhood provides a measure of the relative importance of the social environment. How the household sees itself in relation to that environment is significant, as it has been suggested that residential mobility is sometimes the spatial expression of vertical mobility. The influx of a different socioeconomic class into a neighborhood may be particularly influential in inducing residents to move. The distance variables are proximity to friends and relatives and distance to work. Some studies have rejected job location as an important variable, but other evidence indicates that the length of the journey to work is a significant factor in the relocation of households (Clark and Burt, 1980). As the study under discussion was carried out in Los Angeles, such a variable seems especially appropriate.

The effects of these stressors can be combined in order to calculate an overall measure of residential stress. The *stress value* for an individual household i is expressed as follows:

$$S_i = \text{Md of } [(EOS_j - LS_j) + 7] \tag{11.1}$$

where S denotes the household's median level of stress, EOS is the household's ease of obtaining satisfaction at another location, and LS is its present level of satisfaction. The function is calculated over the j stressors already outlined.

To measure ease of obtaining greater satisfaction elsewhere (EOS), each household was requested to consider how easy or difficult it would be to find a more desirable location elsewhere in Los Angeles. The respondent was asked to use a seven-point attitude scale, going from very difficult (1) to very easy (7), for each of the five characteristics. Similarly, the household was asked to evaluate its present level of satisfaction (LS) on a seven-point attitude scale, going from very dissatisfied (1) to very satisfied (7), for each of the five characteristics. The greatest amount of stress is experienced when the household thinks that it is very easy to find better housing elsewhere while being very dissatisfied with the present location. In this situation the household is most likely to exhibit a strong desire to move. The problem of obtaining negative stress values, as a result of the subtraction involved in equation (11.1), is circumvented by adding seven to each median stress level. The stress scale then varies from 1 (lowest possible stress) to 13 (highest possible stress).

A *modified stress value*, with a weighting factor, was also calculated. The magnitude of this weighting factor is based on the importance of a particular characteristic j to an individual household, as revealed in its ranking of those factors as regards having a satisfactory home. The weighted stress function can be expressed as follows:

$$S_i = \text{Md of } [(EOS_j - LS_j) + 7 + w_j] \tag{11.2}$$

where the notation is the same as in equation (11.1), except for the addition of the weighting factor, w_j.

The data used to calibrate equations (11.1) and (11.2) were obtained from a random sample of 169 households in the city of Santa Monica, within the Los Angeles metropolitan area. The sample was derived by first choosing a simple random sample of 40 equal-sized blocks and then systematically sampling households within those blocks. Of the 169 households, 39 refused to be interviewed, and 21 could not be contacted after three call-backs. The results of the sampling yielded 106 usable interviews. Thirty-seven percent of the respondents lived in houses, and the remainder lived in apartments.

First, the relationship between a household's median stress level and the household's desire to move was investigated, and then the individual stressors were analyzed. The household's desire to move was measured on a seven-point attitude scale going from 1 to 7, with 1 denoting no desire to move and 7 denoting a very strong desire to move. According to the model, higher levels of stress should be associated with a strong desire to move. As the attitude scales provide ordinal-level data, *Kendall's tau* was utilized to examine the relationship between the desire to move and residential stress. Kendall's tau is an appropriate measure of association when both variables are measured at the ordinal level, and its values range from −1.0 for a perfect negative relationship to +1.0 for a perfect positive relationship.

Even though Kendall coefficients are generally thought to be especially appropriate when the data contain a large number of tied ranks or when a large number of cases are classified into a relatively smaller number of categories, it should be recognized that the numerical value of tau can be relatively small even in the presence of a moderately strong relationship. The numerical value of tau depends on the marginal totals of the categories, and we can seldom expect to get a value of tau approaching unity. In the present instance, the correlation between the desire to move and residential stress is 0.384 (Table 11.1). As expected, then, the greater the household's level of stress, the greater the potential mobility exhibited by the household. Note, however, that the weighted stress formulation does not improve the strength of the relationship.

The different amounts of stress produced by each of the five stressors can be most conveniently compared on the basis of the median stress value generated by each stressor (Table 11.2). The differences among the median stress levels are very small, suggesting that, apart from the smog stressor, each of the other four stressors is equally important in generating the overall level of stress. The small difference among the stressors partly explains the lack of improvement in the weighted stress model. Correlations between the stressors and the desire to move give a clearer indication of the relative importance of each stressor. In this case, stress due to the size and facilities of the dwelling unit appears to be the most important factor, and proximity to work seems to be somewhat less important than the other factors (Table 11.3).

Table 11.1 Relationship between the Desire to Move and Residential Stress

Correlation between the household's median stress level and desire to move:	
Kendall tau	0.384
Correlation between the household's weighted median stress level and desire to move:	
Kendall tau	0.343

Source: W. A. V. Clark and M. T. Cadwallader, "Locational stress and residential mobility," *Environment and Behavior*, 5 (1973), Table 3, p. 37.

Table 11.2 Stressors

Stressor	Median Value on the:	
	Stress Scale: 1 to 13	Weighted Stress Scale: 2 to 18
Proximity to friends and relatives	6.091	7.921
Proximity to work	5.417	8.056
Kind of people living in the neighborhood	5.231	8.038
Size and facilities of the dwelling unit	5.094	8.821
Amount of smog in the area	3.000	5.974

Source: W. A. V. Clark and M. T. Cadwallader, "Locational stress and residential mobility," *Environment and Behavior*, 5 (1973), Table 4, p. 37.

Table 11.3 Relationships between the Individual Stressors and the Desire to Move

Correlation between the household's stress due to the size and facilities of the dwelling unit and desire to move:

	Coefficient	Significance
Tau	0.352	0.001

Correlation between the household's stress due to the kind of people living in the neighborhood and desire to move:

	Coefficient	Significance
Tau	0.343	0.001

Correlation between the household's stress due to the distance from friends and relatives and its desire to move:

	Coefficient	Significance
Tau	0.253	0.001

Correlation between the household's stress due to the amount of smog in the area and desire to move:

	Coefficient	Significance
Tau	0.181	0.005

Correlation between the household's stress due to the distance from work and desire to move:

	Coefficient	Significance
Tau	0.082	0.137

Source: W. A. V. Clark and M. T. Cadwallader, "Locational stress and residential mobility," *Environment and Behavior*, 5 (1973), Table 5, p. 38.

Further Modifications

More recent formulations of residential stress have incorporated the duration-of-residence effect by postulating a trade-off between dissatisfaction and inertia. The level of stress, or dissatisfaction, is assumed to increase over time, as the household falls out of adjustment with its present situation. Similarly, however, the resistance to moving, or inertia, is also expected to increase with increasing duration-of-stay. Thus the probability of moving represents a trade-off between stress and inertia, where both components can be expressed as exponential functions of change over time. Despite the analytical appeal of such a conceptualization, however, the black-box nature of the stress and inertia function indicates that the model is best suited to a predictive rather than an explanatory role.

Attempts to further refine the stress–inertia, or stress–resistance, model have suggested that households might be as sensitive to the direction and rate of change of stress as they are to its absolute level at any specific time. Using data from the Canadian city of Saskatoon, Phipps and Carter (1984) were able to show that intention to move was greatest for those households experiencing relatively high levels of stress that had increased over the preceding 2 years. The effects of resistance on mobility were also strong, although somewhat more complex.

It is worth noting at this point that opinions differ about the strength of the relationship between desire to move and actual mobility behavior. Using data from the United States and Canada, Moore (1986) argued that the relationship between the expectation of moving and the actual fulfillment of those expectations is sometimes very limited. Not only do many respondents fail to move in a manner consistent with expectations, but disadvantaged groups are less likely to translate expectation into action than is the population in general. Other studies suggest that short-term migration intentions are an excellent predictor of subsequent behavior and that ambiguities arrive partly because of the failure to distinguish between local and nonlocal moves (McHugh, 1984). Suffice it to say, however, that supply-side constraints obviously play an important role in the relationship between mobility intention and subsequent behavior. In particular, the availability of suitable alternatives to current consumption is crucial.

In efforts to resolve the complex interactions between residential satisfaction, desire to move, and actual mobility behavior, researchers have often employed various kinds of structural equation models. Speare (1974) constructed a path analytical model to explore the interrelationships among five background variables, representing individual or household characteristics, location characteristics, and social bonds; an index of residential satisfaction; and two mobility variables, involving desire to move and actual mobility. In particular, Speare's model proposed that residential satisfaction is the proximate determinant of the decision to consider moving and that satisfaction is determined by the five background variables. That is, the key hypothesis to be tested was that the background variables would operate on the desire to move through the intervening variable of residential satisfaction. Using data from a sample of 700 residents of Rhode Island, Speare provided empirical support for his model.

A comparable study by Landale and Guest (1985), however, suggested that Speare's formulation is only partially appropriate. Using a series of logit models, they supported Speare's contention that subjective satisfaction is a strong predictor of thoughts about moving and that thoughts about moving are good predictors of actual mobility. On the other hand, they found three major problems with Speare's model. First, residential stress, as measured by satisfaction, is not a particularly good predictor of actual mobility, although it does have an indirect effect via thoughts about moving. Second, the background variables affect mobility independently, not just indirectly through satisfaction. Third, the satisfaction variables have little impact in mediating the effects of the background variables on thoughts and behavior concerning mobility.

In a similar vein, Hourihan (1984) developed a path model to explore how personal characteristics cause the perception of neighborhood attributes. Finally, Preston (1984) estimated a series of path models designed to elucidate the relationships between duration-of-residence, personal characteristics, and residential stress. Data were obtained from a survey of 513 elderly residents of the Kansas City SMSA, with each person being interviewed about present living conditions, individual and household characteristics, and residential history. To maximize internal homogeneity, the sample was divided into three groups: white home owners, black home owners, and renters. The results of the analyses suggested that residential stress and inertia are influenced by both years of residence and personal characteristics but that personal characteristics have the larger impact. The strength of the observed relationships differed across the three subpopulations, thus confirming the notion that models of residential mobility should be disaggregated whenever possible.

11.2   The Search for Alternatives

If a household is experiencing high levels of stress and thus a strong desire to move, it must then either modify its present home or begin an explicit search for alternative accommodation. Three interrelated questions are crucial to any understanding of the residential search process (Clark, 1981). First, what information sources are used to find vacant dwellings in the city? Second, how long is the search activity? Third, what is the spatial pattern associated with that search activity? In this section we address each of these questions in turn.

Information Sources

By far the most important sources of information to prospective movers are newspaper advertisements, personal contacts, personal observation of "For Sale" signs, and real estate agents. Rossi (1980:209) has concluded that personal contacts, the second most frequently used medium, after newspapers, are by far the most effective. The use of different sources of information varies, however, according to the type of dwelling unit that is desired. For example, prospective movers more often use newspaper advertisements when searching for an apartment than when seeking a house. Real estate agents are an effective source for house hunters, especially out-of-town buyers (Clark and Smith, 1982). These results are generally substantiated by those of Rossi (1980:209), although he emphasizes the particular significance of personal contacts and direct search when a household plans to rent rather than purchase.

Analyses of the *sequential structure* of information collection suggest that the preliminary stages of search are characterized by a greater dependence on newspaper advertisements, while in the later stages home buyers rely increasingly on real estate agents. In a particularly detailed study of the residential search process, Talarchek (1982) used a questionnaire survey of recent intraurban migrants in the Syracuse, New York, metropolitan area. He detected a two-stage process in the acquisition of information on specific variables. First, households obtain information on neighborhood and locational variables, such as quality of schools, property taxes, and distance to the nearest shopping center. Second, as the residential search progresses, households begin to collect information on the attributes of individual housing units, such as lot size, number of rooms, and floor plan. As might be expected, housing costs seem to be monitored throughout these two stages. While this two-stage process can be identified for the sequential acquisition of information on specific variables, no such pattern appears to exist for the sequential utilization of information sources. Indeed, Talarchek warns that a great diversity of search behavior characterizes the population, even within specific income and household composition groups.

The characteristics of the different information sources are also of some interest. Smith et al. (1982) have conducted an exhaustive examination of newspaper housing advertisements, with the specific aim of comparing the behavior of private sellers and real estate agents. Housing advertisements were sampled over a 20-year period for Santa Barbara, California, and over a 19-year period for Stockton, California. In general, the data provided support for three major hypotheses. First, real estate agents make substitutions between price and locational information in order to achieve a desired level of ambiguity, whereas owners do not appear to make such trade-offs. Second, owners provide relatively more locational information than do real estate agents, while real estate agents are more specific about price. Third, both owners and real estate agents tend to increase the level of price ambiguity when faced with increasing uncertainty concerning bid price distributions.

Real estate agents are perhaps the most complex channel of information transmission, as they act as both recipients and transmitters of information. Prospective buyers pass messages about housing preferences, while real estate agents may either intentionally or unintentionally transform and filter vacancy messages. Various characteristics of real estate agents can affect the nature and extent of their influence on the search process. For example, the ethnicity of the real estate agent contributes a particular set of dimensions to real estate practice. Using a survey of Anglo, black, and Hispanic real estate agents in Denver, Palm (1985) was able to substantiate three research hypotheses. First, sellers tend to select real estate agents of the same ethnic background as themselves. Second, the ethnicity of the real estate agent affects the size and location of territories in which he or she seeks new listings. Third, the ethnicity of the agent substantially shapes his or her network of business contacts.

Length of Search Activity

The length of time spent searching depends on the degree of satisfaction associated with present alternatives combined with the time and money costs that would be incurred by further search. In this context, researchers have constructed a variety of probability models to devise optimal "stopping rules" for housing search activity (Phipps and Laverty, 1983). Most search theories tend to incorporate the idea of a critical utility value that differentiates acceptable and unacceptable alternatives, but this threshold value will change during the course of searching due to learning and preference adjustment. Personality factors are also likely to be involved, with conservative households being prepared to follow a satisficing rather than an optimizing strategy.

Meyer (1980) has outlined a less formal model of decision-making under uncertainty. Although it represents a stopping rule model, in this case the cutoff level is an empirical parameter unique to the individual. Studying hypothetical apartment searches by undergraduate students, Meyer showed that the decision to stop searching is based on three factors: inferences about the distribution of utilities associated with the population of alternatives, the amount of time available for search, and the quality of the particular apartment currently being considered. Efforts to link these stopping rule models with the information acquisition approach have focused on the relationship between the costs of information derived from different sources and the length of search (Smith and Clark, 1980).

Spatial Pattern of Search

Finally, the spatial pattern of search is often remarkably constrained. Huff (1984) has attempted to formalize the constraining influence of *distance* by a series of distance-decay models of residential search. In particular, the intensity of search is expected to decline with increasing distance from certain key reference points in the household's search space, such as current residential location, location of workplace, and the locations of friends and relatives. Nearby possibilities are rejected before search occurs in areas at increasing distances from these various anchor points. The parameters that govern the actual shape of the distance-decay function depend on a combination of the household attributes and the characteristics of the present residence. For example, if a household is dissatisfied with the size of its present residence but is satisfied with the general location, then search activity will tend to be concentrated in the local neighborhood.

Clark and Smith (1982) conducted a detailed empirical analysis of spatial search behavior within the San Fernando Valley region of the Los Angeles metropolitan area. A series of regression equations was constructed, with variables such as the number of different neighborhoods searched and the distance of search serving as the dependent variables of interest. These equations were then estimated using individual data on two distinctive samples: a longitudinal sample of households still actively involved in the search process and a retrospective sample of households who had already bought a house. Despite some noticeable differences between the two groups, the length of residence and number of real estate agents used appeared to perform consistently well as explanatory variables. For example, for the retrospective sample, the length of residence had the highest standardized regression coefficient in the number of areas searched equation. The direction of the relationship was negative, indicating that length of residence acts as a surrogate for information levels. The greater the information is about potential areas, the fewer the areas that have to be actively searched. In contrast, as might be expected, the number of real estate agents used was positively related to the spatial extent of search.

More formal, *theoretical models* of spatial search are still comparatively rare, with the notable exception of those developed by Huff (1982; 1986). Each of Huff's models is designed to generate a search pattern that can be compared with an observed distribution. The generating rules are the assumptions concerning the nature of the search process, and the resulting set of searched vacancies is represented by some kind of spatial pattern. Huff describes residential search behavior as a two-step process consisting of the initial selection of an area in which search is to be concentrated, followed by the actual selection of vacancies within the targeted area. The decision to search in any particular area is assumed to be a function of the household's expected utility from searching in that area, which depends on the relative concentration of potentially acceptable vacancies. Those vacancies located in areas of the city with relatively low expected search costs have a higher probability of being visited than vacancies located elsewhere. Thus, under certain simplifying conditions, the search process can be described as a first-order Markov chain in which the probability of searching in a given area is a function of three elements: the indirect cost of identifying a member of the opportunity set in that area, the direct cost of actually visiting a member of that opportunity set, and the location of the last vacancy visited.

In general, Huff (1986) successfully contended that geographic regularities exist in the spatial pattern of vacancies visited by potential movers and that these regularities primarily arise from two distinct but interrelated sources: the spatial distribution of potential opportunities and the spatial biases in the search strategies employed. Huff was thus able to formally integrate the behavioral approach, with its emphasis on individual choice and decision making, with the constraints approach, which focuses on the structural issues associated with the supply of housing. He effectively demonstrated that these approaches are complementary, in that they highlight different sources of spatial regularity in observed behavior. In sum, his constrained choice set model, area-based search model, and anchor points models exemplarily illustrate a formal, theory-based empirical analysis that goes well beyond descriptive generalizations.

11.3   Neighborhood Evaluation

The third part of the residential decision-making process, the process of neighborhood evaluation and choice, involves two major questions. First, what are the evaluative dimensions across which people assess the relative desirability of alternative neighborhoods? Second, once those evaluative dimensions have been identified, what are the appropriate rules for combining them into an overall utility value for each neighborhood? This section addresses the first of these two questions, and the second question is taken up in the following section.

Evaluative Dimensions

An initial study by Johnston (1973) suggested that neighborhood preferences can be understood as resulting from three underlying cognitive categories, or evaluative dimensions: the impersonal environment, composed mainly of the physical attributes of the neighborhood; the interpersonal environment, composed of the social attributes of the neighborhood; and the locational attributes of the neighborhood. Unfortunately, however, the data for Johnston's study were aggregated across different neighborhoods, thus masking the possibility that different sets of evaluative dimensions might be associated with different individual neighborhoods. With this aggregation problem in mind, the present author (Cadwallader,

1979b) examined, first, whether the same evaluative dimensions are used for different types of neighborhoods and, second, whether the relative importance of these dimensions in determining residential preferences is the same for all neighborhoods. In other words, can the evaluative process be easily generalized, or is it largely context specific?

The data used for this study were obtained, in questionnaire form, from 148 residents of *Madison*, Wisconsin. The original sample size of 255 households yielded 189 completed questionnaires. Of these completed questionnaires, however, 41 were considered unusable, mainly because of missing data. The sample design required the selection of all households within an area containing approximately 10 city blocks, as it was felt that all the subjects should be located in the same neighborhood, thus ensuring some measure of comparability for their ratings of other neighborhoods. In this respect, the subjects represent a statistical population, rather than a sample drawn from such a population. The city blocks were composed of single-family dwelling units and, on average, the subjects had lived at their present addresses for approximately 12 years, although all portions of the duration-of-residence curve were well represented.

Each subject was asked to rate eight Madison neighborhoods on 11 seven-point rating scales. The neighborhoods were chosen by the experimenter with a view toward maximizing ease of recognition on the part of the subjects. If a subject so desired, he or she was shown a map of Madison that outlined the eight neighborhoods. The comparatively long residence in Madison of most of the subjects ensured few problems associated with neighborhood identification.

The subjects were presented with the 11 rating scales, across which the neighborhoods were to be evaluated, in the form of a semantic differential test. The *semantic differential technique* was originally developed by psychologists to investigate the meaning of words, but it has also been used in the analysis of environmental images. The methodological procedure involves presenting the subjects with a set of stimuli, or concepts, that they are required to evaluate across a series of scales consisting of bipolar adjectives. In the present context, the neighborhoods represent the stimuli, while the 11 rating scales represent the bipolar adjectives (Table 11.4). These particular rating scales were chosen partly to maintain comparability with Johnston's study and partly to reflect the selection of attributes used in other studies involving some aspect of residential preferences. Care was taken to randomize the order of both neighborhoods and scales and also to ensure that no regular sequence occurred with respect to whether the right- or the left-hand member of the polar terms represented the positive one.

The data derived from the semantic differential procedure provided a 148 x 11 data matrix for each of the eight neighborhoods. These matrices were subjected to *principal components analyses* (see Section 6.4) in order to identify any underlying evaluative dimensions. Components with eigenvalues greater than 1 were extracted and then rotated to simple structure according to the varimax criterion. For each neighborhood there were three components with eigenvalues values greater than 1, and the highest loading for each scale on these components is reported in Table 11.5. In almost every case the percentage of the total variance accounted for by these three components is fairly low, with the first component accounting for approximately 20 to 30 percent of the total variance. Such low levels of explained variation suggest that the 11 scales are not easily collapsed into significant underlying dimensions.

Table 11.4 Neighborhood Attributes

1	Crowded	—	Spacious
2	Poorly kept yards	—	Well-kept yards
3	Open	—	Private
4	Poor reputation	—	Good reputation
5	Poor-quality housing	—	Good-quality housing
6	Noisy	—	Quiet
7	People dissimilar to me	—	People similar to me
8	Unsafe	—	Safe
9	Inconvenient location	—	Convenient location
10	Poor park facilities	—	Good park facilities
11	Ordinary	—	Distinctive

Source: M. T. Cadwallader, "Neighborhood evaluation in residential mobility," *Environment and Planning A*, 11 (1979), Table 1, p. 395.

The structures of these components were analyzed, however, in order to determine whether they matched those postulated by Johnston (1973). For this purpose, a matrix was constructed to show how many times the highest loading for each scale was associated with the same component as the highest loading for every other scale (Table 11.6). For example, for four neighborhoods, the highest loadings for quiet and privacy were on the same component, and for five of the eight neighborhoods the highest loadings for quiet and spaciousness were on the same component. The major variable groupings in this matrix were then uncovered by means of elementary *linkage analysis*.

Two major groupings emerged. The first contained the scales representing spaciousness, housing quality, distinctiveness, quiet, and privacy, and the second contained the scales representing neighborhood reputation, yard upkeep, safety, type of people, and park facilities. The location variable was not included in either of these typal structures, as its highest loading was not regularly associated with the same component as the highest loading of any other variable. In general, then, these variable groupings are of great interest, as they can be conveniently categorized as representing physical characteristics, social characteristics, and location. As such, they are encouragingly similar to the three evaluative dimensions postulated by Johnston (1973).

Despite this general similarity, however, it is obvious that the factor structures associated with each neighborhood are far from identical (Table 11.5). For this reason, further analysis was pursued to see if any evidence existed for similar neighborhoods being cognized through similar evaluative dimensions. The perceived similarity among the neighborhoods was measured by a *similarity rating*, which was obtained by asking the subjects to take each neighborhood in turn as an anchor neighborhood and to identify the three other neighborhoods that they regarded as most similar to it. The similarity rating for each pair of neighborhoods is shown by the values above the diagonal in Table 11.7, with the highest values indicating the greatest similarity.

Table 11.5 Evaluative Dimensions

Dimension	Middleton			Maple Bluff			Shorewood Hills			Monona		
	I	II	III	I	II	III	I	II	III	I	II	III
Spaciousness	0.72			0.75			0.65			0.75		
Yard upkeep	0.66				0.59		0.73			0.55		
Privacy	0.50				0.72			0.41				0.75
Reputation	0.61				0.49			0.78			0.75	
Housing	0.60			0.78			0.82			0.52		
Quiet	0.75			0.68				0.70		0.78		
People			0.61			0.78			0.85		0.52	
Safety	0.72				0.60				0.41	0.58		
Location		0.74				0.78		0.77				0.56
Park facilities			0.84		0.70			0.65			0.78	
Distinctiveness		0.68		0.86			0.77					0.60
Variance (% of total)	28	16	13	25	19	13	26	25	10	21	18	14

Dimension	Indian Hills			Hilldale			Nakoma			Odana		
	I	II	III	I	II	III	I	II	III	I	II	III
Spaciousness	0.71			0.79			0.71			0.69		
Yard upkeep	0.49				0.60			0.59			0.81	
Privacy			−0.70	0.52					0.87	0.69		
Reputation	0.66				0.83			0.79			0.65	
Housing	0.79			0.60			0.81			0.62		
Quiet	0.53			0.77					0.64			0.65
People		0.51			0.73			0.53			0.76	
Safety	0.62				0.82		0.68					0.77
Location			0.71			0.71		0.68		0.46		
Park facilities		0.86				0.75		0.74				−0.51
Distinctiveness	0.71			0.74			0.69					0.50
Variance (% of total)	30	15	10	23	22	25	26	23	12	18	18	16

Source: M. T. Cadwallader, "Neighborhood evaluation in residential mobility," *Environment and Planning A*, 11 (1979), Table 2, p. 396.

Table 11.6   Associations between Neighborhood Attributes

		1	2	3	4	5	6	7	8	9	10	11
1	Spaciousness	—										
2	Yard upkeep	4	—									
3	Privacy	3	2	—								
4	Reputation	2	6	3	—							
5	Housing	8	4	3	2	—						
6	Quiet	5	3	4	3	5	—					
7	People	0	3	0	4	0	0	—				
8	Safety	4	5	2	4	4	4	2	—			
9	Location	0	1	3	2	0	2	2	1	—		
10	Park facilities	1	2	3	4	1	1	4	1	3	—	
11	Distinctiveness	5	2	2	1	5	4	0	3	3	0	—

Source: M. T. Cadwallader, "Neighborhood evaluation in residential mobility," *Environment and Planning A*, 11 (1979), Table 3, p. 397.

Table 11.7   Matrix of Similarity Ratings and Congruency Coefficients

		1	2	3	4	5	6	7	8
1	Middleton	—	0.01	0.18	0.19	0.02	0.30	0.02	0.22
2	Maple Bluff	0.77	—	0.06	0.02	0.41	0.05	0.35	0.02
3	Indian Hills	0.91	0.88	—	0.16	0.09	0.20	0.07	0.23
4	Hilldale	0.74	0.87	0.82	—	0.06	0.15	0.09	0.29
5	Shorewood Hills	0.86	0.91	0.92	0.85	—	0.04	0.38	0.04
6	Monona	0.88	0.89	0.86	0.84	0.84	—	0.05	0.20
7	Nakoma	0.83	0.94	0.94	0.78	0.90	0.92	—	0.10
8	Odana	0.68	0.55	0.67	0.76	0.59	0.53	0.58	—

Source: M. T. Cadwallader, "Neighborhood evaluation in residential mobility," *Environment and Planning A*, 11 (1979), Table 4, p. 397.

The degree of *factorial similarity* across the neighborhoods was measured by the coefficient of congruence (see Section 6.5). Usually, each component from one factor structure is compared with the components from all other factor structures and then is paired with the one with which it has the highest coefficient of congruence. In the present instance, however, only the first components from each factor structure were compared, as the eigenvalues associated with the remaining components were rather small. The congruency coefficients are shown below the diagonal in Table 11.7.

Inspection of the similarity ratings and congruency coefficients suggests that neighborhoods perceived to be similar are indeed cognized through similar evaluative dimensions. The three neighborhoods of Maple Bluff, Shorewood Hills, and Nakoma especially illustrate this phenomenon. The congruency coefficients among these three are all equal to or greater than 0.90, while these neighborhoods also have the three highest similarity ratings. In sum, although the evaluative dimensions are not the same for all neighborhoods, it does appear that subjects utilize similar dimensions when evaluating similar neighborhoods.

Relative Importance of Neighborhood Attributes

Before considering the relative importance of the neighborhood attributes in explaining neighborhood preferences, it is worth noting the important roles that some of the attributes have been assigned in various theories of urban residential differentiation. We can distinguish, for example, between the historic and structural theories of urban form. The *historic theory*, associated mainly with the name of Burgess, is essentially the spatial manifestation of the filtering process, whereby new houses are built for upper-income families and in time are filtered down to lower-income families (see Section 6.2). An important element of this theory, then, is the assumption that households locate in order to maximize their satisfaction with housing quality.

The *structural theory*, on the other hand, places the emphasis on the trade-off between the demand for accessibility and the demand for land. It is assumed that accessibility behaves as an inferior good, so that as families increase their incomes they prefer to substitute land for accessibility. In other words, it is postulated that lower-income households will locate in order to maximize locational advantages, whereas upper-income households will attempt to maximize their desire for spacious lots.

A third major theory of residential differentiation can be conveniently labeled the *segregation theory*. This theory is implicit in factorial ecological studies, which purport to demonstrate that people attempt to live apart from those unlike themselves and thus minimize the possibility of conflict because of class, generational, racial, and religious or national differences (Davies, 1984). The segregation theory therefore assumes that, when evaluating neighborhoods for residential desirability, households will focus particular attention on the kind of people living in the neighborhoods.

The relative importance of the neighborhood attributes was first examined by analyzing how the subjects rated the attributes on a seven-point scale, going from very unimportant (1) to very important (7). The mean rating is used as the aggregate measure of importance for each attribute, while the standard deviation provides a measure of the level of agreement between subjects (Table 11.8). As expected from the preceding theoretical discussion, the attributes of location and housing quality were both considered very important. The type of people living in the neighborhood was considered somewhat less important, although the influence of this variable might have been muted by the fact that Madison is not characterized by large ethnic or racial differences. Also considered of relatively minor importance was the spaciousness scale; thus some doubt was cast on one of the major ingredients of the structural theory of urban form, although this result could be somewhat misleading, given the strong association between housing and spaciousness (Table 11.6). In general, the standard devia-

Table 11.8 Relative Importance of the Neighborhood Attributes

	Attribute	Mean		Attribute	Mean
1	Location	6.12	7	Spaciousness	5.53
2	Housing	6.06	8	Yard upkeep	5.49
3	Safety	6.02	9	Park facilities	4.87
4	Noise	5.94	10	Reputation	4.82
5	Privacy	5.67	11	Distinctiveness	4.41
6	People	5.55			

Source: M. T. Cadwallader, "Neighborhood evaluation in residential mobility," *Environment and Planning A*, 11 (1979), Table 5, p. 399.

tions indicated that the amount of agreement among the subjects is inversely related to the perceived importance of the attributes.

Further analysis was required, however, to determine whether these attribute rankings remained constant when subjects were asked to think of particular neighborhoods. That is, are the weights attached to the different attributes invariant across neighborhoods? To address this question, a *stepwise regression analysis* was performed for each of the eight neighborhoods. The overall preference rating for the neighborhood, as evaluated by each subject, acted as the dependent variable, while the 11 attributes, or scales, acted as the independent variables.

The particular form of stepwise regression used for this analysis was the backward elimination procedure. This procedure involves computing a regression equation containing all the variables and calculating the partial F value for each variable, treating it as though it were the last variable to enter the equation. The lowest partial F value is then compared with a preselected significance level, in this case 0.10, and if the F value is lower, then the corresponding variable is removed. The regression equation is then recomputed, and the process is continued until no more variables can be removed. The value of this particular procedure lies in the fact that all variables are first included in the equation and then the insignificant variables are removed in the interests of parsimony.

The results of the stepwise regression analysis (Table 11.9) are a little confusing when compared with the original rankings (Table 11.8). The standardized regression coefficients reflect the relative importance of the variables in determining the preference ratings, and it can be seen that the distinctiveness variable appears in more equations than any other, although it previously obtained the lowest ranking. Conversely, the previously important location variable appears in only one of the eight regression equations. Although one should remember that the multiple correlation coefficients are rather small and there is obviously some multicollinearity among the variables, these results clearly indicate the potential hazards involved in attempting to analyze preferences independently of any particular context. By contrast, Preston (1986) suggests that such context effects are comparatively weak and can be reduced by controlling for individual differences in the perceived levels of residential area attributes.

Like the present author, Hourihan (1984) also used semantic differential scales and principal components analysis to uncover the evaluative dimensions used by a sample of residents in Cork, Ireland. For the purposes of constructing a path model of residential satisfac-

Table 11.9 Stepwise Regression Models

Middleton	$Y = 1.32 + 0.31X_1 - 0.17X_6 + 0.32X_9 + 0.16X_{11}$	$R = 0.51$
Maple Bluff	$Y = 1.25 - 0.17X_1 + 0.18X_4 + 0.24X_7 + 0.20X_{11}$	$R = 0.40$
Indian Hills	$Y = 0.96 + 0.37X_4 + 0.40X_{10} + 0.19X_{11}$	$R = 0.52$
Hilldale	$Y = 0.45 + 0.29X_5 + 0.27X_6 - 0.14X_8 + 0.32X_{11}$	$R = 0.66$
Shorewood Hills	$Y = 1.56 + 0.15X_1 + 0.15X_2 + 0.28X_7$	$R = 0.40$
Monona	$Y = 2.76 - 0.18X_2 - 0.27X_3 + 0.24X_4 + 0.36X_{11}$	$R = 0.45$
Nakoma	$Y = 2.65 - 0.32X_2 + 0.39X_4 + 0.30X_7 + 0.17X_{11}$	$R = 0.52$
Odana	$Y = 0.14 + 0.28X_1 - 0.18X_2 + 0.28X_7 + 0.17X_8 + 0.17X_{10} + 0.15X_{11}$	$R = 0.51$

Source: M. Cadwallader, "Neighborhood evaluation in residential mobility," *Environment and Planning A,* 11 (1979), Table 6.
Note: X_1 to X_{11} are in the same order as the neighborhood attributes in Table 11.4.

tion, an initial group of seven components was later reduced to four. These four components summarized the appearance, life-style, quality of life, and stability associated with each neighborhood.

An alternative methodology for capturing these underlying dimensions is available in the form of multidimensional scaling analysis. Based on data derived from a survey of women in Hamilton, Ontario, Preston (1982) concluded that land use, lot size, social character, and housing quality are considered in the evaluation of residential areas. Like principal components analysis, multidimensional scaling is primarily an inductive approach in which the dimensions are subjectively interpreted by the researcher. Unlike the more restrictive orthogonal rotations that have characterized the semantic differential investigations, however, multidimensional scaling allows the evaluative dimensions to be related.

11.4 Neighborhood Choice

Having identified the major evaluative dimensions involved in the residential choice process, we will next explore how the subject ratings associated with those three evaluative dimensions are integrated into an overall utility value for a particular neighborhood. As location, housing, and safety proved to be the three most important attributes, these were used to represent the locational, physical, and social characteristics dimensions, respectively.

Multiattribute Attitude Models

A series of *multiattribute attitude models* (see Section 9.4) was used to determine whether the attributes should be combined in an additive, multiplicative, or weighted additive form (Cadwallader, 1979c). In particular, the following three models were tested:

$$P_i = \sum_{j=1}^{n} A_{ij} \qquad (11.3)$$

$$P_i = \prod_{j=1}^{n} A_{ij} \qquad\qquad (11.4)$$

$$P_i = \sum_{j=1}^{n} A_{ij} W_j \qquad\qquad (11.5)$$

where P_i is the relative attractiveness of neighborhood i; A_{ij} is the attractiveness rating of neighborhood i on attribute j; and W_j is the relative importance of attribute j. These models represent additive, multiplicative, and weighted additive forms, respectively, and all the constituent variables, such as overall neighborhood preference, the attractiveness ratings on each of the three dimensions, and the relative weightings, were measured on the previously defined seven-point scales, which were then averaged across subjects.

Multiattribute attitude models such as these have proved useful in a wide variety of decision-making contexts where a particular spatial alternative can be viewed as a bundle of attributes (Cadwallader, 1981b). The weighted additive model represents a linear compensatory model with weighted components. Such models tend to perform particularly well in those situations where a relatively large variance exists across the weightings. The simple additive model is also a linear compensatory model and is like the weighted additive version, except that the weighting term is omitted. Finally, a multiplicative version of the model was also tested, as evidence exists that subjective judgments often conform to a multiplying rule between factors.

The results for this particular context and sample of subjects indicate that the additive model is more appropriate than its multiplicative counterpart but that there is no difference between the additive versions. The first column in Table 11.10 shows the actual percentage of subjects who preferred each of the eight neighborhoods, while the remaining columns show the percentages predicted by each of the three models. The goodness of fit associated with each model is measured by the index of dissimilarity, which ranges from zero, indicating a perfect fit, to 100.

At first glance all three models appear to provide reasonably good fits, with the two additive formulations being only marginally superior. In this particular situation, however, the maximum possible value of 100 is a little misleading. If we knew nothing about the underlying decision-making process, our best prediction would be that an equal proportion of subjects would rate highest each of the eight neighborhoods. A comparison of this prediction and the observed behavior leads to a value of 7.5 for the index of dissimilarity. Viewed in this light, then, the results suggest that the additive models are genuinely superior to the multiplicative version.

A major reason for the excellent performance of the additive formulations is the fact that whenever the predictor variables are monotonically related to the dependent variable, as in the present instance, the linear model tends to fit well. Also, the predictor variables were genuinely independent, as they had been derived from an orthogonal factor analysis. On the other hand, one major problem with the multiplicative models is that, if any attribute is at a near-zero psychological value, the overall utility value will be very low, irrespective of how high the other attributes might be rated.

Table 11.10 Comparison of Neighborhood Choice Models

	Actual	Additive	Multiplicative	Weighted Additive
Shorewood Hills	16	14	19	14
Nakoma	16	14	16	14
Maple Bluff	13	13	13	13
Hilldale	12	14	16	14
Odana	12	13	13	13
Middleton	11	11	9	11
Indian Hills	11	11	8	11
Monona	9	10	6	10
		$D = 4$	$D = 8$	$D = 4$

Source: M. T. Cadwallader, "The process of neighborhood choice," paper presented at the International Conference on Environmental Psychology, University of Surrey, Guildford, England, 1979.
Note: D refers to the index of dissimilarity.

Nevertheless, it is somewhat surprising that the simple additive model performed as well as the weighted additive model, given the greater information content of the latter. In the present context, however, this situation is perhaps understandable, as the individual indicators for the evaluative dimensions did not vary markedly in relative importance. In those situations where a greater variance in the weightings occurs, one could expect the weighted additive model to perform better than the simple additive model. This issue emphasizes the significance of distinguishing between determinant as opposed to important attributes, as attributes rated as important by subjects will not be pivotal in the decision-making process if the levels associated with such attributes are relatively invariant across the alternatives.

Finally, the simple additive model was retested by incorporating a familiarity variable, as previous research has suggested the importance of familiarity, or information levels, in the formation of preferences (see Section 8.4). This reformulation provided the following equation:

$$P_i = \sum_{j=1}^{n} A_{ij} + F_i \tag{11.6}$$

where the notation is the same as in equations (11.3), (11.4), and (11.5), and F_i is the level of familiarity associated with neighborhood i, as measured by a seven-point rating scale. The index of dissimilarity for equation (11.6) is 3, so the equation is a slightly better predictor than the simple additive model. It should be noted, however, that the utility functions used to evaluate the alternatives might change during the course of the decision-making process (Phipps, 1983).

Other Approaches

Researchers have also offered a number of other approaches to the problem of determining how individual evaluative dimensions are combined into some kind of overall utility value (Ellis and Mackay, 1990; Golledge and Rushton, 1984). The *revealed preference approach*, as the name implies, involves the examination of observed behavior in order to uncover the underlying preference structure (W. Clark, 1982b). Ideally, however, only purely discretionary behavior should be analyzed via revealed preferences. Since desired but unattainable choices are never observed, one runs the risk of confounding preferences and constraints (Desbarats, 1983a). Moreover, models relying on revealed behavior restrict themselves to the domain of experience, and thus one must extrapolate beyond the observed types of spatial alternatives. That is, such models are ineffective in predicting an individual's level of satisfaction for an alternative with attribute levels outside the range of prior experience.

As a result of these problems with the revealed preference paradigm, other researchers have argued for a more experimental approach, whereby simulated choice data are used to present a wider range of alternatives. Such an approach allows the researcher to make observations over repeated, experimentally controlled trials and to estimate parameters that are context free. For example, *information integration theory*, or functional measurement, involves an algebraic model of human information processing (Louviere and Timmermans, 1990b). Individuals assign values to levels of attributes and then combine those values into an overall utility value. Analysis of variance is used to discriminate between the alternative functional forms and thus to uncover the underlying combination rules.

In a similar vein, the *conjoint measurement approach* also requires subjects to evaluate multiattribute alternatives by comparing predetermined levels of supposedly important attributes (Phipps and Carter, 1985). The subjective rankings of various combinations of these experimental levels are then decomposed into preference functions. In conjoint measurement, however, the subjects are simply asked to rank order the stimulus combinations, which is a less demanding task than producing the numerical responses that are used in functional measurement. On the other hand, since conjoint measurement only involves ordinal-level data, its diagnostic value for identifying alternative composition rules is rather weak. In addition, a number of problems are associated with the experimental approach in general. First, the use of factorial designs restricts the number of attributes that can be easily processed by the subjects. Second, specifying appropriate levels of the attributes — deciding how the levels should be spaced — is problematic. Third and most important, decision making under experimental conditions may well differ from decision making in the real world (Timmermans, 1984).

Given these disadvantages of working with experimental data, *discrete choice models* based on observed behavior have become increasingly popular (Wrigley and Longley, 1984; Deurloo et al., 1988; Maier and Weiss, 1991). Researchers have contributed to this literature in at least three major areas: microeconomic theories of consumer behavior, psychological theories of choice, and new statistical methods for analyzing categorical data (Clark et al., 1988; Fischer and Aufhauser, 1988). Discrete multivariate analysis allows a large number of parameters to be estimated, including those for each of the category levels and those describing the interactions among them. These interactions are usually of central concern, although models containing third- or higher-order interactions are often extremely difficult to interpret (Alba, 1987). The most that can be said in such situations is that the association structure is extremely complex.

To date, most decision-making models have been compensatory, in the sense that individuals are assumed to trade off attributes. An alternative perspective, however, suggests that a low value on one attribute cannot be compensated for by a high value on another attribute. Although there are a variety of such noncompensatory choice models, the major ones for discrete choice modeling have been the *elimination-by-aspects* (EBA) models. Such models treat decision-making as a process of elimination and so do not require the simultaneous consideration of every attribute for every alternative. Rather, the important attributes are considered sequentially. At each stage the individual selects a particular attribute, and all the alternatives that do not possess that attribute are eliminated. The decision is thus decomposed into a series of steps, and individuals can eliminate a large number of alternatives after only a few attributes have been considered. The decision maker selects one aspect, or attribute, and then eliminates all alternatives that are unsatisfactory with respect to that single attribute. Then a second attribute is chosen, and so on, until only one alternative remains. In this way the decision maker is not expected to integrate all the pertinent information before arriving at one ultimate decision.

For example, within the context of residential mobility we can hypothesize at least two such steps. First, the household will collect information to help select an appropriate neighborhood. Second, the household will obtain information in order to choose a specific house within that neighborhood. Using data from Syracuse, New York, Talarchek (1982) found that the behavioral patterns of residential search and selection are highly individualized, but that the sequence in which information is acquired does indicate a general two-stage process of the kind just described. When investigating the locational choice process of some new residents in Melbourne, Australia, Young (1984) also found that the elimination-by-aspects model provided an acceptable statistical fit to the data. Such success is not achieved without cost, however, as the elimination-by-aspects approach involves estimating a large number of parameters.

Interaction Effects

A wide range of approaches, therefore, has been used to address the issue of residential choice. No single approach, however, has successfully captured the potential complexity of this process. In particular, insufficient attention has been given to identifying possible interaction effects among the constituent attributes. Such interaction effects were the focus of a study by the present author (Cadwallader, 1992a), in which log-linear procedures were used to calibrate a variety of decision-making models, using the previously described data on Madison, Wisconsin.

Regression analysis is the most convenient methodology for identifying possible interaction effects. If two variables interact in their effect on some response variable, then the effect of either explanatory variable on the response variable depends on the value of the other explanatory variable. That is, the variables X_1 and X_2 interact in the determination of a third variable Y if the effect of X_1 on Y depends on the level of X_2. Similarly, the effect of X_2 on Y

will depend on the level of X_1. For example, a common type of theoretical situation involves the assumption that a given phenomenon will occur when both factors are present, but that it is unlikely to occur whenever either of these factors is absent. If we have three continuous variables, with Y representing neighborhood attractiveness, then such a situation arises when we postulate that Y values will be large only when both X_1 and X_2 are large and that Y values will be small if either X_1 or X_2 approaches zero.

The presence or absence of this interaction effect is usually tested by including product terms as additional variables in a multiple regression model (Allison, 1977). For example, the three-variable case described above would generate the following equation:

$$Y = a + b_1X_1 + b_2X_2 + b_3X_1X_2 \qquad (11.7)$$

where b_3 is the coefficient associated with the product of variables X_1 and X_2. The necessity of including the interaction term can be assessed by using the t statistic to test the null hypothesis that b is zero. If b_3 is significantly different from zero, then the interaction should be included.

A comparable test for interaction effects involves comparing the complete and reduced models by using the F statistic (Agresti and Agresti,1979:383–5). A model with three independent variables, containing all possible first-order interactions, would have the following form:

$$Y = a + b_1X_1 + b_2X_2 + b_3X_3 + b_4X_1X_2 + b_5X_1X_3 + b_6X_2X_3 \qquad (11.8)$$

To test the hypothesis of no first-order interaction, this model would be compared with the following reduced model:

$$Y = a + b_1X_1 + b_2X_2 + b_3X_3 \qquad (11.9)$$

We are thus testing the null hypothesis that the coefficients b_4, b_5, and b_6 are all zero.

The F test statistic can be expressed via the coefficients of multiple determination for the two models, as follows:

$$F = \frac{(R_c^2 - R_r^2) / (k - g)}{(1 - R_c^2) / [n - (k + 1)]} \qquad (11.10)$$

where R_c^2 refers to the complete model and R_r^2 refers to the reduced model; k and g represent the number of independent variables in the complete and reduced models, respectively; and $(k - g)$ and $[n - (k + 1)]$ are the associated degrees of freedom. If the calculated value of this statistic for a given sample is significant when compared with the tabulated F distribution, then we infer that interaction is present. So, while the t test can be used for individual coefficients, the complete versus reduced model F test can be used to test a number of regression parameters simultaneously in order to decide if any of them are nonzero.

In the present example, there are three independent variables, representing the three evaluative scales, for each of the three neighborhoods. The reduced models, those without any interaction terms, all fit the data reasonably well (Table 11.11). As expected, all the coefficients are positive, indicating that an increase in any individual attribute leads to an increase in overall attractiveness. In addition, most of the coefficients are different from zero at the

Table 11.11   Standardized Regression Models

Indian Hills
Attractiveness = 0.139 park facilities[a] + 0.185 distinctiveness[b] + 0.373 reputation[c]
$R^2 = 0.267$

Middleton
Attractiveness = 0.150 distinctiveness[a] + 0.253 spaciousness[c] + 0.315 location[c]
$R^2 = 0.237$

Hilldale
Attractiveness = 0.276 housing[c] + 0.279 distinctiveness[c] + 0.266 quiet[c]
$R^2 = 0.415$

Source: M. Cadwallader, *Migration and Residential Mobility*, University of Wisconsin Press, 1992, Table 6.11.
[a]Significant at 0.10.
[b]Significant at 0.05.
[c]Significant at 0.01.

0.01 significance level. The reduced models were then compared with complete models containing all three first-order terms. The results were inconclusive, as the *t* tests on individual coefficients suggested some interaction effects for Indian Hills and Middleton, but the *F* test described in equation (11.10) was insignificant for all three neighborhoods.

11.5   Log-linear Models of Residential Choice

For a number of reasons, however, the use of multiple regression to analyze statistical interaction can prove problematic (Fischer, 1988). First, some hypotheses concerning interaction cannot be meaningfully tested unless the variables are measured on ratio scales. Second, in equations containing interaction terms, the partial correlations and standardized regression coefficients cannot be interpreted, and thus it becomes misleading to evaluate the relative importance of the main effects and interaction effects (Marsden, 1981). For example, given the functional relationship between the interaction term and its constituent variables, it is not possible to hold X_1 and X_2 constant while investigating the effect of X_1X_2. Third, the use of regression models to identify interaction is especially difficult when measurement error occurs in one or more of the variables. In particular, the reliability of the cross-products term can be low even when the constituent variables have high reliability. For these reasons, the present author (Cadwallader, 1992a) undertook further analysis using a series of log-linear models. These models involved converting the seven-point rating scales into dichotomous categories, indicating either a low or a high value for each neighborhood on each particular scale, as perceived by individual subjects.

Log-linear Models

Regression analysis is not appropriate for analyzing categorical data, as the observations are not drawn from populations that are normally distributed with constant variance. A special class of statistical techniques, called log-linear models, however, has recently been developed for the analysis of such data (Wrigley, 1985). Log-linear models can be used to identify the relationships among a set of categorical variables when the aim is to predict the number of cases in each cell of a multidimensional contingency table. More specifically, a linear model represents the log of the frequencies in each cell as a function of the various combinations of the categorical variables. The smaller the difference between the predicted and observed frequencies, the better the model fits the data. In a sense, log-linear models are similar to multiple regression models; the dependent variable is the number of cases in each cell of the contingency table, and the categorical variables function as independent variables.

A *saturated model*, which contains all main effects and interaction terms, will exactly reproduce all the observed cell frequencies. That is, it will fit the data perfectly. But a log-linear model with as many parameters as there are cells does not constitute a parsimonious description of the relationships among the variables. Rather, models should be identified that contain as few parameters as possible, while still providing an adequate fit to the data. For example, models that do not contain higher-order interaction terms are usually preferable, as such terms are often difficult to interpret. Although it is possible to delete any particular term from a saturated model, most researchers tend to focus on *hierarchical models*. In a hierarchical model, if a term exists for the interaction of a group of variables, then lower-order terms must also exist for all possible combinations of those variables. Thus, if the interaction term ABC is included in a three-variable model, then the terms A, B, C, AB, AC, and BC must also be included.

The likelihood-ratio chi-square statistic can be used to assess the goodness of fit associated with competing hierarchical models. The degrees of freedom for a particular model are calculated by subtracting the number of independent parameters from the number of cells in the contingency table. Small values of chi square indicate a good-fitting model (for a saturated model the chi-square statistic is always zero). Alternatively, in those instances where the observed significance level associated with the chi-square statistic is very small, the model can be rejected. The likelihood-ratio statistic can also be used to assess the contribution of individual terms, as the decrease in value of this statistic when a particular term is added to a model indicates the relative contribution of that term. In other words, two models differing only in the presence of a particular effect can be fit to the data; the difference between the two likelihood-ratio chi-square values, called a partial chi-square, also has a chi-square distribution and can thus be used to test the hypothesis that the effect is zero.

Saturated Models

In the present instance, the saturated model for each of the three neighborhoods can be expressed as follows:

$$\log_e m_{ijkl} = \lambda + \lambda_i^A + \lambda_j^B + \lambda_k^C + \lambda_l^D + \lambda_{ij}^{AB} + \lambda_{ik}^{AC} + \lambda_{il}^{AD} + \lambda_{jk}^{BC} + \lambda_{jl}^{BD}$$

$$+ \lambda_{kl}^{CD} + \lambda_{ijk}^{ABC} + \lambda_{ijl}^{ABD} + \lambda_{ikl}^{ACD} + \lambda_{jkl}^{BCD} + \lambda_{ijkl}^{ABCD}$$

$$(11.11)$$

where m_{ijkl} is the expected frequency of the $ijkl$ cell; λ is the grand mean; A to D are main effects; AB to CD are two-way interaction effects; ABC to BCD are three-way interaction effects; and $ABCD$ is a four-way interaction effect. Such a model would fit the data perfectly, as a saturated model has as many parameters as there are cells in the four-dimensional contingency table. To obtain a more parsimonious model while still maintaining an adequate fit to the observed data, we must eliminate those parameters that are not necessary to describe the structural relationships among the four variables in the contingency table (Wrigley, 1985:173).

With this aim in mind, it is often useful to test whether interaction terms of a particular order are zero. Table 11.12 provides results for the tests that first-, second-, third-, or fourth-order interaction effects are zero for each neighborhood. For example, the test that third-order effects are zero is based on the difference between the likelihood-ratio chi-square for a model without third-order terms and that for a model with third-order terms. If the observed probability level is large, it means that the hypothesis that third-order terms are zero cannot be rejected. In the present case, it can be seen that first- and second-order terms should be included in the final model for each neighborhood, but that the third- and fourth-order interactions are not significantly different from zero at the 0.05 level. The only exception is Hilldale, where the fourth-order interaction term is significantly different from zero.

While the results in Table 11.12 indicate the overall importance of effects of various orders, they do not test the individual terms. In many instances, even if the hypothesis that all second-order terms are zero is rejected, not all second-order terms will be needed for a parsimonious description of the data. In this context, partial chi-squares can be used to assess the contribution of individual terms. A partial chi-square is the difference between the likelihood-ratio chi-square values for two models that differ only with respect to the presence or absence of the effect to be tested. For example, to test the partial association of variables A and B, the complete second-order model would be fitted, and then the same model without the AB interaction term would also be fitted. The difference in the chi-square value for these two models represents the partial chi-square associated with the second-order interaction AB (Wrigley, 1985:197).

In the present instance, partial test statistics were computed for each term in the saturated log-linear model. When the partial values are small, with correspondingly high probability levels, it means that those particular effects are not necessary to adequately represent the data. For Indian Hills (Table 11.13), as expected from the testing of particular k-way effects,

Table 11.12   Tests of Hypotheses That Particular k-Way Effects Are Zero

		Likelihood-ratio chi-square and probability level		
k	d.f.	Indian Hills	Middleton	Hilldale
1	4	154.55 (0.000)	172.84 (0.000)	69.84 (0.000)
2	6	38.45 (0.000)	26.20 (0.000)	107.81 (0.000)
3	4	1.41 (0.842)	1.48 (0.831)	1.71 (0.789)
4	1	3.05 (0.081)	0.09 (0.766)	5.23 (0.022)

Source: M. Cadwallader, "Log-linear models of residential choice," *Area*, 24 (1992), Table 1.

Table 11.13 Partial Chi-Squares for Indian Hills

Effect	Partial Chi-Square	Probability
ARD	0.273	0.601
ARP	0.892	0.345
ADP	0.124	0.725
RDP	0.654	0.419
AR	8.936	0.003
AD	8.567	0.003
AP	2.144	0.143
RD	5.616	0.018
RP	0.402	0.526
DP	0.701	0.403
A	40.740	0.000
R	24.718	0.000
D	67.702	0.000
P	21.394	0.000

Source: M. Cadwallader, "Log-linear models of residential choice," *Area*, 24 (1992), Table 2.
Note: A represents attractiveness; *R*, reputation; *D*, distinctiveness; and *P*, park facilities.

none of the third-order terms were significantly different from zero. Three of the six second-order effects were significant, however, as were the four main effects. Similarly, for Middleton (Table 11.14) the third-order effects were all insignificant, but two of the second-order effects and three of the first-order effects were significantly different from zero. Finally, the partial chi-squares for Hilldale (Table 11.15) indicated the presence of four second-order effects and four main effects.

Stepwise Selection Strategies

Stepwise selection strategies can also be used to identify the most appropriate model. Starting with a saturated model, a *backward elimination* procedure can be used to successively delete the various interaction parameters. For example, if only the saturated model adequately fits the data, then the model selection procedure stops at this point. Otherwise, terms are deleted in a hierarchical fashion. At the first step, the interaction term whose removal results in the least significant change in the likelihood-ratio chi-square statistic is deleted, as long as the observed significance level is larger than the criterion for remaining in the model. In the present context, a significance level of 0.05 was used as the cutoff for remaining in the model. This cutoff meant that for Hilldale the analysis stopped with the saturated model, as the fourth-order interaction could not be deleted.

Table 11.14   Partial Ch-Squares for Middleton

Effect	Partial Chi-Square	Probability
ALS	0.254	0.614
ALD	0.012	0.915
ASD	1.221	0.269
LSD	0.164	0.686
AL	11.599	0.001
AS	6.168	0.013
AD	0.929	0.335
LS	0.020	0.889
LD	0.541	0.462
SD	2.303	0.129
A	47.957	0.000
L	18.327	0.000
S	1.998	0.158
D	104.560	0.000

Source: M. Cadwallader, *Migration and Residential Mobility*, University of Wisconsin Press, 1992, Table 6.14, p. 230.
Note: A represents attractiveness; *L*, location; *S*, spaciousness; and *D*, distinctiveness.

Table 11.15   Partial Ch-Squares for Hilldale

Effect	Partial Chi-Square	Probability
AHD	0.009	0.926
AHQ	0.083	0.774
ADQ	0.841	0.359
HDQ	0.230	0.632
AH	9.032	0.003
AD	5.653	0.017
AQ	13.326	0.000
HD	2.344	0.126
HQ	11.516	0.001
DQ	2.214	0.137
A	5.839	0.016
H	3.052	0.081
D	55.896	0.000
Q	5.057	0.025

Source: M. Cadwallader, *Migration and Residential Mobility*, University of Wisconsin Press, 1992, Table 6.15, p. 230.
Note: A represents attractiveness; *H*, housing; *D*, distinctiveness; and *Q*, quiet.

The final acceptable model for Indian Hills was as follows:

$$\log_e m_{ijkl} = \lambda + \lambda_i^A + \lambda_j^R + \lambda_k^D + \lambda_l^P + \lambda_{ij}^{AR} + \lambda_{ik}^{AD} + \lambda_{jk}^{RD}$$

$$L.R.\lambda^2 = 6.889, \quad d.f. = 8, \quad P = 0.549$$

(11.12)

where $L.R.\lambda^2$ is the likelihood-ratio chi-square statistic, $d.f.$ is the degrees of freedom, and P is the associated probability level. This model contains all four main effects, plus the second-order interactions involving attractiveness and distinctiveness, attractiveness and reputation, and reputation and distinctiveness. It thus contains all the terms indicated by the analysis of partial chi-squares. The absence of any three-variable interaction parameters implies that when two variables are associated that association is unaffected by the level of a third variable. Similarly, the final acceptable model for Middleton contained the following terms:

$$\log_e m_{ijkl} = \lambda + \lambda_i^A + \lambda_j^L + \lambda_k^S + \lambda_l^D + \lambda_{ij}^{AL} + \lambda_{ik}^{AS}$$

$$L.R.\lambda^2 = 7.154, \quad d.f. = 9, \quad P = 0.621$$

(11.13)

Again, there are no third-order effects, but there are two second-order effects, involving attractiveness and location and attractiveness and space, and all four main effects. Also, as with Indian Hills, this is the same model for Middleton that was suggested by the partial chi-square analysis.

Thus far we have concentrated on describing the structural relationships among all four categorical variables. An alternative approach involves treating neighborhood attractiveness as a *response variable*, with the remaining three variables being explanatory. This distinction between response and explanatory variables generates a set of log-linear models that is simply a special case of the general class of log-linear models we have considered thus far (Wrigley, 1985:216–23). In other words, the same principles of parameter estimation, hierarchical structure, and model selection also apply to this more-restricted set of models.

In models that imply a distinction between response and explanatory variables, the structural relationships between the explanatory variables are treated as given and are automatically included in all the models (Fingleton, 1981). For example, in the present context we are analyzing four-dimensional tables in which variable A is the response variable and the remaining three variables, B, C, and D, are explanatory. In any model, then, we must automatically include the interaction BCD and all its lower-order relatives: BC, BD, CD, B, C, and D. That is, these parameters cannot be excluded from any of the models, irrespective of their statistical significance. In this way, all our attention during the process of model selection is focused on the parameters associated with the relationship between the response and explanatory variables.

The hierarchical structure of log-linear models for this situation can be seen in Table 11.16, which describes the results for Indian Hills. The first and simplest model states that the response variable A is independent of the explanatory variables RDP, and it obviously does not provide a very good fit to the data. Rather, in this case the most likely candidates for the best model are models 5, 8, and 13. If more than one model provides an acceptable fit to the data, it has been suggested that the best model is the simplest model that fits the data adequately and to which no single parameter can be added that provides a significant improvement in fit.

Table 11.16   Hierarchical Log-linear Models with Attractiveness as a Response Variable for Indian Hills

Model		Likelihood-ratio Chi-Square	Degrees of Freedom	Probability
1	$RDP + A$	29.976	7	0.000
2	$RDP + AR$	14.289	6	0.027
3	$RDP + AD$	15.294	6	0.018
4	$RDP + AP$	28.293	6	0.000
5	$RDP + AR + AD$	6.209	5	0.286
6	$RDP + AR + AP$	12.632	5	0.027
7	$RDP + AD + AP$	13.001	5	0.023
8	$RDP + AR + AD + AP$	4.125	4	0.389
9	$RDP + ARD$	6.112	4	0.191
10	$RDP + ARP$	12.115	4	0.017
11	$RDP + ADP$	12.910	4	0.012
12	$RDP + ARD + AP$	3.968	3	0.265
13	$RDP + ARP + AD$	3.427	3	0.330
14	$RDP + ADP + AR$	4.108	3	0.250
15	$RDP + ARD + ARP$	3.177	2	0.204
16	$RDP + ARD + ADP$	3.946	2	0.139
17	$RDP + ARP + ADP$	3.327	2	0.189
18	$RDP + ARD + ARP + ADP$	3.054	1	0.081

Source: M. Cadwallader, "Log-linear models of residential choice," *Area*, 24 (1992), Table 3.
Note: Notation is the same as in Table 11.13.

In other words, one can calculate the difference in the likelihood-ratio chi-square for each additional parameter.

As neither model 8 nor model 13 involves a significant improvement over model 5, model 5 is the chosen model. This model contains interactions between attractiveness and reputation and attractiveness and distinctiveness. In its full form this model is represented as follows:

$$\log_e m_{ijkl} = \lambda + \lambda_j^R + \lambda_k^D + \lambda_l^P + \lambda_{jk}^{RD} + \lambda_{jl}^{RP} + \lambda_{kl}^{DP} + \lambda_{jkl}^{RDP} + \lambda_i^A + \lambda_{ij}^{AR} + \lambda_{ik}^{AD} \qquad (11.14)$$

This formulation is different from that shown in equation (11.12), as it is constrained to include the interaction *RDP* and its lower-order relatives. More important, both equations contain the same interactions involving the response variable neighborhood attractiveness. It should be remembered, however, that model 13, which suggests a third-order interaction among attractiveness, reputation, and park facilities, also provides an acceptable fit to the data.

For Middleton (Table 11.17), again a number of models fit the data adequately. In particular, models 5, 8, 14, and 16 are the best models associated with different degrees of freedom. Of these, model 5 is the most parsimonious, and none of the other models exhibits a significant improvement of fit for each additional parameter. In its full form the chosen model is represented as follows:

$$\log_e m_{ijkl} = \lambda + \lambda_j^L + \lambda_k^S + \lambda_l^D + \lambda_{jk}^{LS} + \lambda_{jl}^{LD} + \lambda_{kl}^{SD} + \lambda_{jkl}^{LSD} + \lambda_i^A + \lambda_{ij}^{AL} + \lambda_{ik}^{AS} \tag{11.15}$$

This structure is the same as that in equation (11.13), except that it is constrained to include the third-order interaction between the explanatory variables LSD and its lower-order relatives. Again, however, there is confirmation of the two second-order interactions involving the response variable, that is, attractiveness and location and attractiveness and spaciousness. As

Table 11.17 Hierarchical Log-linear Models with Attractiveness as a Response Variable for Middleton

	Model	Likelihood-ratio Chi-Square	Degrees of Freedom	Probability
1	LSD + A	22.432	7	0.002
2	LSD + AL	9.512	6	0.147
3	LSD + AS	14.738	6	0.022
4	LSD + AD	19.671	6	0.003
5	LSD + AL + AS	2.463	5	0.782
6	LSD + AL + AD	7.702	5	0.173
7	LSD + AS + AD	13.133	5	0.022
8	LSD + AL + AS + AD	1.551	4	0.818
9	LSD + ALS	2.217	4	0.696
10	LSD + ALD	7.699	4	0.103
11	LSD + ADS	12.132	4	0.016
12	LSD + ALS + AD	1.332	3	0.722
13	LSD + ALD + AS	1.542	3	0.673
14	LSD + ADS + AL	0.348	3	0.951
15	LSD + ALS + ALD	1.310	2	0.519
16	LSD + ALS + ADS	0.100	2	0.951
17	LSD + ALD + ADS	0.342	2	0.843
18	LSD + ALS + ALD + ADS	0.088	1	0.767

Source: M. Cadwallader, "Log-linear models of residential choice," *Area*, 24 (1992), Table 4.
Note: Notation is the same as in Table 11.14.

with Indian Hills, however, models for Middleton containing higher-order interactions involving attractiveness also provide acceptable fits to the data.

Finally, for Hilldale (Table 11.18), as implied by the previous stepwise analysis, none of the models provides an exceptionally good fit to the data. Model 8, perhaps the best, contains second-order interactions between attractiveness and each of the explanatory variables. This model is represented as follows:

$$\log_e m_{ijkl} = \lambda + \lambda_j^H + \lambda_k^D + \lambda_l^Q + \lambda_{jk}^{HD} + \lambda_{jl}^{HQ} + \lambda_{kl}^{DQ} + \lambda_{jkl}^{HDQ} + \lambda_i^A + \lambda_{ij}^{AH} + \lambda_{ik}^{AD} + \lambda_{il}^{AQ} \quad (11.16)$$

It is noteworthy that the second-order interactions involving attractiveness were also suggested by the partial chi-squares (Table 11.15). The results for models 12 to 14 again suggest, however, the possibility of third-order effects when attractiveness is treated as a response variable.

The purpose of this discussion has been to address the issue of interaction effects in models of residential choice. Both regression and log-linear approaches were used for this pur-

Table 11.18 Hierarchical Log-linear Models with Attractiveness as a Response Variable for Hilldale

	Model	Likelihood-ratio Chi-Square	Degrees of Freedom	Probability
1	$HDQ + A$	60.863	7	0.000
2	$HDQ + AH$	29.390	6	0.000
3	$HDQ + AD$	41.439	6	0.000
4	$HDQ + AQ$	23.784	6	0.001
5	$HDQ + AH + AD$	19.623	5	0.001
6	$HDQ + AH + AQ$	11.951	5	0.035
7	$HDQ + AD + AQ$	15.329	5	0.009
8	$HDQ + AH + AD + AQ$	6.256	4	0.181
9	$HDQ + AHD$	19.382	4	0.001
10	$HDQ + AHQ$	11.743	4	0.019
11	$HDQ + ADQ$	13.715	4	0.008
12	$HDQ + AHD + AQ$	6.188	3	0.103
13	$HDQ + AHQ + AD$	6.110	3	0.106
14	$HDQ + ADQ + AH$	5.328	3	0.149
15	$HDQ + AHD + AHQ$	6.070	2	0.048
16	$HDQ + AHD + ADQ$	5.310	2	0.070
17	$HDQ + AHQ + ADQ$	5.236	2	0.073
18	$HDQ + AHD + AHQ + ADQ$	5.228	1	0.022

Source: M. Cadwallader, "Log-linear models of residential choice," *Area*, 24 (1992), Table 5.
Note: Notation is the same as in Table 11.15.

pose, although the log-linear models provided a more detailed analysis of the data. As a result of calibrating a variety of models, one can suggest that interaction effects might well play a role in the residential decision-making process, but that such effects will generally involve only lower-order terms. At the very least, however, those researchers undertaking empirical investigations of residential choice should be alert to the possibility of significant interaction effects.

An obvious extension of the approach used here involves the construction of *structural equation models*. When using continuous data, structural equation models allow one to quantify the direct and indirect effects of predetermined variables on endogenous variables (see Section 2.2). Goodman's (1979) suggestions about how causal interpretations might be imposed on systems of discrete variables by the use of log-linear or logit approaches have not met with universal acceptance, however. In particular, the assignment of numerical values to the causal arrows in the path diagram has proved problematic. First, it has been argued that there is no calculus of path coefficients to calculate the magnitude of effects along indirect paths (Fienberg, 1980:120). Second, variables with multiple categories will lead to a variety of estimated path coefficients for a given arrow in the path diagram. Third, the presence of higher-order interaction terms in log-linear models can lead to very complex path diagrams.

Recently, however, considerable progress has been made in the treatment of causal models containing discrete variables. Winship and Mare (1983) present several alternative approaches to the formulation of structural equation models for discrete data and demonstrate that such models can be as analytically flexible as their continuous counterparts. In particular, they show how direct and indirect effects can be calculated by applying Stolzenberg's (1979) methods for the analysis of nonlinear models in general. Although they consider only recursive models with two-category variables, their approach can be extended to systems involving simultaneity and ordered response variables. To date, however, no reliable computer program exists for the maximum-likelihood estimation of structural equations with discrete data that matches the flexibility of LISREL for continuous variables (see Section 10.5).

Chapter 12

The System of Cities

12.1   Urbanization Curves and City Size Distributions

As mentioned in Chapter 2, investigations of the system of cities focus on cities as points in space, generally at either a national or regional level (Simmons, 1986). Attention is paid to both the attributes of cities, such as size and growth rate, and the interaction between them, as in migration. Over time, most countries tend to exhibit a rather systematic pattern of urban population growth, and this pattern is described by the urbanization curve. Within this aggregated picture of urbanization, the urban population is distributed among cities of different sizes, thus generating a particular city size distribution. After discussing the urbanization curve and city size distributions, we will consider how the attributes of urban growth and housing value vary across cities. The chapter concludes with an examination of migration patterns.

Urbanization Curves

The *demographic transition curve* provides a useful context for our discussion of urbanization curves as it summarizes a country's demographic history, as represented by births and deaths. According to this curve, which monitors changing birth and death rates, a country will pass through three phases (Figure 12.1). Initially, during phase I, the population remains fairly stable, as both the birth and death rates are relatively high. Note, however, that the death rate tends to fluctuate somewhat, due to famines, wars, and diseases. This situation describes the demographic experience of preindustrial Europe and contemporary tribes in the Amazon rain forests. During phase II, the death rate drops dramatically due to improvements in nutrition and health care, although the birthrate tends to remain relatively high. This phase

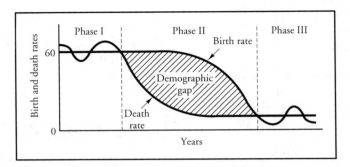

Figure 12.1 The demographic transition curve. (From B. Berry, E. Conkling, and D. Ray, *Economic Geography*, Prentice-Hall, Englewood Cliffs, N.J., 1987, Fig. 3.6.)

involves a period of rapid population growth, as is currently being experienced in much of Africa, Asia, and Latin America.

Phase III is reached when the birthrate falls to match the death rate, and the population stabilizes once again. Unlike phase I, however, it is now the birthrate that tends to fluctuate, mainly as a response to changes in the business cycle. The United States, Canada, Japan, and many European countries provide good examples of this stage of development. It should be noted, however, that the demographic transition curve does not apply equally well to all countries (Berry et al., 1987:45). For example, fertility rates in many less-developed countries are unlikely to decline in the same way they did in Europe. Religious beliefs involving birth control suggest that Latin American countries will continue to experience rapid population growth.

Having discussed how the overall demography of a hypothetical country tends to change over time, we can now turn our attention to the changing level of urbanization. The *urbanization curve*, for a particular country, monitors the changing proportion of the population that lives in cities (Figure 12.2). Once again, we can distinguish three phases (Northam, 1979:65-7). Initially, the curve rises only gradually, but this stage is followed by a period of rapid acceleration during which the country moves from being approximately 25% urbanized to 75% urbanized. Finally, once a large majority of the population has become urbanized, there is a gradual flattening of the curve.

In general, then, the urbanization curve can be described as S-shaped. It is an example of a logistic function and has the property of being asymptotic. That is, we would expect the percent urbanized to approach, but never reach, 100%. More formally, the urbanization curve can be expressed by the following equation:

$$P = \frac{U}{1 + e^{a - bT}} \tag{12.1}$$

where P is the percentage of a country's population that is classified as urban; T is time; U is the upper limit or saturation level; a is the value of P when T is zero; b is the slope; and e is the base of natural or naperian logarithms.

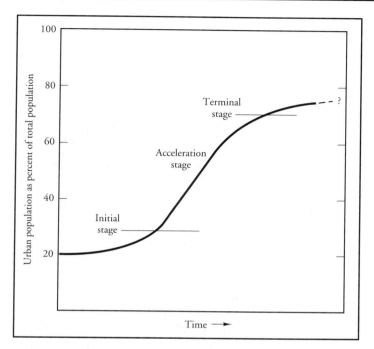

Figure 12.2 The urbanization curve. (From R. Northam, *Urban Geography*, Second Edition, John Wiley & Sons Ltd., New York, 1979, Fig. 4.2, p. 66.)

Why should the urbanization curve be S-shaped, and why should some countries have a steeper curve than others? The answer to the first question lies in the changing economic profile of a country. If the national labor force is partitioned into three sectors, we can imagine a triangle (Figure 12.3). The primary sector involves agricultural and mining activities, the secondary sector refers to manufacturing, while the tertiary, or service, sector represents the white-collar workers in jobs like education, banking, and health. A country in which the jobs are equally distributed across these three sectors would be located at point 1, while a country that is heavily involved in manufacturing would be located at point 2.

Over time, countries tend to evolve through three economic stages, during which they are dominated, in turn, by agricultural, industrial, and, finally, service sector occupations. They thus trace a characteristic trajectory through the triangle (Figure 12.3). This trajectory is matched by the three phases associated with the urbanization curve. While the country is dominated by primary sector activities, the rate of urbanization is gradual. As industrialization unfolds, however, there is a rapid acceleration in urbanization associated with the technology of factory production. Finally, the move toward tertiary sector jobs tends to encourage a flattening of the urbanization curve.

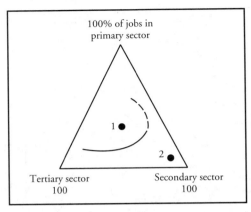

Figure 12.3 The economic triangle. (From P. Haggett, *Geography: A Modern Synthesis*, Second Edition, Harper & Row, Publishers, Inc., New York, 1975, Fig. 13.5a, p. 328. Reprinted by permission of HarperCollins Publishers, Inc.)

The second question, concerning why some countries have steeper urbanization curves than others, is best answered by referring back to the previously described demographic transition curve. In general, less-developed countries tend to exhibit steeper urbanization curves than their industrialized counterparts (Figure 12.4). This difference is partly a matter of timing with respect to the demographic transition curve. The nineteenth-century urbanization associated with contemporary industrialized nations was mainly due to rural–urban migration, while urban growth in currently less-developed nations is fueled not only by migration, but also by high birthrates. For example, in Switzerland, during the late nineteenth century, approximately 70 percent of urban growth was due to rural–urban migration (Haggett, 1983:331). In contrast, in contemporary Costa Rica rural–urban migration only accounts for about 20 percent of the urban growth, while the remainder of that growth is due to the continuing differential between birth and death rates.

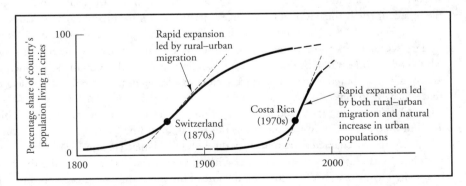

Figure 12.4 Urbanization curves over time. (From P. Haggett, *Geography: A Modern Synthesis*, Revised Third Edition, Harper & Row, Publishers, Inc., New York, 1983, Fig. 14.1, p. 331. Reprinted by permission of HarperCollins Publishers, Inc.)

City Size Distributions

The urbanized population, in a given country, is distributed across cities of varying sizes. For example, in the extreme case, some countries might have a series of cities that are all roughly similar in size. On the other hand, some countries might have one or two very large cities, while all the rest are rather small. In other words, if we rank the size of these cities from largest to smallest, does any regular or systematic pattern emerge? The answer is yes, and the most frequently occurring city size distribution is the *rank-size distribution* (Marshall, 1989). This distribution can be represented by the following equation:

$$P_r = P_1 r^{-q} \tag{12.2}$$

where P_r is the population of the city ranked r, P is the population of the largest city, r is the rank, and q is an exponent. Equation (12.2) can be rewritten as follows:

$$P_r = \frac{P_1}{r^q} \tag{12.3}$$

and it can be simplified further if we assume that $q = 1$. Thus generating the following equation:

$$P_r = \frac{P_1}{r} \tag{12.4}$$

where the notation is the same as in equation (12.2).

An example will help to illustrate the application of equation (12.4). Imagine a country where there are five cities and the largest of these has a population of 100,000 (Figure 12.5). If the rank-size distribution, with an exponent of 1, provides a good fit for this country, then we would expect the second largest city to have a population of 50,000. That is, 100,000 divided by 2. By extension, the third city would be 33,333, and so on. If these five cities are plotted on a graph that has rank on one axis and population size on the other, then we obtain a curvilinear relationship in which all five points fall exactly on the line (Figure 12.5). Equation (12.2),

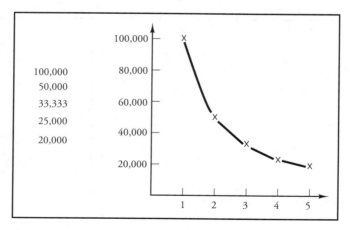

Figure 12.5 The rank-size distribution.

which represents this general relationship, is a power function, so if we take the logarithm of both axes, the relationship will become linear (see Section 3.4):

$$\log P_r = \log P_1 - q(\log r) \tag{12.5}$$

where the notation is the same as in equation (12.2). Note that our exponent q is now simply the slope of the line.

The rank-size relationship turns out to provide a good fit for the United States and a long list of other countries, such as Italy, Poland, Finland, and China. Not all countries display a rank-size relationship, however. For example, some countries have a *primate distribution*, where one very large city dominates all the other cities (Figure 12.6a). England provides a classic example of a primate distribution as London has a population that is nearly seven times as large as that of Birmingham, the second largest city (Champion and Townsend, 1990:161). Austria, Denmark, Mexico, Peru, and Uruguay are examples of other countries that also have primate city size distributions.

Finally, some countries have city size distributions that are best described as *binary* (Figure 12.6b). This kind of distribution has two very large cities, then there is a gap, followed by the normal pattern associated with the lower part of the rank-size distribution. Canada, with Montreal and Toronto, provides a good example of a binary city size distribution, as does Brazil, which is dominated by Rio de Janeiro and Sao Paulo. If there are three or more cities on top of the distribution, then the terms trinary, quarternary, quinary, and so on, can be used. Australia provides an excellent example of a quinary city size distribution, as it is dominated by five large cities: Sydney, Melbourne, Brisbane, Perth, and Adelaide (Holmes, 1987).

Evolution of the U.S. Urban System

In general, three major eras can be identified in the evolution of the U.S. urban system: mercantile capitalism, industrial capitalism, and global capitalism (Johnston, 1982b; Yeates, 1990). *Mercantile capitalism*, which characterized the period between 1790 and approximately

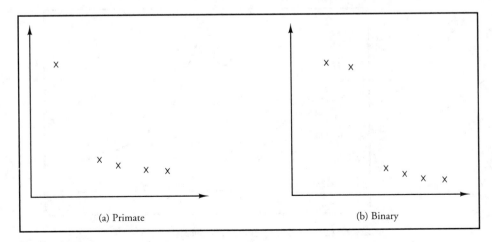

(a) Primate   (b) Binary

Figure 12.6 The primate and binary distributions.

1860, involved the accumulation of wealth through trade. The growth of towns was an integral part of mercantile capitalism, as they provided centers of both consumption and exchange. The largest cities during this era were generally found along the Atlantic coast, with New York, Baltimore, and Boston being especially prominent. Most of the emerging gateway cities to the interior of the country were initially associated with water transportation.

The period of *industrial capitalism*, from approximately 1860 to 1960, was a time of rapid growth for many U.S. cities. Industrial capitalism involved the transformation of the economy from the trading of resources to the processing of those resources and thus focused on the manufacturing of products. In particular, industrial capitalism necessarily led to urban growth as it was associated with the concentration of productive activity that is demanded by the factory system of production. Economies of scale became a major driving force in the location of industry, and it was during this period that the manufacturing belt, extending from New York to Chicago, emerged most clearly. As a result, Chicago, Detroit, Cleveland, and Pittsburgh became major cities during this period, while the west coast cities of San Francisco and Los Angeles also developed into centers of increasing economic power.

Finally, since the 1960s we have entered an era of *global capitalism* that is reflected by a decrease in the international economic dominance of the United States, an increase in the influence of multinational corporations, macro level shifts from the "snow belt" to the "sun belt," and local deconcentration (Scott, 1988; Storper and Walker, 1989). Women have become a vital part of the paid labor force, especially within the service sector of the economy, and recent years have seen the rapid development of new forms of industrial organization. More specifically, we are currently witnessing a period of industrial restructuring with an associated move toward more flexible modes of production and accumulation (Harvey and Scott, 1989). The rigidities of the Fordist regime of capital accumulation, with its emphasis on mass production and the factory system, are being replaced by a regime of flexible accumulation that is characterized by more fluid production and labor arrangements, including the increasing use of subcontracting (Gertler, 1992; Phelps, 1992).

12.2 Urban Growth

We now turn our attention to a more formal explication of the phenomenon of urban growth. Until relatively recently most cities in the United States were experiencing above-average population growth. During the 1970s, however, a new trend emerged, with the term counterurbanization being coined to represent a period of absolute decline or relatively slow growth (Berry, 1988). For example, of the 25 largest metropolitan areas, only Pittsburgh lost population between 1960 and 1970, whereas since then many more have experienced absolute declines (Burns, 1982). These changes were themselves part of a more general redistribution of population from the highly industrialized states in the Northeast to the southernmost tier of states within the South and West. Despite the profound implications of such population shifts in terms of making efficient and equitable decisions concerning the direction of future capital investments or subsidies, remarkably little is known about the actual process of urban growth and decline. Indeed, opinions differ markedly concerning even the degree of volatility in the rate of population change (Borchert, 1983; Gaile and Hanink, 1985).

The traditional approach to urban change suggests that growth or decline is *exogenously determined*. In particular, the export base concept (see Section 7.3) implies that increased demand in the export, or basic sector, of a city's economy will lead to increased employment and purchasing power (Noyelle and Stanback, 1984:29). If this increased purchasing power is spent on goods and services produced within the city, then multiplier effects are set in motion that will generate further growth. Fluctuations in exports are therefore of paramount importance to any urban area, as it is these exports that generate money for the community.

An alternative viewpoint argues that urban growth or decline is primarily controlled by a set of *endogenous mechanisms*. First, it has long been suggested that there is a strong relationship between size and growth (Frey and Speare, 1988:59). Larger cities are thought to possess a comparative advantage due to their greater industrial mix, political power, and fixed capital investment. Second, large cities tend to provide a favorable environment for starting new firms and industries, as the necessary services, skilled labor, and factory space are readily available. Once the new industry matures, however, and the production process becomes more routine, the incubator hypothesis implies that parts of the industry will filter down to smaller, less industrially sophisticated settlements (Nicholson et al., 1981). Third, urban growth is also related to various kinds of agglomeration economies, once again providing larger cities with a comparative advantage.

Finally, a very different perspective on urban and regional growth is provided by neo-Marxist theories of *uneven development*, in which regional growth and decline are considered to be an inevitable manifestation of the inherent crises and instability of capitalism. Indeed, some scholars see uneven development as not only an inevitable consequence of capitalism, but also as a necessary prerequisite for the self-expansion of capital and capital accumulation (Bradbury, 1985). Capital accumulation within a region depends on the rate of profit of the firms located there, and such accumulation cannot occur at an even pace in both space and time (Browett, 1984). Rather, capital tends to accumulate in some places as opposed to, and at the expense of, others.

Behind this pattern of uneven development lies the logic of what Smith (1984:148) has called the "seesaw" movement of capital. Capital moves to where the rate of profit is highest, but the development process itself leads to a decrease in this higher rate of profit. Economic development tends to decrease unemployment, increase wages, and strengthen labor unions, thus lowering the rate of profit and removing the original incentive for development. On the other hand, the lack of capital in underdeveloped regions leads to high unemployment, low wages, and reduced levels of worker organization, precisely those conditions that make an area attractive to capital and susceptible to rapid development. As a result, capital tends to move back and forth, or seesaw, between developed and underdeveloped areas.

It is difficult, however, to imagine any satisfactory explanation of urban growth or decline that does not take migration into account, as most of the variation in growth rates among cities is due to differences in migration (Frey and Speare, 1988:8). Within this context, the present author (Cadwallader, 1991) conducted an empirical analysis of metropolitan growth and decline in the United States. The purpose of the research was threefold. First, urban growth rates for the largest United States Metropolitan Statistical Areas were documented for the periods 1965–1970 and 1975–1980. Second, discriminant analysis was used to identify the major variables that differentiate between growing and declining areas during these two time periods. Third, the interrelationships between migration rates for cities and other variables, such as income, unemployment, taxes, public spending, and housing costs, were explored via a simultaneous equation model.

Patterns of Growth and Decline

The largest 65 metropolitan statistical areas in 1970 were used to document net migration rates for the period 1965–1970, while 64 were used for the period 1975–1980. The reason for this discrepancy is that for the period 1975–1980 Dallas and Fort Worth were merged by the Census Bureau. The net migration rate was determined by subtracting the number of out-migrants from the number of in-migrants and dividing by the population at the end of the time period. For both time periods a substantial number of cities experienced negative net migration rates (Table 12.1). More specifically, 45 percent of the cities had negative net migration rates for 1965–1970, while more than 65 percent had negative migration rates for 1975–1980. In all, 77 percent of the cities experienced a decline in net migration rates between the two time periods.

The *spatial pattern* associated with these net migration rates is shown in Figure 12.7, where the cities have been divided into four categories based on their individual migration experiences during the two time periods. Overall, the regional pattern reenforces the

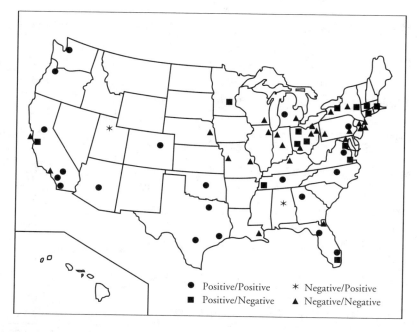

Figure 12.7 Spatial distribution of net migration rates for 1965–1970 and 1975–1980. (From M. Cadwallader, "Metropolitan growth and decline in the United States: An empirical analysis," *Growth and Change*, 22, 1991, Fig. 1, p. 5.)

Table 12.1 Net Migration Rates for 1965–1970 and 1975–1980

	1965–1970	1975–1980		1965–1970	1975–1980
Akron	−0.11	−0.85	Milwaukee	−0.58	−0.82
Albany	0.07	−0.53	Minneapolis	0.29	−0.24
Allentown	0.21	0.07	Nashville	0.08	0.87
Anaheim	4.33	0.47	Newark	−0.67	−1.34
Atlanta	1.45	0.92	New Orleans	−0.51	−0.21
Baltimore	−0.04	−0.27	New York	−0.98	−1.58
Birmingham	−0.45	0.09	Norfolk	0.25	−0.06
Boston	−0.34	−0.86	Oklahoma City	0.16	0.58
Buffalo	−0.63	−1.29	Omaha	−0.39	−0.90
Chicago	−0.64	−1.14	Paterson	−0.39	−1.25
Cincinnati	−0.21	−0.64	Philadelphia	−0.09	−0.61
Cleveland	−0.53	−1.28	Phoenix	2.10	3.56
Columbus	0.11	−0.38	Pittsburgh	−0.60	−0.74
Dallas	1.70	1.32	Portland	1.08	1.33
Dayton	0.09	−0.80	Providence	0.41	−0.36
Denver	1.21	1.09	Richmond	0.60	0.70
Detroit	−0.41	−1.02	Rochester	0.33	−0.81
Ft. Lauderdale	4.16	3.97	Sacramento	0.06	1.18
Ft. Worth	2.03	—	St. Louis	−0.27	−0.69
Gary	−0.65	−0.94	Salt Lake City	−0.33	0.79
Grand Rapids	0.12	0.12	San Antonio	0.50	0.18
Greensboro	0.09	0.35	San Bernardino	1.57	3.46
Hartford	0.77	−0.71	San Diego	2.32	1.69
Honolulu	0.72	−0.97	San Francisco	−0.01	−0.73
Houston	1.68	2.11	San Jose	2.33	−0.27
Indianapolis	−0.05	−0.32	Seattle	1.77	1.12
Jacksonville	−0.13	−0.09	Springfield	0.06	−0.66
Jersey City	−1.44	−1.49	Syracuse	−0.07	−0.69
Kansas City	−0.13	−0.47	Tampa	2.44	3.16
Los Angeles	−0.56	−1.27	Toledo	0.02	−0.44
Louisville	−0.06	−0.78	Washington	1.00	−1.03
Memphis	0.18	−0.27	Youngstown	−0.15	−1.01
Miami	0.78	−0.29			

Source: M. Cadwallader, "Metropolitan growth and decline in the United States: An empirical analysis," *Growth and Change*, Vol. 22, 1991, Table 1, p. 6.
Note: For the period 1975–1980 the Dallas and Fort Worth data were merged by the Census Bureau.

traditional sunbelt–frostbelt dichotomy. The majority of cities with positive migration rates for both time periods are located outside of the heavily industrialized Northeast and Midwest regions of the country, thus supporting Noyelle and Stanback's (1984) contention that manufacturing areas suffered from low employment growth and net out-migration. In particular, Fort Lauderdale, Phoenix, and Tampa have the highest average positive migration rates, thus stressing the importance of retirement centers. In contrast, Buffalo, Jersey City, and New York have the highest average negative migration rates. Those cities exhibiting positive followed by negative migration rates are also concentrated in the frostbelt.

The mean *characteristics* of the growing and declining cities for the two time periods are indicated in Table 12.2. The data were calibrated for 1960 and 1970, as one would expect net migration rates to respond to the characteristics associated with some previous time period. The average income levels are higher for the declining cities, while unemployment rates are higher for the growing cities in 1975–1980. In all cases, however, the differences are relatively minor. A far more substantial difference between growing and declining cities is evident in the data for manufacturing employment, with the declining cities being more heavily oriented toward manufacturing activity. Similarly, housing costs and various kinds of local taxes are uniformly higher for the declining cities. By contrast, local government expenditures are very similar for the two groups of cities, with the growing cities having slightly higher rates for the first period, but slightly lower rates for the second period. Finally, values for the climatic attractiveness index are fairly even, while crime rates are slightly higher for the growing cities. Climatic attractiveness was measured by a combination of variables involving temperature,

Table 12.2 Mean Characteristics of Growing and Declining Cities

	1965–1970		1975–1980	
	Growing (N = 36)	Declining (N = 29)	Growing (N = 22)	Declining (N = 42)
Income per capita	1,875	2,051	3,269	3,436
Unemployment rate	4.61	5.01	4.34	4.04
% employed in manufacturing	24.57	32.18	21.91	28.15
Median housing value	12,777	14,124	17,168	20,200
Property tax per capita	102	122	185	231
Local govt. tax per capita	117	139	224	278
Local govt. spending per capita	279	274	581	625
Education spending per capita	103	99	230	237
Crime rate per 100,000	1,323	1,259	3,462	3,284
Climatic attractiveness	592.34	595.17	607.59	594.50

Source: M. Cadwallader, "Metropolitan growth and decline in the United States: An empirical analysis," *Growth and Change*, Vol. 22, 1991, Table 2, p. 7.

heating- and cooling-degree days, freezing days, zero-degree days, and ninety-degree days, while crime rates were calibrated for seven categories of personal and property crimes.

In general, the relationships between net migration and various other variables were as expected (Table 12.3). Statistically significant negative relationships were found for percent employed in manufacturing, the median value of owner-occupied housing units, per capita property tax rates, and per capita general tax rates, while positive relationships were found for educational expenditures per capita and climatic attractiveness. Somewhat surprisingly, however, net migration was negatively related to per capita income, and positively related to crime rates, although other recent evidence has also suggested that during the 1970s migrants tended to move from high- to low-income areas (Frey and Speare, 1988:75).

Discriminant Analysis

One way to empirically assess which variables play a role in determining urban growth or decline is to use the statistical technique of discriminant analysis (Klecka, 1980). Discriminant analysis involves statistically distinguishing between two or more groups of cases. For this purpose, a set of discriminating variables is chosen that measures characteristics on which the groups are expected to differ. These variables are then weighted and linearly combined in order to form a series of *discriminant functions*, such that the differences between the groups are maximized. The maximum number of functions that can be derived is one less than the number of original groups or equal to the number of discriminating variables, if there are more groups

Table 12.3 Correlation Coefficients between Net Migration Rate and Other Variables

	1965–1970	1975–1980
Income	−0.11	−0.22[b]
Unemployment	−0.07	0.10
Manufacturing	−0.41[a]	−0.45[a]
Housing costs	−0.08	−0.31[a]
Property tax	−0.09	−0.27[b]
General tax	−0.07	−0.31[a]
General spending	0.14	−0.08
Educational spending	0.38[a]	0.02
Crime	0.21[b]	0.22[b]
Climate	0.24[b]	0.05

Source: M. Cadwallader, "Metropolitan growth and decline in the United States: An empirical analysis," *Growth and Change*, Vol. 22, 1991, Table 3, p. 8.
[a]Indicates significant at 0.01.
[b]Indicates significant at 0.05.

than variables. In practice, however, it is often possible to obtain a satisfactory solution by using fewer than the maximum number of functions, similar to extracting factors in factor analysis. In other words, it is assumed that the location of the groups in the m-dimensional variable space can be captured by a few basic composites of those m variables.

Several statistics are available to measure the success of the discriminant functions. The *eigenvalue* associated with a particular discriminant function reflects the relative importance of that function, as the sum of the eigenvalues is a measure of the total variance existing in the discriminating variables. By expressing an individual eigenvalue as a percentage of the sum of the eigenvalues, one can determine the percentage of the total variance accounted for by each discriminant function. The discriminant functions are derived in order of importance, and the procedure can be stopped when the eigenvalues become too small.

The *canonical correlation* is a measure of association between a single discriminant function and the dummy variables that define group membership and thus reflects the function's ability to discriminate among the groups. In other words, the canonical correlation indicates how closely the discriminant function and the groups are related. The discriminant functions can be interpreted by analyzing the standardized discriminant function coefficients. If the sign is ignored, the sizes of these coefficients indicate the relative contribution of the associated variable to each discriminant function. That is, as in multiple regression analysis or factor analysis, they serve to identify the variables that contribute most to the differentiation along each function.

The first discriminant analysis, based on the net migration rates for 1965–1970, used two groups of cities (Table 12.4). To highlight the differences between growing and declining

Table 12.4 Two-group Discriminant Analysis for 1965–1970

	Standardized Coefficients	Means Growing (N = 19)	Means Declining (N = 18)
Income	–0.42	1,979	2,118
Unemployment	0.23	4.56	4.96
Manufacturing	1.23	19.64	32.71
Housing costs	0.35	13,437	15,056
Property tax	–1.20	105	132
General tax	1.48	122	149
General spending	0.55	280	281
Education spending	–0.87	107	96
Crime	0.29	1,508	1,306
Climate	–0.48	620.63	586.83
Canonical correlation = 0.80; cases classified correctly = 95%			

Source: M. Cadwallader, "Metropolitan growth and decline in the United States: An empirical analysis," *Growth and Change*, Vol. 22, 1991, Table 4, p. 9.

cities, two subgroups of extreme cases were chosen. The declining group included all cities with negative migration rates of more than –0.25, while the growing group contained all cities with positive migration rates greater than 0.5. This categorization created a growing group of 19 cities and a declining group of 18 cities. As there were two groups, only one discriminant function was extracted. This function, however, had a canonical correlation of 0.80 and successfully classified 95 percent of the cases. That is, the discriminant function was able to correctly identify 95 percent of the cases as members of the group to which they actually belonged. The standardized coefficients indicated that manufacturing and taxes contributed most to the discriminating function, while spending on education made a somewhat less important contribution to the discrimination process. These results were confirmed by comparing the variable means for each group, which showed that percent employed in manufacturing, property taxes, and general taxes were all higher for the declining group.

A similar analysis was undertaken for the 1975–1980 data (Table 12.5). In this case the declining group consisted of cities with negative migration rates greater than –0.75, while the growing group contained all those cities with positive migration rates. The results were equally encouraging, as the discriminant function had a canonical correlation of 0.82 and successfully classified 90 percent of the cities. The standardized coefficients displayed a similar pattern to those for the previous time period, in that manufacturing, taxes, and spending on education were again the important variables. This time, however, the general tax variable was much more important than any of the other variables, and climatic attractiveness also emerged as an important variable. An examination of the respective group means shows that the percent employed in manufacturing and tax rates were still considerably higher for the declining cities, while climatic attractiveness was greater for the growing cities.

Table 12.5 Two-group Discriminant Analysis for 1975–1980

	Standardized Coefficients	Means Growing (N = 22)	Means Declining (N = 20)
Income	0.06	3,269	3,577
Unemployment	0.33	4.34	4.03
Manufacturing	0.85	21.91	31.06
Housing costs	0.41	17,168	22,400
Property tax	–0.80	185	257
General tax	2.23	224	306
General spending	0.01	581	650
Education spending	–1.26	230	238
Crime	0.03	3,462	3,296
Climate	–0.81	607.59	601.75

Canonical correlation = 0.82; cases classified correctly = 90%.

Source: M. Cadwallader, "Metropolitan growth and decline in the United States: An empirical analysis," *Growth and Change*, Vol. 22, 1991, Table 5, p. 10.

A three-group discriminant analysis was then undertaken for both time periods, with the additional third group containing those cities not included in the previous two-group analysis (Table 12.6). As there were now three groups, two discriminant functions could be extracted. In general, however, the discriminant functions were less successful than in the two-group analysis. For the 1965–1970 data the canonical correlations are 0.68 and 0.56, while 70 percent of the cases were correctly classified. The first function accounted for 66 percent of the total variance, while the remaining variance was associated with the second function. The analysis for 1975–1980 was even less successful, as the second canonical correlation was only 0.21 and only 55 percent of the cases were correctly assigned by the model. Furthermore, the first function accounted for nearly 97 percent of the total variance, leaving an insignificant role for the second function. As in the two-group analysis, however, the standardized coefficients emphasized the importance of the manufacturing, tax, and educational spending variables for both time periods. In general, then, the two- and three-group discriminant analyses suggested that manufacturing activity, local tax rates, and spending on education were particularly important discriminators between growing and declining areas.

Simultaneous Equation Models

The next step in the analysis was to model the interrelationships between the characteristics of cities and their growth rates. Whenever reciprocal relationships are anticipated, as is the case here, simultaneous equation models are appropriate (see Section 10.3). Such models have already been used in the context of migration (Cadwallader, 1992b) and allow the

Table 12.6   Three-group Discriminant Analysis

	Standardized Coefficients			
	1965–1970		1975–1980	
Income	−0.04	0.25	−0.17	1.23
Unemployment	0.29	0.09	0.08	0.37
Manufacturing	1.01	−0.61	0.93	0.12
Housing costs	0.29	0.43	0.54	0.07
Property tax	−0.18	0.69	−0.93	0.97
General tax	0.49	0.11	2.02	−2.15
General spending	0.75	−0.59	0.26	0.90
Education spending	−1.08	−0.18	−1.12	−0.59
Crime	0.35	0.37	0.20	0.17
Climate	−0.53	−0.05	−0.68	−0.01
% Variance accounted for	66.26	33.74	96.77	3.23
Canonical correlation	0.68	0.56	0.76	0.21
Cases classified correctly	70%		55%	

Source: M. Cadwallader, "Metropolitan growth and decline in the United States: An empirical analysis," *Growth and Change*, Vol. 22, 1991, Table 6, p. 11.

researcher to specify a number of endogenous variables. In the present instance, net migration, property taxes, and spending on education were treated as endogenous variables, while income, housing costs, manufacturing, climate, and crime were exogenously determined. Besides treating migration as an endogenous variable, property taxes and spending on education were postulated to be reciprocally related, as local government taxing and spending decisions are simultaneously determined (Loehman and Emerson, 1985). Of the original variables, unemployment and general spending were not included in the simultaneous equation models as they did not perform well in terms of either the zero-order correlation coefficients (Table 12.3) or the discriminant analyses (Tables 12.4, 12.5, and 12.6).

As there were three endogenous variables, three equations were estimated for each time period (Table 12.7). A necessary, although not sufficient, condition for the identifiability of a structural equation within a given linear model is that at least $k - 1$ of the variables should be excluded from that equation, where k is the number of structural equations. In this case, therefore, at least two variables should be excluded from each equation. Consequently, the educational spending and property tax equations are overidentified, while the migration equation is exactly identified. Note that the structure of the models is the same for both time periods in order to facilitate comparison.

The structural parameters contained within simultaneous equation models can be estimated in a variety of ways. The two-stage least-squares estimates for the present equations and

Table 12.7 Simultaneous Equation Models

1965–1970										
Education spending =	40.64	+	$8.25X_1$	+	$0.27X_3$	+	$0.01X_4$			
	(17.22)		(10.58)		(0.44)		(0.03)			
Property tax =	−53.79	−	$23.90X_1{}^a$	+	$1.33X_2{}^a$	+	$0.29X_5{}^b$			
	(38.80)		(7.65)		(0.42)		(0.16)			
Net migration =	0.82	+	$0.01X_2$	−	$0.01X_3$	−	$0.04X_6{}^b$	+	$0.001X_7$	+ $0.0002X_8$
	(1.53)		(0.03)		(0.01)		(0.02)		(0.002)	(0.0003)
1975–1980										
Education spending =	−68.48	+	$12.76X_1$	+	$0.02X_3$	+	$0.09X_4$			
	(243.33)		(20.64)		(0.94)		(0.13)			
Property tax =	−117.75	−	$22.25X_1{}^b$	+	$0.99X_2{}^a$	+	$0.52X_5{}^a$			
	(46.62)		(10.09)		(0.19)		(0.16)			
Net migration =	−0.143	+	$0.02X_2{}^a$	−	$0.02X_3{}^a$	−	$0.03X_6$	+	$0.002X_7$	+ $0.0001X_8$
	(1.63)		(0.01)		(0.01)		(0.02)		(0.001)	(0.0002)

Source: M. Cadwallader, "Metropolitan growth and decline in the United States: An empirical analysis," *Growth and Change*, Vol. 22, 1991, Table 7, p. 12.
[a]Indicates significant at 0.01.
[b]Indicates significant at 0.05.
X_1 is net migration, X_2 is spending on education, X_3 is property taxes, X_4 is income, X_5 is housing costs, X_6 is employment in manufacturing, X_7 is climatic attractiveness, and X_8 is crime rate.

the associated standard errors revealed a fairly similar pattern across both time periods. In neither case were the relationships for the first equation significant, but for the property tax equation all the coefficients were significantly different from zero at the 0.05 probability level, and four of them were significant at the 0.01 probability level. Moreover, the directions of the relationships were as expected. Property taxes were negatively associated with net migration, but positively related to educational spending and housing value.

Finally, the net migration equation performed considerably better with respect to the 1975–1980 data than the 1965–1970 data. For the earlier time period only the manufacturing variable had a significant coefficient at the 0.05 level. By contrast, for the later time period, two of the variables were significant at the 0.01 probability level. Again, the directions of the relationships were as expected, given the previous analyses, as net migration was positively related to spending on education and climatic attractiveness, but negatively related to property taxes and percent employed in manufacturing.

To summarize, the spatial pattern associated with these migration rates tended to reenforce the traditional sunbelt–frostbelt dichotomy, as the majority of cities with positive migration rates for both time periods were located outside of the heavily industrialized Northeast and Midwest regions of the country. An examination of the characteristics associated with declining cities indicated that they were more heavily dependent on manufacturing activity than the growing cities and that they also exhibit higher housing costs and local taxes. Two- and three-group discriminant analyses supported these conclusions, with manufacturing activity, local tax rates, and spending on education being particularly important discriminators between growing and declining areas. The interrelationships between the characteristics of cities and their growth rates were then more formally modeled via the use of simultaneous equations, with net migration, property taxes, and spending on education being treated as endogenous variables. The estimated coefficients and their associated standard errors suggested that the property tax equation was more successful than the one for spending on education and that the equation for net migration worked better for the 1975–1980 data than for the previous time period.

12.3 Intermetropolitan Housing Value Differentials

As with population growth or decline, the value of housing also varies from city to city in the United States. Remarkably little is known, however, about the determinants of such differentials. Analyses of urban housing markets have generally been concerned with intraurban variation (Follain and Jimenez, 1985) and the change in house prices over time (Engelhardt and Poterba, 1991). The lack of interest in intermetropolitan differences is unfortunate, as such differences substantially affect real incomes. Indeed, the existence of significant interregional housing value differentials can impede a nation's macroeconomic performance, as migrants will be discouraged from relocating to resource-rich regions (Barlow, 1990; Forrest, 1991).

To date, there have been very few attempts to systematically explain intermetropolitan housing value differentials. A notable exception is provided by the work of Ozanne and Thibodeau (1983), who analyzed the sources of intercity rent and house price variation in the United States. Their regression results indicate greater success in explaining rent variations than in explaining housing price variations. In particular, the rent equation explained

88 percent of the variation, while the house price equation explained 58 percent. In the latter equation only three coefficients were statistically significant at the 0.10 level.

Using data from the United Kingdom, Nellis and Longbottom (1981) investigated the price of new privately owned dwellings. They concluded that income is the most important determinant. Similarly, in a study of Canadian intercity house price differentials, Fortura and Kushner (1986) also emphasize the importance of demand factors such as income. More specifically, they found that a one percent increase in household income raises house prices by just over one percent. In their regression model the explanatory variables combined to account for 83 percent of the variation in the price of housing. Three of the seven estimated coefficients were significant at the 0.05 level.

Within this context, the purpose of this section is to present an empirical analysis of housing values for the largest United States Metropolitan Statistical Areas in 1960, 1970, and 1980 (Cadwallader, 1993). First, correlation analysis is used to explore the relationships between housing value and other selected variables. Second, discriminant analysis is used to identify the major variables that differentiate between cities exhibiting high and low housing values. Third, the interrelationships between housing value and other variables are investigated via a simultaneous equation model.

The period from 1960 to 1980 is of particular interest, as it was a time of rapidly appreciating prices in the U.S. housing market. The median value of owner-occupied housing units rose from $12,000 in 1960 to $17,100 in 1970, an increase of just over 42 percent (Adams, 1987:98). The increase during the 1970s was even more dramatic, however, as the median value rose to $51,300, an additional jump of 200 percent (Adams, 1987:98). This exceptionally rapid increase in house prices during the 1970s was at least partly due to the peak of the post World War II baby boom entering the housing market, but the increase was unevenly distributed across even the major cities. For example, in Los Angeles the median value of owner-occupied housing units rose from $24,285 in 1970 to $87,400 in 1980, representing an extraordinary increase of 260 percent (Adams, 1988:124). During the same time period, the parallel rise in Chicago was 167 percent, while that in New York was 126 percent.

Correlation Analysis

Data were culled from a variety of sources for the 65 largest metropolitan statistical areas in the United States. Income was used because of its importance in previous studies involving housing value (Nellis and Longbottom, 1981; Fortura and Kushner, 1986), with the conventional wisdom being that high average incomes should increase demand and thus drive up housing values. The local level of unemployment has also been found to be an important determinant of housing value, with higher levels of unemployment being associated with lower housing values (Champion et al., 1988). Percentage of the labor force involved in manufacturing provides a further measure of the city's economic structure. Those cities with a significant proportion of their labor force in the tertiary sector tend to be among the fastest growing cities, while declining cities are more heavily oriented toward manufacturing activity (see Section 12.2). As a result, we would expect a negative relationship between level of manufacturing activity and housing value.

Local taxes and local public spending have long been considered important determinants of property values. Using data for communities in northeastern New Jersey, Oates (1969) regressed property value on a series of variables, including annual expenditure per pupil in public schools and local property taxes. He found significant relationships between property values and local taxes and public spending, but warned of simultaneity bias, as expenditure and taxes also depend on each other, as well as on property value. Similarly, Cebula (1983), using data for cities in the United States, found a positive relationship between per capita expenditure on public education and housing value. Finally, research on interurban variation in migration and wage rates (Graves, 1980; Roback, 1982) indicates that households are prepared to trade off income and house prices for certain urban amenities, such as a favorable climate or a low crime rate. According to this argument, we would expect housing value to be positively related to climatic attractiveness and negatively related to crime rate.

Initially, zero-order correlation analysis was used to explore the relationships between housing value and the other variables involved in the study (Table 12.8). For the most part the relationships are as expected. For all three years there is a strong positive correlation between income and housing value. There are also strong positive correlations associated with local government spending, local taxes, and climatic attractiveness. Unemployment, manufacturing, and crime only exhibit significant relationships to housing value in 1980. As expected, unemployment and manufacturing are negatively related to housing value, but crime has an unexpected positive relationship.

Discriminant Analysis

Discriminant analysis was used to further explore these relationships. A separate discriminant analysis was run for 1960, 1970, and 1980. In each case the cities were divided into three groups based on housing values. To highlight the differences between cities with low and high housing values, the two subgroups of extreme cases were used in the analysis. For 1960

Table 12.8 Correlation Coefficients between Housing Value and Other Variables

	1960	1970	1980
Income	0.601[a]	0.743[a]	0.668[a]
Unemployment	0.003	0.091	−0.325[a]
Manufacturing	0.072	−0.043	−0.253[b]
Taxes	0.583[a]	0.646[a]	0.191[c]
Spending	0.318[a]	0.324[a]	0.193[c]
Crime	−0.019	0.173	0.388[a]
Climate	0.349[a]	0.461[a]	0.605[a]

Source: M. Cadwallader, "Inter-metropolitan housing value differentials: the United States, 1960–1980," *Geoforum*, 24 (1993), Table 2. Reprinted with kind permission from Elsevier Science Ltd.
[a]Indicates significant at 0.01.
[b]Indicates significant at 0.05.
[c]Indicates significant at 0.10.

each of these groups contained 22 cities, with a mean housing value of $10,700 for the lower group and $16,600 for the higher group. For 1970 the two groups contained 22 and 23 cities, respectively, with associated means of $14,600 and $24,400. For 1980 the two groups contained 21 and 22 cities, with means of $39,300 and $75,700.

The discriminant function for 1960 has a canonical correlation of 0.729 and successfully classifies 88 percent of the cases (Table 12.9). That is, the discriminant function was able to correctly identify 88 percent of the cities as members of the group to which they actually belonged. The standardized coefficients indicate the importance of taxes, manufacturing, and income. The discriminant function for 1970 performs even better, with a canonical correlation of 0.762 and 91 percent of the cases classified correctly (Table 12.10). Again, income, taxes, and manufacturing have the highest standardized coefficients. By 1980 the picture has changed somewhat (Table 12.11). The canonical correlation is 0.772, and the percentage of cases classified correctly is still a respectable 83 percent, but this time the most important three variables are crime, income, and unemployment. For all three years the income variable plays an important role, and in each case the variable means show that income is highest for the group of cities with high housing values.

Simultaneous Equation Models

The next step in the analysis was to model the interrelationships between the characteristics of cities and housing value. In the present instance, housing value, taxes, and spending are treated as endogenous variables, while income, manufacturing, unemployment, crime, and climatic attractiveness are exogenously determined. Housing value is treated as an endogenous variable because it is the variable that is of major interest in the present analysis. Taxes and spending are also endogenously determined, in this case because of their reciprocal rela-

Table 12.9 Discriminant Analysis for 1960

	Standardized Coefficients	Means	
		Low	High
Income	0.331	1746.045	2215.524
Unemployment	0.236	4.464	4.824
Manufacturing	−0.421	25.441	27.829
Taxes	0.820	98.591	157.619
Spending	0.011	251.864	320.143
Crime	−0.281	1341.864	1322.476
Climate	0.020	563.045	635.905

Canonical correlation = 0.729; cases classified correctly = 88%

Source: M. Cadwallader, "Inter-metropolitan housing value differentials: the United States, 1960–1980," *Geoforum*, 24 (1993), Table 4.
Reprinted with kind permission from Elsevier Science Ltd.

Table 12.10 Discriminant Analysis for 1970

	Standardized Coefficients	Means	
		Low	High
Income	0.667	3133.762	3714.478
Unemployment	0.242	3.800	4.348
Manufacturing	−0.320	26.371	25.757
Taxes	0.497	200.857	319.304
Spending	−0.181	526.952	697.609
Crime	−0.028	3086.143	3571.000
Climate	0.151	546.190	655.348

Canonical correlation = 0.762; cases classified correctly = 91%

Source: M. Cadwallader, "Inter-metropolitan housing value differentials: the United States, 1960–1980," *Geoforum*, 24 (1993), Table 5. Reprinted with kind permission from Elsevier Science Ltd.

Table 12.11 Discriminant Analysis for 1980

	Standardized Coefficients	Means	
		Low	High
Income	0.589	7291.150	8465.273
Unemployment	−0.334	7.435	6.055
Manufacturing	0.154	28.645	23.591
Taxes	0.089	438.750	529.955
Spending	0.035	1360.100	1599.682
Crime	0.699	5943.500	7595.682
Climate	0.183	549.850	656.091

Canonical correlation = 0.772; cases classified correctly = 83%

Source: M. Cadwallader, "Inter-metropolitan housing value differentials: the United States, 1960–1980," *Geoforum*, 24 (1993), Table 6. Reprinted with kind permission from Elsevier Science Ltd.

tionship. Not unexpectedly, there is considerable evidence that local government taxing and spending decisions are simultaneously determined.

As there are three endogenous variables, three equations are estimated for each of the three time periods (Table 12.12). The taxes and spending equations are overidentified, while the housing value equation is exactly identified. Note that the structure of the models is the

Table 12.12 Simultaneous Equation Models

1960						
Taxes =	-27.79	$-$ $0.33X_3$	$+$ $1.34X_1$	$+$ $0.04X_4$		
	(55.51)	(0.38)	(1.48)	(0.06)		
Spending =	79.53	$-$ $0.03X_2$	$+$ $1.77X_1$	$-$ $1.24X_5$		
	(163.40)	(1.85)	(2.61)	(1.65)		
Housing =	28.01	$+$ $0.22X_3$	$+$ $0.42X_2^b$	$-$ $2.79X_6$	$-$ $0.08X_7$	$+$ $0.01X_8$
	(33.04)	(0.21)	(0.20)	(3.43)	(0.81)	(0.05)
1970						
Taxes =	-317.46	$+$ $0.14X_3$	$-$ $0.26X_1$	$+$ $0.16X_4^a$		
	(87.75)	(0.09)	(0.47)	(0.05)		
Spending =	433.91	$+$ $2.77X_2^a$	$-$ $1.97X_1^c$	$-$ $6.31X_5^a$		
	(88.87)	(0.72)	(1.11)	(1.73)		
Housing =	-20.26	$+$ $0.25X_3$	$+$ $0.22X_2$	$-$ $16.02X_6^c$	$-$ $0.07X_7$	$+$ $0.16X_8^c$
	(54.68)	(0.21)	(0.31)	(8.79)	(0.10)	(0.09)
1980						
Taxes =	-341.24	$+$ $0.01X_3$	$-$ $0.24X_1$	$+$ $0.12X_4^a$		
	(222.26)	(0.15)	(0.19)	(0.03)		
Spending =	1107.30	$+$ $0.70X_2$	$+$ $0.35X_1$	$-$ $7.84X_5^c$		
	(260.88)	(0.65)	(0.33)	(4.22)		
Housing =	-529.01	$+$ $0.13X_3$	$+$ $0.93X_2$	$-$ $23.27X_6$	$+$ $0.08X_7$	$+$ $0.90X_8^a$
	(493.64)	(0.92)	(0.88)	(16.55)	(0.79)	(0.25)

Source: M. Cadwallader, "Inter-metropolitan housing value differentials: the United States, 1960–1980," *Geoforum*, 24 (1993), Table 7.
[a]Indicates significant at 0.01.
[b]Indicates significant at 0.05.
[c]Indicates significant at 0.10.
X_1 is housing value, X_2 is taxes, X_3 is spending, X_4 is income, X_5 is manufacturing, X_6 is unemployment, X_7 is crime, X_8 is climate.

same for all three time periods in order to facilitate comparison. It should also be noted that income is not directly involved in the housing value equation, but is indirectly involved via its relationship with taxes.

As previously noted, the *structural parameters* contained within simultaneous equation models can be estimated in a variety of ways. The two-stage least-squares estimates for the present equations and the associated standard errors reveal some similarities and differences across the three time periods. The relationships specified by the tax equation are not significantly different from zero for 1960, but for both 1970 and 1980 there is the expected positive coefficient for income. The spending equation fits best for 1970, when taxes, housing, and manufacturing all have significant coefficients. There are no significant relationships for 1960, in the spending equation, and only manufacturing has a significant coefficient for 1980. The housing value equation also fits best for 1970, with significant coefficients associated with both unemployment and climate. Taxes are significantly related to housing value for 1960 and climate is significantly related for 1980. It is noteworthy that the directions of the significant relationships are the same for all three time periods, thus indicating some degree of system stability where significant relationships have been identified.

To summarize, the results of the discriminant analyses suggest that there is a strong relationship between income and housing value and that local taxes, percent employed in manufacturing, unemployment levels, and crime rates also play a role in explaining housing value differences. The simultaneous equation models suggest a series of recursive relationships among housing value, local government taxes, and local government spending, while the estimated coefficients and their associated standard errors provide further evidence for the role of income. The relationship between income and housing value is not unproblematic, however, as it could well be argued that in the short run it is housing value that partly determines the income level of in-migrants.

Further work is required in a number of areas. First, the present analysis does not incorporate the effect of demographic pressures on housing values. For example, those cities experiencing high rates of new household formation are likely to experience escalating house prices. Second, the current analysis also ignores the effects of changes in other parts of the housing market. A shortage of rental units, for example, can lead to an increase in demand for homeownership. Third, the analysis does not take into account the influence of supply variables, such as the price of nonresidential land and the role of local development controls. Finally, the present analysis does not explicitly consider the dynamic nature of the housing market, whereby factors such as change in unemployment, as opposed to the absolute level of unemployment at a particular point in time, might play a role in determining housing value.

12.4   Patterns of Migration

Many of our current ideas about interregional migration stem from the pioneering work of E. G. Ravenstein. More than 100 years ago, Ravenstein published two influential papers in the *Journal of the Royal Statistical Society*, describing his laws of migration (Ravenstein, 1885, 1889). These laws, or perhaps more accurately generalizations, were inductively derived from place-of-birth data published in the British censuses of 1871 and 1881, together with similar data from North America and Europe. Among his more important generalizations were the following: the majority of migrants move only a short distance; migrants do not proceed directly to their ultimate destination, but get there via a series of steps; each migration stream tends to generate a compensating counterstream; and the major causes of migration are economic. These simple statements about migration flows have generated a remarkable amount of empirical research, and many of Ravenstein's hypotheses have since been confirmed.

Spatial Patterns

Many scholars have substantiated Ravenstein's generalization that migration behavior is affected by distance (Stillwell, 1991). More specifically, the *friction-of-distance* effect has frequently been observed; this means that the number of migrants from any given region decreases with increasing distance. The form of this relationship between the amount of migration and distance from the origin has usually been described by a power function (see Section 3.4). The power function, in this particular context, can be expressed as follows:

$$M = aD^{-b} \qquad\qquad (12.6)$$

where M is migration, D is distance, and a and b are empirically derived constants. The negative b coefficient indicates that migration is an inverse function of distance.

Some researchers have suggested that the negative exponential function (see Section 3.4) is more appropriate for representing the relationship between migration and distance. Such a function would be expressed as follows:

$$M = ae^{-bD} \qquad\qquad (12.7)$$

where the notation is the same as in equation (12.6), and e is the base of natural logarithms. Regardless of the precise functional form, the weight of evidence indicates that the role of distance has decreased over time. There is considerable doubt, however, about whether the individual effect of distance can ever be satisfactorily isolated. It has been argued that distance-decay parameters reflect a complex combination of spatial structure, involving the size and configuration of origins and destinations in a spatial system and intrinsic interaction behavior, thus precluding any simple interpretation of such parameters (Fotheringham, 1981, 1991; Eldridge and Jones, 1991).

In a causal sense, distance is merely being used as a surrogate for other, less easily measured variables. In particular, the fact that migration decreases with increasing distance has been attributed to the notion that distance serves as a proxy for the psychic costs of movement. In some situations, however, distance might reflect information flows or the uncertainty about employment and income prospects in other regions. Given this range of interpretations, it is not surprising that distance has been measured in a variety of ways. Mileage, time, and cost distance are all appropriate under different circumstances, and researchers have also attempted to use the concept of cognitive distance in migration models.

Ravenstein's pioneering research on migration is also responsible for the more recent interest in chain or *stepwise migration*. Ravenstein suggested that cities grew by attracting migrants from the surrounding area and that those migrants were then replaced by others from even remoter regions. It can thus be argued that the historically important rural to urban migration flows are actually composed of a series of discrete steps. That is, people first move from rural areas to small towns, then from small towns to larger cities, and so on. Such a sequence might occur as a result of information flows, whereby new opportunities present themselves as a migrant moves upward through the urban hierarchy. These new opportunities might also be accompanied by changing aspiration levels.

Besides stepwise flows, a significant amount of *return migration* can also be observed at the interregional level (Long, 1988:136; Rogers and Belanger, 1990). Return migration describes the movement of people to regions where they have previously lived, and approximately 25 percent of all migrants in the United States can be designated as returnees (DaVanzo and Morrison, 1981). In particular, return migration has often reversed the direction of traditional migration streams. For example, the net outflow of black migrants from the South is now exceeded by a counterstream, two-thirds of which represents returnees (Cromartie and Stack, 1989; McHugh, 1987). Ravenstein had mentioned the notion of such counterstreams, but it is unclear whether he was referring to returnees or merely to flows of migrants moving in the opposite direction to the dominant stream.

Despite its widespread occurrence, however, return migration has been a rather neglected phenomenon. The main reason for this comparative neglect is that the data requirements for the analysis of return migration are particularly severe (Alexander, 1983). Data on migration are usually only available on a discontinuous time scale, and many moves go unregistered in conventional statistics that measure migration by comparing an individual's residential location at two points in time. As always, the level of temporal resolution influences the quality of observed data, and this is especially true for return migration, as such moves often occur soon after a previous move. Besides these data problems, research on return migration has also been hampered by the lack of an appropriate conceptual framework. The challenge is to formulate an approach that can account for migration sequences where the emphasis is on a series of moves rather than on an individual event (Dierx, 1988).

DaVanzo and Morrison (1981) have made some progress toward this goal by specifying a framework that highlights the role of location-specific capital and imperfect information. Location-specific capital comprises all the factors that tie a person or household to a specific place. These factors include concrete assets plus more intangible variables, such as job seniority, an established clientele, personal knowledge of the area, community ties, and friendship networks. Such considerations suggest that a migrant will favor a previous area of residence as a potential destination due to the location-specific capital there. In addition, the propensity to return will decrease with increasing length of absence, since location-specific capital tends to depreciate over time. Consistent with this latter proposition, DaVanzo and Morrison found that family heads are most prone to return within a year and that the probability of moving back decreases thereafter.

The notion of imperfect information suggests that a move can sometimes turn out to have been an unwise decision, thus generating a category of discouraged migrants whose employment or income expectations did not materialize. These unfulfilled expectations may be the result of imperfect information, leading to an overly optimistic assessment of the labor market conditions at the destination. A return move might then be the most appropriate corrective action, as the migrant would not want to repeat the mistake. Support for his line of reasoning is provided by two kinds of empirical evidence. First, the unemployed are especially prone to make return moves, particularly when the migration interval is short. Second, the less-educated segments of the population, who tend to possess only fragmented information about national employment opportunities, are also more likely to return to familiar surroundings.

Finally, it has been argued that return migrants can have significant consequences for regional growth and development (King, 1986:18–27). First, returning migrants sometimes have sources of accumulated capital, which are then invested in new enterprises and thus generate further employment and capital in the region. Second, returning migrants often possess newly acquired labor skills and innovative entrepreneurial attitudes. On the other hand, evidence also suggests that return migrants do not necessarily assume positions of social or psychological leadership (Townsend, 1980). Rather, the locals who have stayed within the region often tend to take on positions of industrial and political responsibility.

Migration Fields

Migration fields represent a useful way of summarizing different kinds of migration flows; a migration field refers to the overall pattern of flows associated with a particular place (Roseman and McHugh, 1982). An in-migration field is thus a functional, or nodal, region that delineates the origins of immigrants to an area, while an out-migration field characterizes the destinations of emigrants from that area. Such fields can be represented formally and abstractly by using such concepts as vector flows and linkage analysis (Slater, 1984). In particular, the technique known as primary linkage analysis can be used to identify individual migration fields (Haggett et al., 1977:485–90). Consider Figure 12.8, which depicts a hypothetical flow matrix for ten different regions. The values in the cells of the matrix indicate the number of migrants moving between various pairs of regions during a particular time period. For example, 14 people moved from region 2 to region 3, and 12 people moved from region 3 to region 2.

The hierarchical order of a region is measured by the total number of people moving to that region, so region 7, with 207 migrants, is the first region, region 3 is the second region, and so on. The hierarchical structure of the various regions is determined by the largest outflow to a higher-order region. For regions 1, 3, and 7, the largest outflow is to a low-order region, so these three regions form the terminal points of the graph. The remaining seven regions are then assigned, either directly or indirectly, to one of these terminals. For example, the largest outflow from region 2 is to region 1; thus an arrow connects those two regions. Similarly, the largest outflow from region 8 is to region 5, and the largest outflow from region 5 is to region 3. The resulting graph identifies three distinct subgraphs, or migration fields.

Clayton (1977a, 1977b) has used linkage analysis to explore interstate migration flows in the United States for the periods 1935–1940, 1949–1950, 1955–1960, and 1965–1970 (Figure 12.9). The number of terminal nodes identified for each of these time periods is six, eight, four, and four, respectively. Those identified for the last two periods—California, Florida, New York, and Virginia—are also present in both earlier periods. This reduction in the number of terminal nodes reflects the increasing concentration around a few major destinations. In general, then, the structure of the migration system contains two major components. First, a series of local linkages connect small clusters of contiguous states, such as Washington–Idaho–Montana, Minnesota–Dakotas, and Texas–Oklahoma–Louisiana. Second, there is a series of national-level linkages, such that Washington, Minnesota, Texas, and all their dependencies are linked to California. Over time, the national-level linkages appear to be increasingly dominant, especially with respect to California and Florida.

In many respects the study of migration fields can be regarded as a problem in formal and functional regionalization (see Section 2.1). Within this context, general field theory provides an attractive framework for capturing the interdependency between the characteristics of places and the flows between them (Schwind, 1975; Slater, 1989). General field theory involves a spatial system that comprises places, the attributes of those places, and the interac-

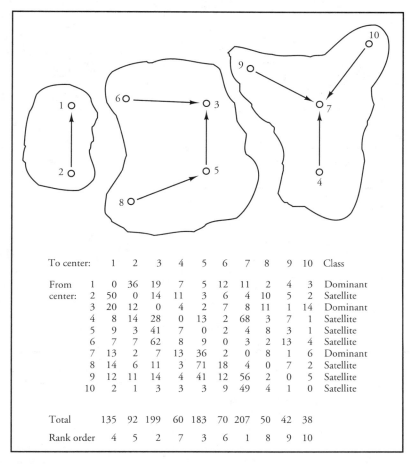

Figure 12.8 Hypothetical migration fields based on interaction data. (From M. Cadwallader, *Migration and Residential Mobility*, University of Wisconsin Press, Madison, Wisconsin, 1992, Fig. 2.3.)

tions among them. While groups of places with similar characteristics form uniform regions, functional regions are based on similar patterns of interaction. The essential postulate of general field theory is that these two types of regions are isomorphic; there is a mutual equilibrium in structure and behavior. Any change in the characteristics of places will change the pattern of migration, and vice versa.

Schwind (1975) has used general field theory and the associated technique of canonical analysis (see Section 8.4) to explore interstate migration in the United States. The migration data covered the period 1955–1960, and the attribute data consisted of 31 variables describing the demographic, economic, social, and climatic characteristics of the individual states. Nine migration fields were extracted by the canonical technique, and eight of them were significant. Four of the fields reflected relatively closed migration systems, with strongly sym-

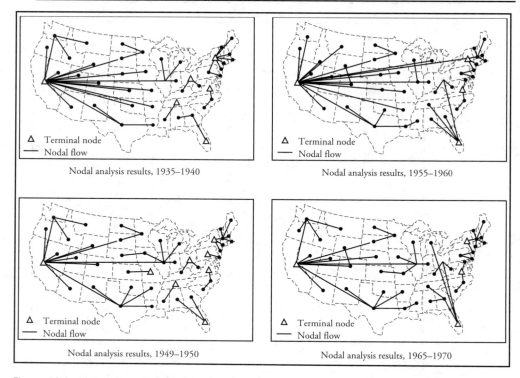

Figure 12.9 Linkage analysis for interstate migration in the United States. (From C. Clayton, "Interstate population migration process and structure in the United States, 1935–1970," *Professional Geographer,* 29, 1977, Figs. 1–4.)

metrical, or reciprocal, origin–destination patterns. The remaining fields were more open, with asymmetrical origin–destination patterns. For example, California received large numbers of migrants from the Midwest, while its own out-migrants were more oriented toward Oregon and Washington. Thus climate was particularly associated with asymmetrical patterns, while industrial labor force characteristics tended to be associated with reciprocal movements.

Chapter 13

Interregional Migration

13.1   A Conceptual Framework

To date there have been essentially two major approaches to the analysis of migration behavior. The first, or *macro approach*, is concerned with explaining migration in terms of measured characteristics of the socioeconomic and physical environments, such as income, unemployment, and climate. This literature is firmly rooted within the rubric of neoclassical economics, with due attention being given to interregional wage differentials and investment in human capital. The second, or *micro approach*, attempts to explain migration within the context of the psychological decision-making process and is concerned with how individuals choose between alternatives. The perception and evaluation of potential destinations and concepts such as place utility are subsumed within the general framework of choice behavior developed by psychologists.

This philosophical dichotomy within migration research has been discussed by White (1980). He refers to objective philosophy and cognitive philosophy and notes that this distinction is somewhat analogous to the behaviorist–cognition dichotomy in psychology. In a broader context, Golledge and Stimson (1987:2) make a similar distinction when they discuss the differences between the structurally oriented approach and the process-oriented or behavioral approach. While both these approaches have been partly successful, however, it is also true that a synthesis of the two approaches might afford greater insight than either approach on its own. To argue thusly is to suggest that these two lines of inquiry are complementary rather than competitive (Golledge, 1980). In many instances the same types of explanatory variables are used in both, but it is the methodological perspective that differs most markedly.

Despite the increased use of subjective, or cognitive, variables when explaining migration patterns, however, no explicit attempt has been made to formulate a theoretical framework that shows the interrelationships between the macro and micro perspectives. While existing schemas tend to be primarily taxonomic (W. Clark, 1982a), more recent exhortations for theory development have suggested that a synthesis of the micro and macro approaches is likely to be the most rewarding in terms of providing a unified yet flexible theoretical framework for investigating migration behavior (Woods, 1982:157). In many respects, this line of reasoning parallels the structure–agency debate (see Section 1.4), with its consideration of the interaction between large-scale socioeconomic landscapes and individual decision-making. Within this context, the present author has constructed a theoretical schema that provides an example of how macro- and micro-level work on migration might be most usefully integrated (Cadwallader, 1989a). Such a framework can be used to locate a variety of research questions that have previously only been related in a rather ad hoc fashion. In addition, the gaps in our present knowledge and the potential areas of future research can be easily identified.

The proposed framework contains four major sets of relationships (Figure 13.1). First, the objective variables O_h, O_i, and O_j can be combined in various ways to explain overt migration behavior M. Thus the first link represents the traditional macro or aggregated approach to modeling migration patterns. Second, the objective variables are transformed, through the individual cognitions of potential migrants, into their subjective counterparts S_h, S_i, and S_j. Third, the subjective variables are combined to form an overall measure of attractiveness U that allows potential migrants to choose between alternative destinations. Fourth, subject to certain constraints, the individual utility functions are translated into overt behavior. Note that the links below the broken line represent the behavioral perspective, which attempts to shed light on the factors that intervene between the objective variables and migration behavior.

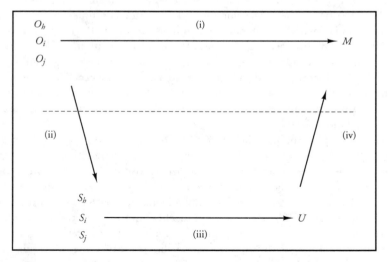

Figure 13.1 A conceptual framework for linking macro- and micro-level approaches to migration. (From M. Cadwallader, "A conceptual framework for analysing migration behavior in the developed world," *Progress in Human Geography,* 13, 1989, Fig. 1.)

This schema is similar to that used in psychology, in which the two major conceptions of an organism are explicated. The top part of the diagram would represent a behaviorist view, in which a set of physical stimuli impinges on the organism to produce an overt response, thus generating a physical law that relates the observable response to the observable stimuli. A more cognitive conception is contained within the lower part of the diagram, where unobservables within the organism become a legitimate part of the conceptualization. A psychophysical law is used to describe the ways in which the physical stimuli produce subjective representations within the organism, a psychological law describes how these subjective values are combined to produce an implicit response, and a psychomotor law addresses the relationship between this implicit response and the overt response. Louviere (1982) has commented on the appropriateness of this overall framework as a paradigm for research into spatial behavior in general.

The explication of this framework involves an examination of the following four relationships:

$$M = f(O_h, O_i, O_j) \qquad (13.1)$$

$$S_h = X(O_h) \qquad (13.2)$$

$$U = \phi(S_h, S_i, S_j) \qquad (13.3)$$

$$M = Z(U) \qquad (13.4)$$

where M is overt migration behavior; O_h, O_i, and O_j are objectively measured attributes of regions or neighborhoods; S_h, S_i, and S_j are their subjectively measured counterparts; and U is the overall attractiveness associated with any region or neighborhood.

Most explorations of equation (13.1) have involved the estimation of single-equation regression models, whereby a set of regional characteristics, such as wage rates and unemployment levels, is used to predict migration rates for various kinds of spatial units, such as states or cities. Distance has tended to play a prominent role in such models, although there is considerable debate about whether a purely friction-of-distance effect can ever be satisfactorily isolated (Sheppard, 1984). These single-equation models are often unsatisfactory, in that they fail to reflect the mutual interaction between the characteristics of regions and their associated migration patterns. For this reason, a simultaneous equation approach is often preferable.

Following the lead of psychophysics, the transformation represented by equation (13.2) is usually expressed by using some kind of power function. For example, it is generally agreed that the relationship between cognitive distance and physical distance is best represented by a power function (see Section 8.3) and is thus similar to the psychophysical law relating the strength of an external stimulus to an individual's impression of the subjective intensity of that stimulus. This relationship has been shown to hold for a variety of directly perceivable stimuli, such as brightness and loudness, as well as for stimuli that cannot be directly perceived, such as the relationship between perceived status and annual income.

A growing number of researchers have begun to investigate explicitly the interrelationships between the objectively and subjectively measured attributes of places (Pacione, 1982). Knox and MacLaran (1978), in a study based on Dundee, Scotland, assessed the compatibility between objective and subjective indicators of well-being. Although there was some overlap between the objective and subjective indicators, only one-third of the correlations between the two were statistically significant, and one-sixth of the correlations were negative. These results indicate, then, that the relationships between objective and subjective variables are exceedingly complex. The correspondence between objective and subjective measures of the same phenomenon can vary from a very strong relationship to a comparatively weak one, and in the present absence of appropriate transformations, it has been suggested that subjective attributes should be regarded as independent sources of explanation in migration models, not merely as dependent variables of objective attributes (Todd, 1982).

Personality traits are among the factors that have been investigated in order to help disentangle the interrelationships between objective and subjective variables. For example, Walmsley's (1982) evidence indicates that introverts and extroverts differ in their residential preferences for different parts of eastern Australia. Apparently, introverts prefer inland, temperate, and settled areas, whereas extroverts prefer wilderness areas, coastal resorts, and large cities. The author himself notes, however, that the extremely small sample size prevents his drawing any general conclusions.

The integrative function involved in equation (13.3) is especially difficult to disentangle, as one is attempting to algebraically describe the combination rule by which individuals act as if they integrate various items of information in order to evaluate some stimuli. A variety of integrative functions are possible, and the following three are representative:

$$U_{ak} = \sum_{l=1}^{n} S_{alk} \tag{13.5}$$

$$U_{ak} = \sum_{l=1}^{n} S_{alk} W_{lk} \tag{13.6}$$

$$U_{ak} = \prod_{l=1}^{n} S_{alk} \tag{13.7}$$

where U_{ak} is the overall attractiveness of region a for individual k; S_{alk} is the attractiveness rating of region a on attribute l for individual k; and W_{lk} is the relative importance of attribute l for individual k. These equations represent additive, weighted additive, and multiplicative forms, respectively. Much of the recent experimental work suggests that multiplicative forms, or related algebraic forms, are the dominant integrative function associated with spatial choice, although in the case of residential preferences there is perhaps insufficient evidence to disconfirm additivity. Unlike the multiplicative version, the additive and weighted additive formulations are examples of compensatory models, in which low ratings on one attribute can be compensated for by high ratings on another attribute.

Finally, equation (13.4) represents the translation of preferences into overt behavior. The often-found disparity between preferences and behavior could simply imply that people are inconsistent in their behavior or that currently available techniques are inappropriate vehicles for eliciting true attitudes (Pipkin, 1981). More important, however, a series of *constraints* is in operation that inhibits the behavioral manifestation of underlying preferences. Such constraints make it exceedingly difficult to isolate underlying preferences by observing overt behavior. Rather, migration patterns are a complex balance of preferences and constraints. Indeed, the observed discrepancies, or residuals, associated with the relationship between preferences and behavior enable us to speculate on the nature and extent of the constraints in any specific situation.

The constraints that operate within the context of migration involve obstacles that produce attitude-discrepant behavior in three major ways: by restricting the opportunity set, by influencing the formation of preferences, or by preventing choice actualization (Desbarats, 1983b). These constraints can be either individual or institutional in their origin. For example, at the individual level, a potential migrant's income will obviously restrict where he or she can live, regardless of underlying preferences. Similarly, the location of the individual's workplace is a constraining influence on residential choice (Reitsma and Vergoossen, 1988; Long, 1988:236). Individuals are also constrained by the amount of information they possess concerning potential destinations, as choice set formation is a function of location. The expected relationship between information levels and migrational search behavior has been well documented (Clark, 1982c).

At the institutional level, the urban housing market provides an excellent laboratory for investigating the constraints that are generated by various kinds of institutions. Real estate agents, banks, and savings and loan associations all regulate the flow of information and mortgage money into the housing market. In addition, both national and local governments seek to influence migration patterns as a means of achieving various policy goals (Clark, 1983). Such influence is usually indirect, involving policies designed to create a more desirable distribution of population. For example, tax incentives can be used to attract business and industry to less-developed regions. Similarly, the distribution of new highways and defense contracts can also exert an indirect influence on migration patterns (Ellis et al., 1993). More direct governmental intervention is involved in such projects as the provision of public housing in cities.

The balance between choice and constraints has been most fully explored within the context of ethnic migration and segregation (Jackson and Smith, 1981). Social geographers, in particular, have engaged in a prolonged debate concerning the role of voluntary and discriminatory forces in the maintenance of ethnic neighborhoods. It is exceedingly difficult to explain segregation purely as a result of discrimination, but the current focus on a simple dichotomy between choice and constraints is probably inadequate to unravel the complex interplay between social and physical space. It has been suggested that the choice–constraint concept might be more usefully formulated within a Marxist framework, whereby a broader societal analysis would take precedence over the present empirical emphasis (Brown, 1981).

Returning to the overall framework, the behavioral component of this conceptual schema, as represented by equations (13.2), (13.3), and (13.4) is especially susceptible to a sequential series of analyses. For example, if we have a set of three objectively measured variables with subjective counterparts, and we are postulating an additive integrative function, then we might expect the following succession of relationships:

$$S_{ahk} = a_1 O_{ahk}^{b_1} \tag{13.8}$$

$$S_{aik} = a_2 O_{aik}^{b_2} \tag{13.9}$$

$$S_{ajk} = a_3 O_{ajk}^{b_3} \qquad (13.10)$$

$$U_{ak} = S_{ahk} + S_{aik} + S_{ajk} \qquad (13.11)$$

$$M_{ak} = a_4 + b_4 U_{ak} \qquad (13.12)$$

where the notation is the same as in equations (13.1) to (13.7).

It should be noted, however, that a variety of assumptions is associated with this theoretical structure. First, it is assumed that the development of aggregated models relating overt behavior and objectively measured variables might obscure significant intermediate relationships. Second, it is assumed, as in much contemporary choice theory, that potential migrants do not evaluate regions, neighborhoods, or houses per se, but rather their cognitions of the varying levels of attributes associated with those alternatives. Third, it is assumed that the levels of the independent variables that are used to generate an overall evaluation are some function of the observed levels of those same variables. Fourth, it is assumed that individuals maximize utility; that is, when faced with a choice between alternatives possessing different overall values, they are most likely to choose the alternative with the highest value.

13.2   The Determinants of Migration

Many variables are either directly or indirectly linked to migration (Liaw, 1990; Shaw, 1985). Notions of equilibrium derived from neoclassical economics suggest that in any situation characterized by *income differentials*, labor will tend to migrate from low- to high-income areas (Clark and Gertler, 1983; Tabuchi, 1988). As a result of this movement, the supply of labor is increased at the destination and decreased at the origin so that over time, due to supply and demand considerations, income levels will tend to equalize throughout the system. Such an argument assumes that there are no major barriers to mobility and that people desire to maximize income. It is noteworthy, however, that a number of studies have suggested that destination income has a greater influence on migration flows than does origin income (Shaw, 1985:144). This relationship is partly conditioned by cost-of-living differentials, as there is empirical evidence that variations in the cost of living have a significant impact on migration patterns (Cebula, 1980a; Renas and Kumar, 1981). Finally, when considering the relevance of the income variable, it should be remembered that only wage earners will respond to interregional differences. This variable would presumably have less impact in predicting the migration flows of the retired sector of the population. Also, it would seem that short-distance moves would be less reflective of income differentials than would long-distance moves.

Employment opportunities are also related to migration patterns (Odland, 1988; Barff, 1990). Theoretically, the greater the unemployment in a region, the greater the rate of out-migration. Conversely, in-migration should be negatively related to unemployment levels. Some studies, however, have indicated that employment levels, especially in origin regions, do not account for much of the variation in migration rates (Shaw, 1985:145). One reason for this finding can be found in the use of aggregated data, as the unemployed constitute a relatively small proportion of the total population. High unemployment rates are likely to be of less significance to those individuals who already have jobs. Using micro-level data, DaVanzo

(1978) has been able to substantiate this argument by showing that the unemployed are more likely to move than the employed. Indeed, Herzog and Schlottmann (1984) provide evidence that the premove unemployment rate of migrants is as much as three times that of nonmigrants. In sum, high unemployment rates tend to encourage the out-migration of those who are unemployed, but exert little influence on the majority of the population.

The probability of obtaining employment elsewhere is often the key issue when one considers the influence of unemployment rates. The calculation of such probabilities requires information on the number of job vacancies and the number of people seeking those jobs. Data on job vacancies are generally not available, and using unemployment levels as a measure of expected job competition entails a number of problems (Isserman, 1985). First, unemployment rates ignore those workers who are not actively seeking jobs but who would quickly do so if jobs became more widely available. Second, unemployment rates do not take into account the aging of the labor force in different regions and thus do not reflect potential openings due to retirement. As a result, various surrogates for the probability of gaining employment have been used. For example, Plaut (1981) used the ratio of vacancies to unemployment.

In a similar vein, the employment profile of a region tends to be related to the net migration of that region. In particular, growth in the secondary and tertiary sectors of an economy, rather than in agricultural employment, has been found to be positively related to levels of in-migration. Evidence about the relationship between migration and urbanization is less clear, however. Until recently, migrants have had a strong tendency to move from nonmetropolitan to metropolitan areas of the United States. During the 1970s, however, a reversal of this trend occurred, and the growth rates for nonmetropolitan areas exceeded those of metropolitan areas (Greenwood, 1981:91).

Researchers have suggested a variety of reasons to explain this *migration turnaround* (Fuguitt, 1985; Wilson, 1987). First, the economies of scale that characterized the process of urban growth have often turned into diseconomies, and the social costs of urban disamenities have begun to exceed urban–rural wage differentials. Second, an increasingly affluent retirement-age population has been moving in large numbers to small retirement communities in Florida and the Southwest. Third, the increased demand for domestic energy sources, especially coal, has induced growth in rural regions. Fourth, the decreasing importance of manufacturing employment has led to the declining attraction of urban areas.

Education is generally found to be positively related to in-migration (Shaw, 1985:147). At the aggregate level, a number of studies have succeeded in showing that educational differentials account for differential migration rates between places. In particular, the well educated are much more likely to make long-distance moves than are their more poorly educated counterparts, although there is less of a distinction for short-distance moves. In other words, the correlation between education and migration appears to become stronger as the migration distance increases. This differential is due to the fact that the better educated are dealing with a labor market that is national rather than local in scale, and they also have better information concerning job opportunities.

The propensity of labor force members to migrate tends to decrease with increasing *age* (Nakosteen and Zimmer, 1980), as older people have a shorter expected working life over which to realize the advantages of migration. Moreover, family ties and job security are also likely to be more important for older people, thus further decreasing their incentive to migrate. As a result, adults in their early twenties have the highest migration rates, and migration decreases monotonically beyond that age. On the other hand, the event of retirement, usually between the ages of 65 and 75, often precipitates a long-distance move, especially toward a more congenial climate. So, although migration rates generally decrease with increasing age, there is often a slight upturn in rates for those approaching retirement age.

The provision of *local government services* also has an impact on migration patterns (Cebula, 1980b; Harkman, 1989; Charney, 1993). High welfare payments act as a strong attraction, especially to nonwhites, as welfare is an important source of income for many families. However, the direction of causality between migration and welfare can run both ways; this suggests that both migration and welfare should be treated as endogenous variables in a system of simultaneous equations. Migration is also influenced by local governmental expenditure on services such as education, but as with welfare payments, a reciprocal relationship exists between expenditure and migration. Moreover, one can also expect local governmental expenditures and local taxes to be simultaneously determined, the level of one depending on the level of the other (Loehman and Emerson, 1985).

Quality of life and amenity variables have become increasingly common ingredients of migration models (Cushing, 1987a; Knapp and Graves, 1989). In the context of metropolitan migration, Porell (1982) used principal components analysis to reduce individual quality of life variables into a smaller number of underlying components, representing climate, natural recreational amenities, social amenities, crime, air pollution, and health. When comparing the quality of life variables with their more traditional economic counterparts, Porell found that neither set of variables seemed important as determinants of out-migration. On the other hand, both quality of life and economic factors were significant determinants of in-migration. In this context, it is likely that economic determinants become more important than quality of life considerations in eras of national economic stagnation and high unemployment (Clark, 1983:52).

A number of recent migration studies have used *climatic variables* as indicators of quality of life. Most of these studies use some kind of temperature measurement (Ballard and Clark, 1981), although a wide variety of indices have been constructed (Cushing, 1987b). After testing a variety of climatic indicators in the context of net migration into SMSAs, Renas and Kumar (1983) concluded that people generally prefer areas that have moderate climates, rather than extremely hot or cold climates.

Climatic variables have most often been invoked within what Graves (1980) has termed the amenity-oriented equilibrium approach to explaining migration. In this approach, a city is envisioned as supplying a particular climatic bundle. For equilibrium, or spatial indifference, to obtain, attractive climatic bundles can be traded off against less attractive economic attributes. That is, it can be argued that income and unemployment differences across cities do not reflect real utility differentials, but rather the compensation required for differences in location-fixed goods, such as climate. Thus, one would not necessarily expect income differences to generate migration, since those differences might merely reflect compensation for climatic differences. On the other hand, rational individuals might use migration to trade off income for increased quality of life consumption. If such a long-run equilibrium between economic and quality of life bundles is to be accepted as plausible, however, it must be accepted that the present system is in a state of significant disequilibrium. Several cities, such as San Francisco and Atlanta, presently offer favorable economic incentives in addition to an attractive quality of life, while other cities, such as Buffalo and Syracuse, offer neither (Porell, 1982). Recent evidence suggests that such variables as jobs and wages are considerably more important than location-specific amenities in explaining metropolitan migration rates of employed individuals (Greenwood and Hunt, 1989).

Finally, a variety of studies have argued that migration is a cause of economic change as well as a reaction to it (Vanderkamp, 1989). In particular, a number of regional development theorists have suggested that migration contributes to increased inequality in regional per capita income. The major contributor to this increased inequality is the fact that migrants do not typically represent the population in general, and so the high-income regions that gain in population as a result of migration also experience a favorable change in their population composition. In general, the empirical evidence supports the notion that migration tends to increase rather than decrease regional economic differences. In particular contexts, however, migration may have a variety of effects on local income levels (Clark, 1983:55). For example, out-migration could reduce the labor supply and thus increase wages or, alternatively, the loss of high-wage earners might decrease average wages in the short run.

Migration also influences unemployment levels and, contrary to expectations, out-migration can worsen rather than improve the unemployment problems in depressed areas. Unemployed persons take with them a certain level of expenditure when they migrate, thus decreasing the purchasing power and ultimately the level of employment in the region. Irrespective of the detailed manifestations in particular regions at particular points in time, evidence clearly indicates that migration rates are both a cause and an effect of income and unemployment variables (Chalmers and Greenwood, 1985).

These interrelationships suggest that the overall relationship between migration and capital can be very complex (Clark, 1983:86–100). In general, however, it can be argued that labor migration should lag behind the movement of capital. That is, capital can react faster than labor to the changing map of economic opportunities. Researchers need to take such leads and lags into account when modeling the empirical interrelationships between labor and capital. In the short run the movement of capital can initiate migration, but in the long run capital and migration might well readjust to each other as part of the larger macroeconomic system. The empirical evidence on this issue, at the level of interstate labor movements, indicates that indeed lags do occur between migration and capital. In particular, Clark and Gertler (1983) suggest that most states experience one- to two-year lags. Only in very rare instances does prior migration initiate a spatial flow of capital.

Such empirical evidence tends to compromise the neoclassical economic viewpoint that migration and capital growth are initiated separately, but in reaction to similar macroeconomic variables (Milne, 1993). Rather, capital often determines migration because of its ownership of the means of production and its right to initiate and terminate employment. This uneven distribution of power in employment relationships, although partly mitigated by unionization, casts some doubt on the assumptions of free will and independence that characterize neoclassical conceptions of labor migration. Indeed, it can be argued that in those situations where a firm has the luxury of choosing from a wide variety of potential locations the firm will locate in order to use a relatively vulnerable labor force. That is, two sets of variables will come to dominate the process of locational choice: labor costs and the degree of local worker organization (Walker and Storper, 1981).

13.3   Models of Interregional Migration

Statistical models can be used to help sort out the relative importance of the variables that influence migration. One of the most enduring migration models is the *gravity model*, whereby the amount of migration between any two regions is expected to be directly proportional to the product of their populations and inversely proportional to the distance between them (Haynes and Fotheringham, 1984). In symbolic form it is usually represented as follows:

$$M_{ij} = a\frac{P_i^{b_1}P_j^{b_2}}{D_{ij}^{b_3}} \tag{13.13}$$

where M_{ij} is the amount of migration between regions i and j; P_i and P_j are the populations of regions i and j; D_{ij} is the distance between regions i and j; and a, b_1, b_2, and b_3 are constants that reflect the relative weightings of the constituent variables. If the variables are expressed in logarithmic form, multiple regression analysis can be used to estimate the coefficients:

$$\log M_{ij} = \log a + b_1 \log P_i + b_2 \log P_j - b_3 \log D_{ij} \tag{13.14}$$

where the notation is the same as in equation (13.13).

Flowerdew and Salt (1979) used the gravity model to explore the migration flows between a set of 126 Standard Metropolitan Labor Areas in Great Britain. Fitting a gravity model of the form described in equation (13.14), they derived a multiple coefficient of determination of 0.527, thus implying that over 50% of the variation in migration flows could be statistically accounted for by the gravity model formulation. All the coefficients were significantly different from zero and had very low standard errors. As expected, the coefficients associated with both population variables were positive, while the coefficient associated with the distance variable was negative.

Single-equation Regression Models

In recent years a wide variety of single-equation multiple regression models (see Section 4.1) have been fit to migration data (Flowerdew and Amrhein, 1989; Owen and Green, 1989). For example, Renas and Kumar (1983) estimated a number of models involving net migration rates for Standard Metropolitan Statistical Areas in the United States. One of their models took the following form:

$$M_i = a + b_1 I_i + b_2 IC_i + b_3 C_i + b_4 CC_i + b_5 U_i + b_6 E_i + b_7 WT_i + b_8 ST_i \tag{13.15}$$

where M_i is the net number of migrants into SMSA i between 1960 and 1970 expressed as a percentage of the 1960 population; I_i is the median family income in SMSA i in 1969; IC_i is the annual rate of change of median family income in SMSA i between 1959 and 1969 expressed as a percentage; C_i is a measure of the cost of living in SMSA i in 1969; CC_i is the annual rate of change in the cost of living in SMSA i expressed as a percentage; U_i is the unemployment rate in SMSA i in 1960; E_i is the median school years completed in 1960 for the population aged 25 years old and over in SMSA i; WT_i is the absolute deviation of the mean January temperature from 65 degrees Fahrenheit for SMSA i; and ST_i is the absolute deviation of the mean July temperature from 65 degrees Fahrenheit for SMSA i.

The multiple coefficient of determination for this model was 0.69, and six of the coefficients were significantly different from zero at the 0.05 probability level (Table 13.1). These six coefficients also had the expected signs. Net migration was negatively related to cost of living, unemployment, and the two temperature variables. On the other hand, there was a positive relationship between net migration and level of education and between net migration and change in income. In this latter context it is interesting to note that absolute income level was not significantly related to migration. One weakness of this model, however, is that some of the explanatory variables were measured for 1960 while others were measured for 1969.

McHugh (1988) provided a second example of a single-equation regression model. In this case the dependent variable of interest was a measure of interstate migration among blacks in the United States. The following model was tested:

$$M_{ij} = a + b_1 P_i + b_2 P_j + b_3 PB_i + b_4 PB_j + b_5 D_{ij} + b_6 I_i \qquad (13.16)$$
$$+ b_7 I_j + b_8 M_i + b_9 M_j + b_{10} W_i + b_{11} W_j$$

where M_{ij} is the number of blacks five years of age and older residing in state j in 1980 and in state i in 1975; P_i is the population of state i in 1985; P_j is the population of state j in 1975;

Table 13.1 Regression Statistics for the Renas and Kumar Model of Net Migration Rates for SMSAs in the United States

	Coefficients
Constant	28.436
Income	0.001
Change in income	4.862[a]
Cost of living	−0.007[a]
Change in cost of living	2.256
Unemployment	−2.757[a]
Education	2.741[a]
Winter temperature	−0.467[a]
Summer temperature	−0.716[a]
	$R^2 = 0.69$

Source: S. Renas and R. Kumar, "Climatic conditions and migration: An econometric inquiry," *Annals of Regional Science*, 17 (1983), Table 1.
[a]Indicates significant difference from zero at the 0.05 probability level.

PB_i is the percentage of the population that is black in state i in 1975; PB_j is the percentage of the population that is black in state j in 1975; D_{ij} is the distance in highway miles between the principal city of black population in state i and the principal city of black population in state j; I_i is the per capita income for blacks in state i in 1979; I_j is the per capita income for blacks in state j in 1979; M_i is the number of blacks in the armed forces in state i in 1980; M_j is the number of blacks in the armed forces in state j in 1980; W_i is the average monthly payment for Aid to Families with Dependent Children in state i in 1978; and W_j is the average monthly payment for Aid to Families with Dependent Children in state j in 1978.

Estimates for the model were generated via least-squares regression analysis using a double logarithmic specification. Data involved migration flows between the contiguous 48 states, plus the District of Columbia, providing 2352 individual migration flows in all. The multiple coefficient of determination for this particular model was 0.777, and all but one of the beta coefficients was significantly different from zero at the 0.001 level, while that for W_j was significantly different from zero at the 0.05 level (Table 13.2). As expected, all four population

Table 13.2 Regression Statistics for the McHugh Model of Interstate Migration among Blacks in the United States

	Beta Coefficients
Population i	0.326[a]
Population j	0.308[a]
Percentage black i	0.259[a]
Percentage black j	0.168[a]
Distance ij	−0.247[a]
Black per-capita income i	0.062[a]
Black per-capita income j	0.091[a]
Blacks in armed forces i	0.127[a]
Blacks in armed forces j	0.222[a]
Welfare i	−0.037[b]
Welfare j	−0.068[a]
	$R^2 = 0.777$

Source: K. McHugh, "Determinants of black interstate migration, 1965–70 and 1975–80," *Annals of Regional Science*, 22 (1988), Table 3.
[a]Indicates significant difference from zero at the 0.001 probability level.
[b]Indicates significant difference from zero at the 0.05 probability level.

variables had significant positive relationships with black migration, while the distance variable displayed the traditional negative relationship. The variables related to the military appeared to be important determinants of black migration, but income and welfare levels seemed to play only minor roles.

In addition to isolating the important variables to be included in such single-equation regression models, researchers have also focused attention on identifying the most appropriate *functional form*. The two most popular specifications involve the linear formulation, in which migration is assumed to be a linear function of the explanatory variables, and the log-linear formulation, in which all the variables are logarithmically transformed. The log-linear formulation has been especially common, as it generally provides a higher coefficient of determination, and the elasticity of migration with respect to an explanatory variable can be obtained directly from the estimated coefficients. A variety of other functional forms, such as semi-log, reciprocal, or log reciprocal, can also be utilized (Goss and Chang, 1983).

Most recently the polytomous logistic model has become a popular vehicle for modeling migration (Shaw, 1985:184). Here it is assumed that the migration decision involves choosing among a finite number of mutually exclusive potential destinations. The utility of a particular destination and thus its probability of being chosen depend on both the attributes of the destination and the attributes of the individual making the decision. The dependent variable can be represented as the ratio of the probability of moving to destination j and the probability of staying in origin i. Estimating the effects of destination and personal characteristics involves using log transformations to produce a log-linear model (see Section 11.5).

Even when an appropriate functional form is identified, however, single-equation migration models are often inadequate. First, their use precludes the possibility of establishing causal links between the explanatory variables themselves, and thus they disregard indirect effects. Second, they do not allow for any kind of feedback effect, or reciprocal causation, between the constituent variables. These more complex interrelationships can be explored by using structural equation models, including both path analysis and systems of simultaneous equations.

Path Models of Migration

Path analysis is a useful way to analyze the interrelationships between a set of explanatory variables and migration, as it allows one to estimate the magnitude of the linkages between variables, which then provide information about the underlying causal processes (see Section 4.4). In the example described here (Cadwallader, 1985), the causal model is recursive, since there are no feedback effects between variables. As a result, ordinary least-squares regression techniques can be used to obtain the path coefficients, although all the normal regression assumptions should be met.

The term *path coefficient* refers to the standardized regression coefficient, or beta coefficient. As Wright (1985) pointed out, however, the standardized and unstandardized coefficients should be seen as complementary, since they provide different kinds of information. In the context of migration models, when comparing models for different subgroups or parts of the country, it seems advisable to use unstandardized estimates, since they will be unaffected by the different variances in the same variable that might occur for different subgroups or regions. On the other hand, if one wishes to compare the relative importance of variables within a particular subgroup or regional context, as in the present instance, then the standardized coefficient is the more appropriate, as it adjusts for the different measurement scales associated with the different variables.

The causal model to be analyzed via path analysis is represented by the following set of structural equations:

$$X_4 = b_{41}X_1 + b_{42}X_2 + b_{43}X_3 + e_4 \tag{13.17}$$

$$X_5 = b_{51}X_1 + b_{52}X_2 + b_{53}X_3 + b_{54}X_4 + e_5 \tag{13.18}$$

$$X_6 = b_{61}X_1 + b_{62}X_2 + b_{63}X_3 + b_{64}X_4 + b_{65}X_5 + e_6 \tag{13.19}$$

where X_1 is the percentage of the civilian labor force employed in agriculture, forestry, and fisheries; X_2 is the percentage of the population living in urban settlements; X_3 is the median number of school years completed by all adults aged 25 years and over; X_4 is the percentage of males 16 years and over in the civilian labor force who are unemployed; X_5 is median income; X_6 is net migration; b_{41} to b_{65} are the path coefficients; and e_4, e_5, and e_6 are the disturbance terms, or residual path coefficients.

The data to calibrate this model were collected for all 42 State Economic Areas in the Upper Midwest (Michigan, Minnesota, and Wisconsin). State Economic Areas were chosen as the basic units of analysis because they have greater internal homogeneity, in socioeconomic characteristics, than do either states or counties. The data for all the variables were obtained from the State Economic Area Reports. In particular, the major variable of interest, net migration, was measured by subtracting the amount of out-migration from the amount of in-migration and dividing by the population, using the migration flow data in the State Economic Area Reports for each State Economic Area. This particular measure of migration was used as the central variable in the analysis because it conveniently combines information concerning both in- and out-migration flows, although it should be noted that there is no such individual as a "net migrant" (Rogers, 1990). In other circumstances, where the focus is on distinguishing between the causal factors involved in generating out-migration as opposed to in-migration, more disaggregated measures of migration flows are desirable. Net migration was obtained for two time periods, 1955–1960 and 1965–1970, in order to explore any temporal variation.

The standardized regression coefficients associated with the structural equations represent the path coefficients (Figures 13.2 and 13.3), and the path coefficient associated with the disturbance term, incorporating the combined effect of all unspecified variables, is the square root of the unexplained variation in the dependent variable under consideration. For comparative purposes, statistical significance is reported both at the 0.10 and 0.05 levels, although it should be remembered that the significance level per se says nothing about the strength of the relationship but merely indicates the amount of risk associated with assuming that the relationship exists.

The constituent path coefficients can be conveniently discussed by distinguishing between the effects of the exogenous and endogenous variables on migration. Focusing first on the direct influence of the three *exogenous variables* (agricultural employment, the percentage urban population, and education), we can see that the percentage urban population is negatively related to net migration in both 1955–1960 and 1965–1970. Similarly, education is positively related to net migration in both time periods, which indicates that higher educational levels are related to higher levels of migration-induced regional growth. The behavior of the agricultural employment variable is less systematic, however, as it is negatively associated with migration in 1965–1970 but not significantly related in 1955–1960.

Second, with respect to the *endogenous variables*, unemployment is negatively related to migration in both time periods, which implies that high levels of unemployment are not associated with large net influxes of migrants. As expected, income levels are positively associated with net migration in 1955–1960. This relationship is negative in 1965–1970, however, although it is not significantly different from zero at the 0.10 level.

Overall, the pattern of statistically significant path coefficients is remarkably similar for the two time periods. The only differences are that the percentage urban population is not significantly related to unemployment in 1955–1960 or income in 1965–1970, agricultural employment is not significantly related to migration in 1955–1960, and income is not significantly related to migration in 1965–1970. It is somewhat of a surprise that education is not significantly related to income in either of the two time periods, although it should be remembered that the percentage urban population, agricultural employment, and unemployment were all being controlled for.

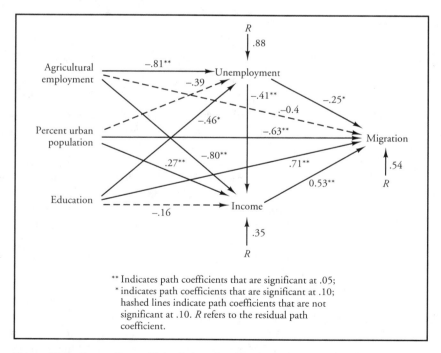

Figure 13.2 The path coefficients for migration in the Upper Midwest, 1955–1960. (From M. Cadwallader, "Structural equation models of migration: An example from the Upper Midwest USA," *Environment and Planning A*, 17, 1985, Fig. 1.)

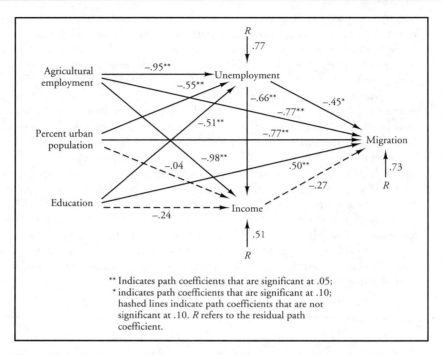

Figure 13.3 The path coefficients for migration in the Upper Midwest, 1965–1970. (From M. Cadwallader, "Structural equation models of migration: An example from the Upper Midwest USA," *Environment and Planning A*, 17, 1985, Fig. 2.)

The total effect of each causally prior variable on the succeeding variables can be readily partitioned into direct and indirect components (see Section 4.4). The direct effect is represented by the path coefficient linking two variables, whereas the indirect effect is calculated by multiplying the path coefficients associated with the intervening variables. For example, besides the direct path between agricultural employment and migration, there are three indirect paths, one via unemployment, a second via unemployment and income, and a third via income. As previously noted, the total effect between any two variables in a recursive model is not necessarily the same as the correlation coefficient between those two variables, as the latter term can also include association due to the correlation between predetermined variables and spurious association caused by joint dependence on a prior variable, as well as direct and indirect effects (Alwin and Hauser, 1981).

In the present instance, many of the indirect effects run counter to the direct effect and thus reduce the total effect (Table 13.3). For example, the large negative direct effect of the percentage urban population on migration is partly offset, in both 1955–1960 and 1965–1970, by positive indirect effects. Similarly, the negative direct effect of agricultural employment in 1965–1970 is partly offset by a positive indirect effect, although in 1955–1960 both the direct and indirect effects of this variable on net migration are only marginal. In contrast, the positive direct effect of education is compounded in both time periods by the indirect

Table 13.3   Effects of Each Variable on Migration in the Upper Midwest

	Direct Effect	Indirect Effect	Total Effect
	1955–1960		
Agricultural employment	−0.04	−0.04	−0.08
Percentage urban population	−0.63	0.32	−0.31
Education	0.71	0.14	0.85
Unemployment	−0.25	−0.22	−0.47
Income	0.53	—	0.53
	1965–1970		
Agricultural employment	−0.77	0.52	−0.25
Percentage urban population	−0.77	0.16	−0.61
Education	0.50	0.20	0.70
Unemployment	−0.45	0.18	−0.27
Income	−0.27	—	−0.27

Source: M. Cadwallader, "Structural equation models of migration: An example from the Upper Midwest USA," *Environment and Planning A,* 17 (1985), Table 1.

effects via unemployment and income. It is thus important, in any migration context, to be able to disentangle these different kinds of influences.

Simultaneous Equation Models of Migration

In this section we will consider how simultaneous equation models (see Section 10.3) have been used to study migration. The present author (Cadwallader, 1985) calibrated a simultaneous equation model of net migration in the upper Midwest, using the same data as for the previously described path model. The particular model that was estimated is shown in Figure 13.4 and can be represented by the following structural equations, which indicate the expected relationships:

$$X_6 = a - b_{64}X_4 - b_{65}X_5 - b_{62}X_2 + e_6 \tag{13.20}$$

$$X_4 = a - b_{46}X_6 - b_{45}X_5 - b_{41}X_1 + e_4 \tag{13.21}$$

$$X_5 = a - b_{56}X_6 - b_{54}X_4 + b_{53}X_3 + e_5 \tag{13.22}$$

where X_1 is the percentage of the civilian labor force employed in agriculture, forestry, and fisheries; X_2 is the percentage of the population living in urban settlements; X_3 is the median number of school years completed by all adults aged 25 years and over; X_4 is the percentage of males 16 years and over in the civilian labor force who are unemployed; X_5 is median income;

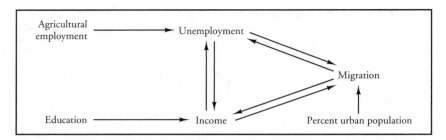

Figure 13.4 A simultaneous equation model for migration in the Upper Midwest. (From M. Cadwallader, "Structural equation models of migration: An example from the Upper Midwest USA," *Environment and Planning A*, 17, 1985, Fig. 3.)

X_6 is net migration; a is a constant; the b terms are estimated coefficients; and e_4, e_5, and e_6 are the disturbance terms.

Note that reciprocal links are postulated between migration, unemployment, and income. These three variables are thus treated as endogenous variables, whereas agricultural employment, the percentage urban population, and education are treated as predetermined variables. The endogenous variables are to be explained by the theory or model, whereas the predetermined variables are treated as given and are used to provide explanatory power to the model. In general, the predetermined variables can either be lagged values of the endogenous variables or exogenous variables, which are lagged or nonlagged variables that are considered to be separate causes of the endogenous variables. In the present situation the predetermined variables consist entirely of nonlagged exogenous variables.

In particular, the following relationships are postulated. First, with respect to equation (13.20), it is expected that net migration will be negatively related to unemployment, in that areas experiencing relatively high unemployment are likely to experience relatively heavy out-migration (Greenwood, 1981:158). On the other hand, net migration should be positively related to median income, as one would expect migrants to move out of relatively low income areas and into relatively high-income areas. Also, based on the results of the previously reported path analysis, we might expect migration to be negatively related to the percentage of the population living in urban settlements.

Second, with respect to equation (13.21), unemployment should be negatively related to net migration, as the in-migration of labor will increase the demand for local goods and services, thus increasing employment (Greenwood, 1981:150). In addition, migration to a region tends to induce investment in that region, which further increases the demand for labor. The relationship between unemployment and income should be negative, with the highest levels of unemployment being associated with low-income regions and, given the results of the path analysis, the relationship between unemployment and the percentage employed in agriculture should also be negative.

Third, with respect to equation (13.22), it is expected that median income will be negatively related to net migration, as out-migration will tend to increase wage levels for those remaining, whereas in-migration will tend to decrease them (Greenwood, 1981:150). The relationship between income and unemployment is expected to be negative, with regions experiencing high unemployment levels tending also to be low-income regions. Finally, a positive relationship is expected between income and level of education. This relationship is specified as running from education to income and not vice versa, because the time lags involved in the relationship between income change and education change are likely to be greater than those between education change and income change, since the educational effects of income growth tend to accrue to the offspring (Greenwood, 1981:161).

With the use of two-stage least-squares analysis, the estimated coefficients in the structural equations are largely as expected for both 1955–1960 and 1965–1970 (Table 13.4). The signs of the coefficients in the *migration equation* are as predicted, with net migration being negatively related to unemployment and the percentage urban population and positively related to income. The coefficients for the unemployment and the percentage urban population variables, however, are not significantly different from zero at the 0.10 probability level. Other migration studies have also found insignificant coefficients with respect to the unemployment variable (Greenwood, 1981:174). It has been suggested that, when explaining migration, variables relating to job turnover are more relevant than unemployment rates, since potential migrants tend to be more concerned about the rate at which hiring for new jobs is taking place.

The coefficients in the *unemployment equation* for both 1955–1960 and 1965–1970 are exactly as predicted, except that for 1965–1970 the net migration variable is not statistically

Table 13.4 Estimated Structural Equations for Migration in the Upper Midwest

Variable	Structural Equation			
	1955–1960			
Migration =	-33.699	$- 3.281X_4$	$+ 0.016X_5^a$	$- 0.503X_2$
	(22.431)	(2.640)	(0.011)	(0.398)
Unemployment =	25.640	$- 0.255X_6^b$	$- 0.003X_5^b$	$- 0.307X_1^b$
	(5.723)	(0.105)	(0.001)	(0.072)
Income =	$-21,665.00$	$- 466.168X_6$	$+ 310.508X_4$	$+ 2342.500X_3^a$
	(18,410.700)	(460.947)	(369.514)	(1657.840)
	1965–1970			
Migration =	-29.604	$- 1.701X_4$	$+ 0.007X_5^a$	$- 0.338X_2$
	(20.875)	(2.208)	(0.005)	(0.263)
Unemployment =	32.290	$- 0.190X_6$	$- 0.003X_5^b$	$- 0.487X_1^b$
	(9.914)	(0.226)	(0.001)	(0.178)
Income =	$-38,284.000$	$- 442.335X_6^a$	$+ 546.931X_4$	$+ 3656.560X_3^b$
	(18,908.200)	(306.229)	(434.288)	(1472.710)

Source: M. Cadwallader, "Structural equation models of migration: An example from the Upper Midwest USA," *Environment and Planning A,* 17 (1985), Table 2.
[a]Indicates significant at 0.10.
[b]Indicates significant at 0.05.

significant at the 0.10 probability level. All the other coefficients are significantly different from zero at the 0.05 probability level. Agricultural employment and income are negatively related to unemployment in both time periods, and migration also has a significant negative relationship in 1955–1960. This latter finding indicates that high rates of net migration tend to be associated with low levels of unemployment.

Finally, the *income equation* contains the expected relationships with respect to net migration and education. That is, income is negatively related to net migration, since out-migration tends to increase wage levels for those remaining, and vice versa, whereas a positive relationship exists between income and level of education. The only exception to this general pattern is that the effect of net migration is statistically insignificant in 1955–1960. Rather surprisingly, however, income is positively related to unemployment for both time periods, although the coefficients are not significantly different from zero.

Of particular interest are the estimated parameters for the postulated reciprocal relationships among unemployment, income, and migration. There do not appear to be exceptionally strong feedback relationships between unemployment and migration, although migration was a statistically significant determinant of unemployment in 1955–1960. Similarly, income is a statistically significant determinant of unemployment, but unemployment seems to have far less effect on income. Finally, at least for 1965–1970, income and net migration are reciprocally related, as income is a statistically significant determinant of migration and migration is also a statistically significant determinant of income. More specifically, high-income levels attract migrants, but as net migration increases, income levels decrease.

13.4   Dynamic Approaches

During recent years scholars have increasingly tried to construct migration models that reflect the dynamic nature of migration streams (Odland and Bailey, 1990; Rogerson, 1990). For example, temporal change has been formally analyzed using *Markov chain models*. Migration transition probability matrices can be constructed to describe the probability of moving from one region to another during some specified time period. Clark (1986b), for example, reported a transition probability matrix in which the values represented the probability of moving to some other region in the United States, based on data for the period 1975–1980. Thus, there was a 0.0153 probability that an individual living in the South in 1975 had moved to the North Central region by 1980 (Table 13.5). As one would expect, the highest values were along the diagonal, indicating the probabilities of people remaining in their regions of origin.

Although Markov models are attractive, in that the transition probabilities reflect both the relationships between regions and the stochastic nature of migration decisions, they are limited by a number of rather restrictive assumptions. First, the transition probabilities are assumed to be temporally stable. Second, the population is assumed to be homogeneous, and this implies that everyone obeys the same transition matrix. Third, the Markov property suggests that the probability of migrating between two areas is solely dependent on current location, not on previous behavior. More recently, researchers have attempted to construct models

Table 13.5 Transition Probabilities for Regional Population Shifts in the United States

	1975				
	Northeast	North Central	South	West	Total
Northeast	0.9358	0.0064	0.0097	0.0069	0.2188
North Central	0.0098	0.9380	0.0153	0.0165	0.2629
South	0.0381	0.0332	0.9592	0.0274	0.3310
West	0.0163	0.0224	0.0159	0.9492	0.1873
Total	0.2273	0.2695	0.3215	0.1817	1.0000

Source: W. Clark, *Human Migration*, Sage (Beverly Hills), 1986, Table 3.6d.

of change that relax the assumption of stationary transition matrices by explicitly incorporating heterogeneous populations and changing preferences.

Perhaps the most impressive use of the Markov approach has been in multiregional demographic analysis (Rees, 1983; Rogers and Willekens, 1986), which is characterized by the demographer's concern for precise measurements on disaggregated populations. A multiregional life table requires information concerning the probability of individuals surviving to certain ages, measures of fertility, and estimates of mobility. Such data allows one to address the issues of how individuals born in one region will redistribute themselves across other regions and at what stage in the life cycle such movement is most likely to occur. In this way, changing patterns of population distribution can be related to differential growth rates across regions over time. Increasingly, however, multiregional demography has begun to introduce socioeconomic factors that can be used to explain the values taken by the model parameters. As a result, the lack of behavioral content in Markov models indicates that they can be profitably combined with a regression-based approach, which explicitly incorporates a set of explanatory variables (Rogerson, 1984).

An alternative approach for capturing the dynamics of migration flows is also available in the form of various *time-series techniques* (Chatfield, 1984). For example, one can develop moving average models and autoregressive models. In the moving average model the value of an individual variable at a particular point in time is generated by the weighted average of a finite number of previous and current random disturbances, or shocks. By contrast, in an autoregressive model the current observation is generated by a weighted average of past observations, together with a random disturbance in the current period. Mixed autoregressive moving average models can be developed for stationary time series, and certain types of nonstationary time series can be differenced in order to produce a stationary time series, thus allowing the development of a general integrated autoregressive moving average model. Within the context of migration, this autoregressive integrated moving average process, known as ARIMA, suggests that the amount of migration from one time period to the next is a complex function of previous migration plus current and previous random disturbances.

The implications of these time-series models for migration have only just begun to be explored. Markovian models of interregional migration are essentially autoregressive. Similarly, demographic forecasting models that specify the determinants of population change by accounting for births, deaths, and migration are also autoregressive and assume the random disturbance term to be negligible. Most geographic processes are stochastic rather than completely deterministic, however, as a variety of unknown variables, or an exogenously generated random element, may affect the dependent variable. Consequently, when a variable like migration is treated in dynamic terms, one is observing an underlying stochastic process that can be treated as some form of ARIMA model.

G. Clark (1982a) has investigated gross migration flows using ARIMA models and suggests that both autoregressive and moving average processes are at work. In particular, Markovian, or autoregressive, models are apparently inappropriate for many rapidly growing regions. Such regions contain a sizable moving average component, implying that in-migration is a volatile process, more influenced by exogenous and rapid shocks than by previous patterns of in-migration. In contrast, average and decelerating growth areas seem better approximated by autoregressive models. Future efforts along these lines are likely to incorporate spatial autocorrelation by using space–time autocorrelation functions. The prior assignment of weights reflecting the effects of contiguous interregional migration flows, however, remains problematic. The usual assignment of weights according to distance-decay principles might be unwise, as regions are not necessarily highly integrated with their immediate neighbors, at least in the context of migration (G. Clark, 1982b).

Distributed lag models allow the incorporation of temporal information in the explanatory variables. In this case, migration acts as the dependent variable, while some explanatory variable, such as unemployment level, is measured over a series of prior time periods. The equation can then be specified as follows:

$$Y_t = a + b_0 X_t + b_1 X_{t-1} + b_2 X_{t-2} + b_3 X_{t-3} + e \tag{13.23}$$

where Y_t is a measure of migration at time period t, X_t is a measure of unemployment at time period t, X_{t-1} is unemployment in the previous time period, and so on. The use of such lagged variables allows one to take into account the fact that the relationships between variables such as unemployment and migration are not always instantaneous and often involve time lags.

Bearing in mind the usual assumptions of the multiple regression model, one can use ordinary least-squares procedures to estimate the coefficients, at least in principle. In practice, however, certain problems associated with the distributed lag model tend to preclude the direct application of ordinary least-squares analysis (Katz, 1982:181). First, the estimation of an equation with a large number of lagged explanatory variables will be compromised by multicollinearity, leading to imprecise parameter estimates. Second, a lengthy lag structure will require the estimation of a large number of parameters, often severely reducing the degrees of freedom. Third, the creation of lagged variables reduces the number of observations, which are often in short supply in time-series models.

Some of these difficulties can be resolved if one is prepared to make some a priori specifications concerning the form of the distributed lag and thus impose some restrictions on the lag weights. For example, in the Koyck geometric lag model, the weights associated with the lagged explanatory variables are all positive and decline geometrically with time. Such a condition reduces the number of parameters to be estimated; although the weights never become zero, beyond a certain time the effects of the explanatory variable are negligible. While the geometric lag model can be very useful, it is somewhat limited by the fact that it postulates a declining set of lag weights. A more general formulation is provided by the Almon polynomial distributed lag model, which merely assumes that the lag weights will follow certain patterns and that these patterns can be approximated by fitting polynomial functions. Often a third- or fourth-degree polynomial will be sufficient, although higher-order polynomials can be specified when sufficient data are available.

Where the data allow, Fourier analysis and spectral analysis can be used to describe fairly complex temporal changes. *Fourier analysis* involves modeling a time series by fitting sine and cosine functions. A cosine series represents a regular wave with a specific length from crest to crest. Sine waves can be added to allow for the fact that the series may not begin at the top of a crest. *Spectral analysis* can be viewed as an extension of Fourier series analysis, in that it does not use exact wavelengths but instead involves frequency bands of specified width; thus it is more appropriate for the kind of fluctuations encountered in migration flows, which are seldom exactly periodic. Cross-spectral analysis can be employed when one is interested in the relationships between two or more time series.

Event history analysis represents a particular type of dynamic approach that allows one to model the interrelationships between specific life course events, such as leaving school, marriage or divorce, and migration. In general, an event is some qualitative change that occurs at a particular point in time, rather than a gradual change in some quantitative variable (Davies, 1991). Thus the occurrence of an event implies a relatively sharp disjunction between what precedes and what follows, and an event history is a longitudinal record of when events happen to a sample of individuals (Allison, 1984). There is no single method of event history analysis, but rather a variety of related techniques developed in such disparate disciplines as sociology, biostatistics, and engineering. Whereas biologists and engineers have tended to emphasize methods for single, nonrepeatable events, such as death, sociologists have tended to emphasize repeated events, such as job change. All three traditions, however, have focused on regression-type models where the occurrence of an event depends on some linear function of explanatory variables. Sandefur and Scott (1981), for example, were able to show that the traditional inverse relationship between age and migration is primarily due to the effects of family life cycle and career variables.

Dynamic data on mobility, however, are often much more readily available for discrete time intervals, as in *panel studies*, than for the continuous time mobility histories employed in event history analysis (Clark, 1992). Panel data involve successive waves of interviews, using a particular sample of individuals or households (Sandefur et al., 1991). Panel surveys became especially popular during the late 1970s as a means of monitoring the effectiveness of federally funded social experiments, but only more recently has information on geographic mobility been available from such sources. Great care should be taken when utilizing panel data, however, as different experimental designs concerning the number and spacing of waves can lead to different substantive conclusions (Sandefur and Tuma, 1987). For example, monitoring the mobility of individuals during a single short interval can generate different interpretations than monitoring the same group of individuals during a single long interval. Furthermore, measuring explanatory variables at the end of a particular time period can be misleading, especially if those variables are likely to change over time due to migration. Finally, panel studies suffer what has been called the initial condition problem (Davies and Pickles, 1985). That is, data from panel studies interrupt the process of interest at some intermediate point, rather than starting at the beginning.

Chapter 14

Urban Planning

14.1 State Intervention

In Chapter 2 it was suggested that one way of understanding the evolution of cities is to think of the urban future as being a response to the present pattern of socioeconomic forces plus the intervention of the state and other factors outside the urban system itself (Figure 2.1). It is now time to consider explicitly the role of state or government intervention and the nature of urban planning. After reviewing various theories of the state, in order to appreciate fully the role of the state in planning, we consider the planning process itself. The problem of urban deprivation and the cycle of poverty is then discussed, followed by a brief overview of possible urban spatial plans.

Theories of the State

There are many theories of the state (see also Section 1.1), and a variety of typologies have been produced (Johnston, 1980). Most of the typologies focus on the function of the state, and Clark and Dear (1981) identify six particular characterizations of the state that have been developed by a number of authors. First, and perhaps most simply, the state has been viewed as a *supplier of public goods and services*. The provision of public goods is regarded as an allocative function of government, and consideration is given to the optimal rules for distributing public goods. In particular, the state is seen as providing those goods and services that the majority of the population require but are unable to provide for themselves.

Second, the state is also seen as a mechanism for *regulating and facilitating* the operation of the market economy. The state intervenes in the private market to ensure that the market creates the best possible allocation of resources and to achieve the approximately equilibrium conditions that are characterized by high employment and low inflation. Within this context, the state also tries to maintain the rules of the free-market economy by enacting antitrust and antimonopoly legislation.

Third, part of the state's behavior is designed to adjust market outcomes in order to achieve some kind of policy goal. In this way the state engages in what might be called *social engineering*, which involves subjective judgments about what society ought to be rather than what it is. Although still accepting the marketplace as the major means of distribution, the state attempts to redress socioeconomic imbalances and to protect the interests of certain disadvantaged or minority groups. More specifically, the state tends to set minimum thresholds with respect to standards of living and provides the facilities to ensure that nobody falls below these designated thresholds.

Fourth, the state acts as an *arbiter* in disputes between different interest groups in society. In this context the state has a number of options. For example, it can try to act as a neutral umpire, it can try to be rational and achieve some optimal outcome that benefits all the parties involved in the dispute, or it can be elitist and merely reflect the interests of the ruling power groups. This characterization of the state emphasizes the power vested in the state and its ability to use that power to further its own ends.

Fifth, the *instrumentalist* view of the state regards the state as being simply an instrument of the business elite. This viewpoint is perhaps best exemplified by the work of Miliband (1977), who explored the "conspiracy" between the ruling class and the state. He argues that the social backgrounds of the top decision makers in the government, judiciary, and law enforcement agencies ensure that the interests of major business, or capital, will always receive a sympathetic hearing. Thus the link between capital and the major state institutions tends to maintain the status quo of the socioeconomic and political system.

Sixth and finally, a *structuralist* perspective would argue that the functions of the state are determined by the structure of society itself, rather than by a few key people within that society. The state responds to the prevailing balance of class forces within society and attempts to alleviate persistent class contradictions. In this respect, the state is not an autonomous entity, but rather a reflection of the balance of power among competing social classes at any particular time.

Although it is convenient to distinguish among these different viewpoints, it is obvious that the categories are not mutually exclusive; some of them may subsume others. For example, both the structuralist and instrumentalist positions can be derived from Marxist theory. What is most important, however, is that all six approaches contain both useful insights and inherent weaknesses. Their different theoretical perspectives are not easily reconciled, and it is best to adopt a position of theoretical pluralism.

Forms of Urban Government

These various theories of the state are partly manifested at the local level by different forms of urban government. Most city governments fall into one of three categories: the mayor–council form, the council–manager form, and the commission form. In the *mayor–council* form of government, both the mayor and the city council are directly elected by the public. The mayor is regarded as the chief executive, while the council is the legislative body.

Unlike the mayor, the council members usually have other jobs and can devote only a limited amount of time to their council duties. As shown in the organizational chart (Figure 14.1), the mayor typically has the power to appoint the individual department heads.

It is conventional to distinguish between strong and weak mayor–council forms of government. A mayor–council form of government is considered to be *strong* if it has some or all of the following characteristics. First, the mayor can veto ordinances passed by the council. Second, the mayor has significant power over the city budget, such as the right to submit an executive budget or to have veto power over individual items in the budget. Third, the mayor has the power to appoint or remove department heads or city commissioners without council approval. Fourth, the mayor has a four-year term of office with the possibility of reelection for many terms and enjoys the support of powerful local interest groups and an effective political organization.

On the other hand, a *weak* mayor–council system exists in those situations where many department heads, such as city treasurer, city assessor, or city attorney, are either elected directly by the public or are appointed by the city council rather than by the mayor. Such situations are characterized by a diffusion of power and responsibilities, and individual department heads often have conflicting political philosophies. The power of the mayor also tends to be rather weak if he or she can only be elected to a comparatively short term of office, for example two years, and if other layers of local government, such as counties or special districts, have significant authority.

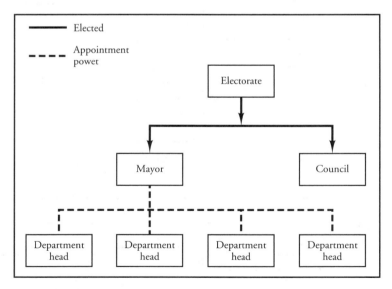

Figure 14.1   Mayor–council form of government. (From *City Lights: An Introduction to Urban Studies* by E. Barbara Phillips and Richard T. LeGates, Fig. 10.1, p. 262. Copyright © 1981 by E. Barbara Phillips and Richard T. LeGates. Reprinted by permission of Oxford University Press, Inc.)

As one might expect, there has been considerable debate over the merits of council versus mayoral power. Most council members are elected from individual districts within the city, so it is often argued that the mayor should assume increased power and responsibility, as he or she is elected on a citywide basis and presumably represents the general interests of the entire city. On the other hand, others argue that a strong mayor can lead to the overpoliticization of local decision making, whereby public policy is designed to benefit those interest groups who have direct access to the mayor. The decentralization of power tends to benefit those economic, racial, or ethnic minorities who are represented by council members elected from their own neighborhoods.

Under the *council–manager* form of government, which is common in many medium-sized American cities, the mayor has much less power and authority than in a mayor–council government (Phillips and LeGates, 1981:262). The most important individual figure in this form of government is the city manager, who is appointed by the city council (Figure 14.2). In most instances the city manager is responsible to a directly elected city council and has the power to appoint and remove department heads. The manager is responsible for the day-to-day running of the city, and although the council retains general legislative power, the manager generally prepares the city budget and develops policy recommendations for the council's action. Also, the combination of professional expertise and access to detailed information tends to give city managers considerable power.

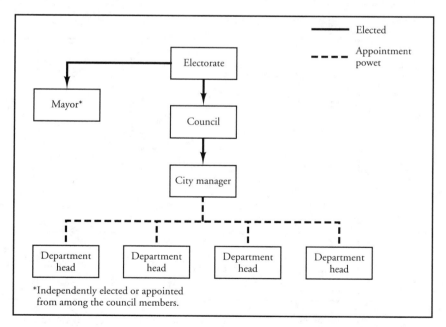

Figure 14.2 Council-manager form of government. (From *City Lights: An Introduction to Urban Studies* by E. Barbara Phillips and Richard T. LeGates, Fig. 10.2, p. 263. Copyright © 1981 by E. Barbara Phillips and Richard T. LeGates. Reprinted by permission of Oxford University Press, Inc.)

Finally, under the *commission* form of urban government, the voters elect a relatively small number of commissioners, who function as both legislators and executives (Figure 14.3). In other words, there is no separation of power similar to that between the council and a mayor. The commission form was introduced in the early 1900s, but it has not been widely adopted. In particular, this form of government suffers from the fact that there is no executive leader, and thus it is very difficult to coordinate individual policy initiatives. One of the commissioners usually serves as mayor, but his or her function is largely ceremonial.

Public Participation

The idea of direct citizen input into urban government, and planning in general, received its greatest stimulus during the early 1960s. The introduction of public participation was viewed as a means by which some of the power could be transferred from the bureaucracy to the people. In addition, it was argued that local government officials should benefit from the increased flow of information and thus be in a position to make more rational decisions. A more cynical view, however, would suggest that the formal recognition of public participation simply legitimizes the activity of planners, while giving the public the impression that their opinion counts (Knox, 1982:209). Also, the socioeconomic characteristics of those citizens who do become involved in the local decision-making process suggest that it is local business interests that are likely to benefit most.

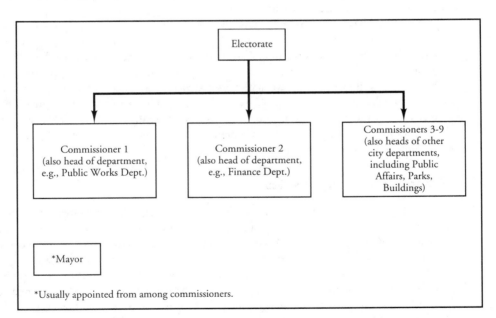

Figure 14.3 Commission form of government. (From *City Lights: An Introduction to Urban Studies* by E. Barbara Phillips and Richard T. LeGates, Fig. 10.3, p. 264. Copyright © 1981 by E. Barbara Phillips and Richard T. LeGates. Reprinted by permission of Oxford University Press, Inc.)

In any event, it is not hard to imagine why local officials have been less than enthusiastically receptive to the notion of extensive public participation in the planning process. Most bureaucracies try to preserve money and time by considering a minimum number of alternatives and restricting participation to a small number of decision makers. Also, planners tend to feel that too much public participation undermines their professional expertise and that the different interest groups will tend to cancel each other out anyway.

Perhaps one of the most effective forms of public participation falls under the heading of advocacy planning. *Advocacy planning* involves the use of experts by neighborhood organizations and other local groups to ensure that their interests and needs are articulated in the technical language of professional planners. There are really two types of advocacy planning. In one type an advocate directly represents a particular client or neighborhood on a specific issue, while the other type involves an advocate working on his or her own initiative to help a particular group of citizens, but no formal arrangements are involved. In the latter instance, the advocate proceeds solely on the basis of his or her own ideological judgment, so the principle of advocacy planning explicitly recognizes that planning goals are value statements that are not objectively verifiable, and so public decision making must reflect the will of the people. In other words, conflicting interest groups must be directly represented, as planners cannot truly arrive at objective solutions. Advocate planners disavow the view that planning is merely the application of technical expertise to society's problems and suggest that the advocate must identify his or her position with respect to a particular issue and then represent the interests of a clientele that has similar views.

Most advocacy planning has tended to focus on the issues of urban renewal and freeway construction through low-income neighborhoods, and although this kind of approach has had some success in blocking, or at least modifying, the official plans, it has been criticized on a number of grounds. In particular, critics have identified three key problems: a neighborhood's susceptibility to the advocate's personal values, the exclusion of neighborhood members from the more technical processes of plan preparation and data manipulation, and the focus of advocacy planning on short-term issues rather than on long-term goals. These first two points suggest that the advocate planner can sometimes assume, either knowingly or unknowingly, the role of community manipulator.

Location of Public Facilities

Public input would seem to be especially appropriate in the context of locating public facilities, such as waste treatment plants, parks, museums, art galleries, libraries, and public hospitals (Kirby, 1982). The allocation of public facilities requires not only the computation of demand levels, but also the calculation of *externality* effects, which summarize the effects of a locational decision on those not directly involved with the facility (Palm, 1981:243). A positive externality, or spillover effect, occurs when the residents of an area gain from a particular locational decision, such as a new public park. A negative externality, however, will lower the quality of life in an area, such as when a new freeway divides an existing community and increases the amount of noise and air pollution. It is in the context of these negative externalities that most conflicts tends to occur, and there is an important difference between accessibility and proximity. *Accessibility* to a public facility is generally regarded as a positive asset, but *proximity* refers to the effect of being close to a facility that is not directly utilized, thus imposing a cost on those households involved.

In essence, the public facility locational decision is inherently a *political act*, representing a compromise between various interest groups (Lake, 1987). The political nature of the decision occurs for three main reasons. First, the lack of clear theory for locating public facilities ensures that the decision is the subject of various political pressures, although much recent work has been devoted to the question of maximizing accessibility. Second, the lack of detailed information concerning potential demand and budget restrictions does not encourage the development of a purely criteria based decision-making process. Third, because financial resources are limited and priorities must be established, decisions concerning the location of public facilities are unavoidably forced into the political arena.

A good example of the problems associated with locating public facilities is provided by *mental health care* establishments. Such establishments generate two types of externalities. First, there are the tangible effects, which involve clearly identifiable and usually quantifiable impacts. For example, most residents fear that community mental health centers result in declining property values. This situation can often result in a self-fulfilling prophecy syndrome, although there is in fact some evidence that property values are unrelated to the siting of such facilities. The intangible effects, on the other hand, are associated with such things as the fear for personal safety, the stigma attached to mental illness, and the dislike of loitering clients.

The perceptions of the neighborhood residents themselves play an extremely important role in the identification of externality effects. In a household survey conducted in metropolitan Toronto, it was found that a strongly neutral group of respondents did not anticipate any impact on their neighborhood from the introduction of mental health facilities (Dear et al., 1980). In general, respondents who claimed to be aware of the local mental health care facility were relatively more tolerant in their estimation of the neighborhood impact of such facilities. Also, the spatial extent of the negative externality field appeared to be remarkably confined. As proximity to a potential facility increased, so did the perceived undesirability of that facility, but the most negative responses occurred within one block of a facility location; beyond six blocks there was a far more tolerant attitude.

Planners appear to have developed three major conflict-avoidance strategies when dealing with community opposition to plans for mental health facilities. The simplest involves finding locations where no community opposition is expected or where controversial facilities would generally be unnoticed (Dear and Wittman, 1980). Such locations are usually found within areas of the inner city that are characterized by rental accommodation and transient residents. If such locations are not appropriate, one of two other approaches tends to be adopted. Either a community is educated and coerced into accepting a facility before it is actually introduced, or a facility is set up without prior warning, in the hope that it will go unnoticed until it can be demonstrated that it does not generate any harmful spillover effects.

14.2   The Planning Process

Berry (1973) has postulated an urban policy model (Figure 14.4) and suggested a sequence of four modes of planning that are variants of this general model. In essence, one can distinguish

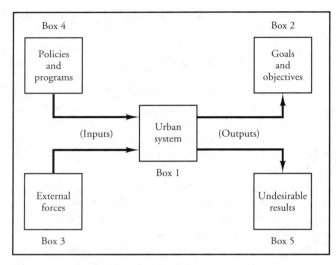

Figure 14.4 Urban policy model. (Reprinted from *The Human Consequences of Urbanization: Divergent Paths in the Urban Experience of the Twentieth Century* by B. J. L. Berry, Fig. 17, p. 173. Copyright © 1973 by St. Martin's Press, Inc. Reprinted by permission of the publisher.)

between two categories of inputs, external forces and policies and programs, that produce change in the urban system. This change can generate two different types of outputs: undesirable results or desirable results; that is, either problems or the goals or objectives that are being sought. Box 1 represents the existing urban system that has to be acted on to achieve the goals contained in box 2. In general, urban planning can either be consciously directed toward achieving these long-range goals, or it can simply react to the problems or crises represented by box 5. Box 3 contains the exogenous forces that influence the urban system, but whose causes lie outside the influence of the urban policymaker. Such forces include the behavior of the birthrate and the gross national product, which often affect the growth of urban areas far more than the policies of individual city planning departments. Finally, box 4 represents the policies and programs generated by government planning agencies. These programs can be likened to levers that are pulled in order to cause desired changes in the urban system.

Four Modes of Planning

Within this general framework, Berry identified four modes or styles of planning: ameliorative problem solving, allocative trend modifying, exploitive opportunity seeking, and normative goal oriented. The simplest and perhaps most common form of planning is the *ameliorative problem-solving* form, in which nothing is done until the problems reach crisis proportions. This strategy is very present oriented, and little thought is given to long-run goals or objectives. American local authorities are often forced into this essentially reactive mode of planning because of their reliance on elected officials with short terms of office, their restricted budgets, and their limited legal jurisdictions (LaGory and Pipkin, 1981:276).

The *allocative trend-modifying* form of planning is more future oriented and uses projections of existing trends to forecast problems that will arise in the future. Based on the prediction of these future problems, available resources are allocated to promote the most desirable outcomes. In this scenario, the regulatory mechanisms are devised to modify and make the best of existing trends. Traffic forecasting demand models provide an excellent example of this strategy of predicting the future and then trying gently to modify it.

The *exploitive opportunity-seeking* planning style does not identify future problems, but seeks out new growth opportunities. These new growth opportunities are identified by imaginative leaders in both the public and private sectors of the economy. As well as planners themselves, the individual actors are corporate managers, real estate developers, and industrialists. Whereas the trend-modifying approach seeks to make the best of existing trends, the opportunity-seeking approach merely aims to maximize profits while it can, with less concern for the future.

Finally, *normative goal-oriented* planning is explicitly future oriented and seeks to identify a desired future state for the urban system. Specific goals are set in accordance with the kind of future that is desired, and plans are implemented to guide the system toward these goals. This approach is more long term in nature and assumes that consensus can be reached with respect to what an ideal urban system should be.

In reality, planning policy in most countries represents a mixture of these four major types. However, normative goal-oriented planning is really possible only in those countries that have a centralized form of government, with sufficient control over the different sectors of the economy. Thus this style of planning is characteristic of countries with strong central governments, such as Sweden and Great Britain. At the other extreme, countries such as the United States and Canada tend to exemplify the ameliorative problem-solving and the allocative trend-modifying forms of planning.

Stages in the Planning Process

A variety of schematic summaries of the various stages in the planning process have been devised (Herington, 1989). The one presented here (Figure 14.5) is a simplified composite of these and, like them, falls within the genre of systems planning. Fundamental to the concept of systems planning is the idea of interaction between two types of systems: the controlling system, represented by planners themselves, and the urban system, which it seeks to control. In other words, just as cities and regions can be usefully conceptualized in terms of interacting systems, so can planners themselves, thus creating a planning system.

The first stage in the overall planning process is the actual *decision to plan*. Planning, as we understand it today, is of relatively recent origin and should not be taken for granted. The decision to plan requires moving from a society dominated by laissez-faire principles to one that accepts the need for at least a certain amount of state intervention. Such a decision is a major step, as it involves the restriction of individual freedoms, such as the right to do what one likes with one's own property. In this respect, the need for planning and the definitions of its roles and purposes should be kept under constant review by all sections of society.

Having made the decision to plan, the next step is to *understand the operation of the present urban system*. The system must be represented in summary form, and that representation can be either purely descriptive, without providing an explanation of the underlying processes at work in the system, or causal, identifying the constituent cause-and-effect relationships (Oppenheim, 1980:2). The bulk of the present book has been devoted to describing a number of theories and models that contribute to our understanding of the

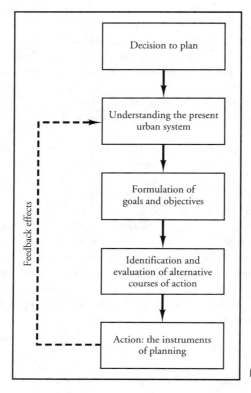

Figure 14.5 Planning process.

present urban system. For the most part these models have been partial rather than general, in that they are concerned only with one identifiable subsystem, such as the housing market, within the overall urban system.

Batty (1978) has suggested that the models used in this stage of the planning process can be categorized according to a number of major characteristics. First, in terms of the temporal dimension, they range from static, through quasi-dynamic, to fully dynamic. Because of the unavailability of true time-series data, however, most urban models are static in nature, which creates a serious problem, as planning is concerned with change over time. Second, models have been constructed at different levels of both sectoral and spatial aggregation. Sectoral aggregation involves aggregating different occupational or ethnic groups, for example, while spatial aggregation involves aggregating different spatial units, such as blocks and census tracts. Third, from a technical perspective, models incorporate either linear or nonlinear forms of relationships, with linear models usually being regarded as special forms of nonlinear models. Because linear mathematics has been more thoroughly developed than its nonlinear counterpart, most urban models are linear, although a pursuasive argument can be made that most real-world relationships are nonlinear. Finally, models entail either direct or indirect solution procedures. Direct procedures involve some kind of immediate solution, whereas indirect methods are sequential and involve iterative or simulation techniques.

Having attempted to understand the nature of the urban system as it is presently consti-tuted, usually through the construction of appropriate urban models, the next step is to *for-mulate a series of goals and objectives*. A hierarchy of policy decisions can be identified. For example, a first-order policy decision might be to contain urban development at a particular level. A second-order policy decision might involve decisions concerning concentration versus dispersal or a few large nuclei versus several smaller ones. Third-order decisions would then get down to the specifics of land use patterns, residential densities, and transportation net-works.

One can also distinguish between goals and objectives. Goals are typically rather vague and general, and progress toward a particular goal requires the attainment of certain more pre-cise objectives. For example, a stated goal might be to increase the opportunities for outdoor recreation in the city, while the associated objectives might be to double the acreage of parks within the next 10 years and to acquire 5000 acres of river and lakeside land within the next 20 years. Similarly, a goal of providing a convenient pattern of major shopping centers might be matched with the particular objective of ensuring that the average distance of households from their nearest major shopping centers is no more than 5 miles.

In most instances, goals represent general areas of concern. One can identify political goals, such as the maintenance of efficient and democratic government; economic goals, such as accessibility to goods and services; social goals, such as safety; and environmental goals, such as environmental quality. There is always the thorny problem, however, of who should be responsible for identifying the particular goals toward which an urban system is directed. The broad goals for society are generally a matter for the politicians, but politicians are often pre-occupied with acute short-term issues (Hall, 1992). Somehow, then, goals must be formulated via cooperative effort involving direct participation by the public, perhaps through public opinion polls and referendums, local politicians, and professional planners.

Once the goals and objectives have been defined, *alternative courses of actions for achiev-ing those goals must be identified and evaluated*. In particular, the proposed alternative courses of action must be related specifically to the original goals and objectives. For example, alterna-tives related to the physical form of the city, such as whether it should be linear, circular, or polynuclear, might have very little to do with stated objectives concerning employment opportunities and accessibility to shopping centers. Once a set of feasible alternatives has been suggested, however, three major principles of program evaluation should be adhered to: first, various impacts of a program over time should be clearly identified; second, the seriousness of these impacts, both good and bad, should be estimated; and third, the costs of the program should be considered in the light of available funds.

Three relatively formal methods for evaluating alternative courses of action, or planning programs, have been established in the planning literature. The most popular methodology can be generally labeled as *cost–benefit analysis*, whereby the anticipated benefits to be gener-ated by a given program are compared to its anticipated costs (Schofield, 1987). The costs and benefits are usually itemized in money terms, and it is assumed that the best alternative is the one that provides the greatest quantity of economic benefits in relation to economic costs. One obvious problem of this approach, however, is that many planning elements, such as a beautiful landscape or building, cannot be evaluated in financial terms. In other words, notions of social costs and benefits must also be included in the overall analysis.

After particular courses of action have been chosen, they are implemented, or put into practice, using the *instruments of planning*. Scott (1980:61) has suggested that there are three major ways that urban governments, or planning agencies, have intervened in urban affairs: by using fiscal policies, land regulation policies, and development policies. One major instrument of fiscal policy has been the property tax, which is ubiquitous in the United States. Also, monetary variables are manipulated by controlling the prices of certain urban goods and services, such as legislating limits on housing mortgage rates, and rent controls. Finally, urban governments often directly subsidize or provide grants for mass transit systems, low-income housing programs, and industrial location incentives. Many would argue, however, that such subsidies are merely palliative and simply contribute to the inefficient production of a variety of urban services.

One of the most popular types of land regulation policy has been the use of zoning ordinances, which we discussed in the context of land use and land value theory (see Section 3.3). Other legal restrictions on the use of land include such devices as subdivision controls, construction codes, and building height limitations. Often, however, such restrictions have been lifted in those situations where some kind of political pressure has been brought to bear. For example, exceptions are sometimes made to land use zoning ordinances in order to placate a particular interest group. In particular, land use restrictions are often relaxed in those situations where they threaten to depress land prices.

The final type of policy instrument is represented by the direct development or redevelopment of targeted areas by government agencies (Church and Hall, 1989). Examples of such development projects include public housing construction, urban renewal, and the initial development of industrial estates. These projects all play important roles in terms of guiding the urban land market and thus shaping the spatial patterns of land use and land value.

The effect of manipulating these policy instruments is then monitored in order to calibrate their influence on the urban system. Sometimes the *feedback effects* are in the anticipated, positive direction, but often there are unanticipated side effects. These unanticipated side effects are comparable to those sometimes produced by surgical intervention in the case of hospital patients. The overall planning process has a certain trial-and-error element to it, and the systems approach allows it to be viewed in terms of a functioning system, with internal feedback effects that can be induced by manipulating certain exogenous factors. In particular, the systems model of planning contains two major assumptions, one explicit and the other more implicit (Hall, 1979:5). First, it is explicitly stated that this is a scientific approach to planning, which assumes that the urban system can be understood to such a degree that it is possible to forecast accurately both the direct and indirect effects of government intervention on the operation of that system. Second, and more implicitly, it is assumed that the urban planner can act in a value-free manner, with all the detachment of a physical scientist.

As can be imagined, these assumptions have been rather harshly criticized by some authors, especially in terms of what is seen as the inability to place the process of urban planning within the overall activity of the capitalist state (Scott, 1980:231). As a result of this and other criticisms of mainstream planning theory, a series of alternative approaches has been suggested. Most of these alternatives focus on the role of public participation in the planning process and emphasize the fact that planning can be viewed as a political act of redistribution.

In this latter context, however, the existing distribution of income and resources is sometimes simply reinforced by the distribution of public goods.

The Comprehensive City Plan

On the most local scale, much of the planning effort is devoted to producing and implementing a comprehensive city plan (Branch, 1985). Such plans are intended to provide a blueprint of how the city should develop, and they are long range in the sense that they often project up to 20 years into the future. Most of the following objectives tend to be incorporated into these plans. First, there should be provision for an orderly physical growth and development of the city. Second, land should be allocated among various alternative uses in order to maximize its potential benefits. Third, facilities should be provided to satisfy the various needs of a society with increasing amounts of leisure time. Fourth, an attempt should be made to provide a range of neighborhood environments. Fifth, economic development should be promoted in an effort to ensure sufficient employment opportunities for all sectors of the labor force. Sixth, plans should be made to ensure the provision of certain public services, such as transportation and eduction facilities.

Generally, the overall plan is divided into a number of components, or elements (Northam, 1979:474–6). *Population studies* are used to project the future demand for city services. Independent projections are made for various subgroups, based on age, income, and sex, so that the demand for particular types of services, such as transportation for the elderly, can be anticipated. These projections are also disaggregated for different parts of the city, as future demands will obviously vary across different types of neighborhoods.

Housing studies are conducted in conjunction with the population studies to ensure that sufficient amounts of different types of housing will be available in the future. For example, there should be an appropriate mix of single-family and multifamily dwelling units, and attention should be paid to the spatial distribution of different kinds of housing. Often, the private sector tends to be aimed at providing high-income housing, so the future need for low-income housing must be accurately assessed, including the feasibility of subsidizing low-income housing projects.

Aspects of the economic structure of the city are considered in *economic studies*. These studies project future demand for different parts of the labor force and try to anticipate the need for particular specialized skills. Job retraining programs might be advocated to generate these skills or attempts made to attract appropriate migrants from other cities. The interrelationships between the local economy and the state and national economies are also explored in order to assess the impact of national fluctuations. The economic base and input–output analysis procedures described in Chapter 7 are important methodological tools in this context.

Land use studies are a key element of the overall planning process, and an inventory of existing land use is essential to the development of a city plan. Trends in land use and current land use densities serve as guides to future requirements, and land regulation policies are implemented to minimize the effects of negative externalities. Particular attention is paid to problems that might develop in the rural–urban fringe, where the competition for land between urban and agricultural uses is the most intense.

Finally, all comprehensive city plans involve a series of *transportation studies*. The obvious relationship between land use and traffic flow is utilized to estimate future traffic demands, often using some form of gravity model (see Section 9.1). Consideration must also be given to the future provision of mass transit systems and parking facilities for private automobiles. For example, the future use of the downtown area by private automobiles can be manipulated, to a certain extent, by the structure of parking costs.

14.3   Urban Deprivation

A number of urbanists have conceptualized a cycle of poverty that is particularly associated with the central areas of North American cities. In this section we discuss one such conceptualization (Johnston, 1982b:253–61), especially as it relates to the interrelationships between different manifestations of poverty and the concept of multiple deprivation. We will then describe and evaluate the idea of tackling these problems by means of area-based positive discrimination. The section concludes by exploring some alternative explanations of the cycle of poverty and by briefly examining the related phenomenon of the urban homeless.

The Cycle of Poverty

Johnston's (1982b) view of a cycle of poverty is represented in Figure 14.6. This diagram shows how the problems are linked in a cumulative fashion, thus creating a chain of situations that is difficult to break. *Poverty* is experienced by those members of society who are unable to find employment or who are excessively poorly paid. The greatest single item of

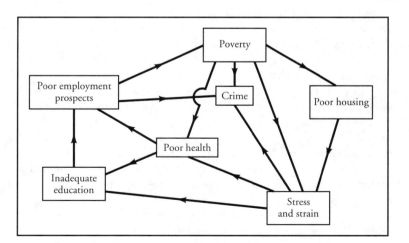

Figure 14.6 Cycle of poverty. (Reprinted from *The American Urban System: A Geographical Perspective* by R. J. Johnston, Fig. 9.2, p. 257. Copyright © by St. Martin's Press, Inc. Reprinted by permission of the publisher.)

expenditure is generally housing, so those with the lowest incomes tend to live in *poor-quality housing*, both because they cannot afford high rents and because they are unable to qualify for mortgages. Often, the rents demanded are even higher than what might be suggested by prevailing market conditions because of the oligopolistic control of the housing supply by a few, very powerful landlords.

Poverty, especially through its relationship to poor housing, creates certain *stresses and strains* for the households concerned. This stress often generates certain kinds of chronic *health problems* that are related to poor diet, cold, and exposure to vermins. Sometimes severe mental illness can result from the problem of trying to cope with inadequate resources, and interpersonal disputes are accentuated by the high-density living. Palm (1981:245) has pointed to the association between poverty and particular health problems and points out that health care facilities are often unequally distributed and discriminatory.

The cycle of poverty is further strengthened by the *inadequate educational facilities* that are characteristic of many inner-city areas. The school buildings are often obsolete, and the financial resources available to local school boards are relatively meager compared to their suburban counterparts. It is also true that unhealthy children are less likely to reach their full potential than are those who are well fed and housed. This relative deprivation in terms of educational opportunities tends to translate into *poor employment prospects*. Without certain qualifications, most of the school graduates are destined to end up in low-skilled, poorly paid jobs that provide little opportunity for advancement. The employment problem is compounded by the fact that in recent years many types of industry have been moving to the suburbs (see Section 7.2), thus generating high levels of unemployment in inner-city areas.

A central element in this cycle of poverty is the incidence of *crime* (Smith, 1989; Herbert, 1993). Two types of crime appear to be particularly prevalent in this context. First, there are high rates of crimes against property, or economic crimes, that tend to be stimulated by high rates of poverty. Second, the stresses and strains associated with high unemployment also tend to be associated with crimes against the person, such as spouse beating and child abuse. Unemployment also tends to encourage the formation of youth gangs and associated turf rivalries.

Area-based Positive Discrimination

Given the spatial concentration of poverty and associated problems, it was perhaps natural that solutions should be sought in terms of providing help on a local basis. Area-based policies were introduced to complement existing welfare programs, which did not provide long-term help for the most deprived segments of the population and were unable to break the cycle of poverty in inner-city areas. Much of the initial impetus for this strategy was provided by the Ford Foundation, which during the 1950s was especially concerned with the problems of metropolitan government and urban renewal. In particular, the foundation started a gray-areas program, which was specifically aimed at decaying neighborhoods in the central city. A series of demonstration projects were undertaken in Boston, New Haven, North Carolina, Oakland, Philadelphia, and Washington. Similar experimental community action programs were instigated by the President's Committee on Juvenile Delinquency. In both cases the importance of education and employment opportunities were stressed, and resources were concentrated into a few demonstration or experimental projects. It is probably fair to say, however, that these projects were not conspicuously successful, with public involvement remaining rather minimal and few jobs being produced.

These initiatives were later adopted by the federal government in its War on Poverty program during the Kennedy administration and the associated Economic Opportunity Act and Office of Economic Opportunity. A series of community action programs stressed the need to coordinate services and to encourage the participation of the poor. The provision of employment opportunities was seen as a primary goal, and significant contributions were made in terms of creating jobs for the urban poor. The Model Cities program, set up in 1966, was also based on the area projects. Grants and technical assistance were provided to help communities establish demonstration programs that would reduce social and economic disadvantages. The program eventually involved about 150 cities, although the emphasis was placed on physical blight and urban renewal rather than on the poor themselves.

Despite some limited success, however, there is a growing feeling that such area-based programs do not get to the root of the poverty problem, in that it is unrealistic to abstract the problems of particular neighborhoods from their wider social context (MacLaran, 1981; Knox, 1989). Hamnett (1979) and others have criticized what they call the fetishism of space, whereby spatial causes are inferred from spatial manifestations. They argue that deprived neighborhoods are merely the setting for a number of deprived persons who happen to reside there and that the problems experienced by such persons arise from nonareal causes. In other words, if the problems are not areally caused, area-based solutions are unlikely to eliminate those problems. Similarly, Eyles (1979) argues that, although area effects can intensify or compound individual deprivations, they are not the source of those deprivations, which lie in the broader socioeconomic structure of capitalist society.

Alternative Explanations

Rather than simply focusing on the cycle of poverty and area-based solutions, it is advisable, then, to think in terms of a variety of possible explanations of urban deprivation (Knox, 1982:207). In this context, there appear to be at least five major candidates (Hamnett, 1979). First, the idea of a *culture of poverty* suggests that problems arise from the internal pathology of deviant groups, who suffer from a lack of opportunity and aspiration. Consequently, strategies to improve the situation should concentrate on social education in general rather than on particular projects with respect to housing and income maintenance.

Second, the concept of *transmitted deprivation* has been invoked to suggest that social maladjustment is transmitted from one generation to the next. The emphasis is on the relationships among persons, families, and groups. Thus the home environment is seen to be the major culprit rather than the problems of low wages and poor housing. For adherents of this idea, the solution lies in using professionals such as social workers to help parents improve the home environment.

Third, it has been suggested that the problems of the socioeconomically deprived can be blamed on *institutional malfunctioning*. In this scenario it is argued that problems have arisen due to the failure of planning and administrative bureaucracies. Such bureaucracies have tended to generate a series of separate departments concerned with education, housing, and the like, and so have been unable to fashion a coherent public policy that is capable of addressing the interrelated problems that together lead to multideprived persons and households. The solution to institutional malfunctioning apparently lies in a more coordinated approach to rational social planning.

Fourth, it has been argued that problems of urban deprivation simply arise from the *unequal distribution of resources and opportunities*. The underprivileged are unable to obtain their fair share of society's resources, as they are not represented or allowed to participate fully in the political process. The solution to this inequitable distribution of resources is seen to lie in positive discrimination policies and greater public participation in the planning process.

Finally, Marxist theory suggests that problems of urban deprivation can be ultimately explained in terms of the overall socioeconomic structure of capitalist society. In particular, problems arise from *class conflict*, whereby certain class divisions are necessary to maintain an economic system based on private profit. The key concept here is inequality, and the proposed solution lies in the redistribution of power and political control. Although these explanations of deprivation are by no means mutually exclusive, it is certainly clear that while policies of positive discrimination might make sense in the context of some explanatory frameworks, they are regarded as nothing more than short-term cosmetic devices by those who are persuaded by the class conflict model of explanation.

The Urban Homeless

In its extreme form, the downward spiral associated with the cycle of poverty often results in homelessness. The term homeless, which refers to those individuals or families without permanent shelter, has traditionally been associated with skid row and substance-abuse problems (Hoch, 1991). In addition, a large and growing group of homeless individuals includes the mentally ill. Indeed, the deinstitutionalization of mentally ill patients, which began in the late 1960s, has created a series of service-dependent inner-city areas (Dear and Wolch, 1987; C. J. Smith, 1988:105–19). These individuals are often unable to take advantage of job retraining schemes and thus remain part of the chronically unemployed.

Homelessness has also grown dramatically, however, due to the increasing number of individuals or families who are simply unable to afford housing. This last group has been especially affected by the economic restructuring of many U.S. cities that has involved the suburbanization of low-skilled jobs (Law and Wolch, 1991). Thus, although at the individual level prolonged homelessness is often the result of a series of personal misfortunes, these misfortunes are not independent of larger structural forces. The problem has been further exacerbated by the increasing fiscal burden that has fallen on the local welfare state and by the continued reliance on charitable organizations. Finally, while community attitudes toward the homeless can be supportive in the abstract, there is no doubt that shelters and other homeless facilities are not viewed as attractive forms of land use when they are located in one's own neighborhood (Dear and Gleeson, 1991).

14.4   Possible Urban Spatial Plans

In terms of urban form, a variety of possible spatial plans, or patterns, has been suggested. One of the earliest conceptions of an ideal city was formulated by *Ebenezer Howard*. Howard tried to incorporate the best of rural and urban living in a series of potentially self-sufficient communities of approximately 30,000 people that would be surrounded by permanent green belts. These cities were called *garden cities*, and he emphasized the need for relatively low population densities. The overall city would be about 1000 acres, of which the central 5 acres was designed to include commercial and entertainment activities. Beyond the central area were a series of residential rings, including a grand avenue, with an industrial area just within the surrounding green belt. Two garden cities were built near London, Letchworth and Welwyn, and garden cities became fashionable in the United States after 1910. But many of these cities, like Forest Hills Gardens, New York, which was designed by Olmsted and financed by the Russel Sage Foundation, became affluent suburbs for commuters rather than self-contained communities (Phillips and LeGates, 1981:408).

Another visionary in terms of ideal urban form was *Frank Lloyd Wright*. He planned a hypothetical city called Broadacre that encouraged a completely low density urban spread. Each home would be surrounded by an acre of land to grow crops on. These homes should then be connected by superhighways, thus creating an automobile-based city, complete with suburban shopping centers. A major drawback of Wright's semiagrarian conception, however, was that it paid insufficient attention to the economic base of the city. Industry was to be decentralized to a series of scattered nodes, thus inhibiting the development of economies of scale (see Section 7.1).

The Swiss-born architect *LeCorbusier* had very different ideas from Wright, in that he favored a very centralized city, with business activity concentrated downtown in multistory office buildings. In terms of the residential areas, most families would live in multistory apartment complexes, surrounded by open parklike spaces. Thus LeCorbusier advocated very high overall population densities, while leaving the bulk of the land unbuilt. His impact is especially clear in numerous multistory projects in Europe, where slums were cleared to make way for high-rise residential developments with open space between them. These high-rise housing projects, however, were often criticized as being worse than the slums they replaced, in that they tended to inhibit community feeling and the supervision of children's playtime activity (LaGory and Pipkin, 1981:295).

The more recent planning literature tends to have been concerned with three basic forms of plan: the circular, the linear, and the sectoral. The *circular* plan is based on a set of ring and radial roads, which focuses on a central business district (Figure 14.7). Many cities have tended naturally to follow this type of radial-concentric growth, with individual neighborhoods developing in the interstices between each radial and ring road. Green belts have been suggested as a major technique for controlling the outward expansion of such cities, although the urban sprawl often begins just outside the green belt. Besides being unable to contain the spatial spread of cities, green belt policies have also been criticized because they favor higher-income families, who tend to live at the edge of the city.

The *linear* plan was proposed as a planning framework for London and is characterized by a central corridor containing the major areas of employment and economic activity (Figure 14.8). Residential areas are located along roads that run at right angles from the central corridor, and these are also connected by a circumferential highway. This plan provides for an orderly arrangement of land use, while maintaining the potential for industrial economies of scale along the central corridor. Open land is preserved between the residential areas, and the form of the plan is ideal for towns located on major rivers such as the Thames.

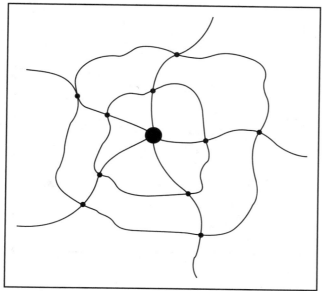

Figure 14.7 Circular plan.

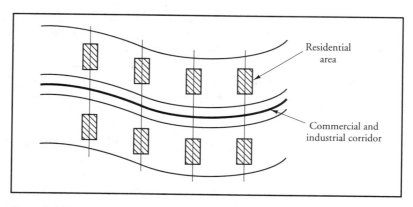

Figure 14.8 Linear plan.

Finally, the *sectoral* or radial plan has a downtown nucleus with a series of radial corridors (Figure 14.9). Along these corridors are various-sized urban subcenters, with their own array of goods and services. Rather like central place theory (see Section 5.2), a hierarchy of service centers is postulated, with the downtown providing certain highly specialized goods and services for the whole urban area, while the subcenters cater to the more local, less specialized demands. Open space is preserved between the radial corridors, and the corridors themselves are especially conducive to the establishment of mass transit systems. One drawback of this formulation, however, is the relative difficulty of moving from one radial to another and the associated potential for downtown congestion.

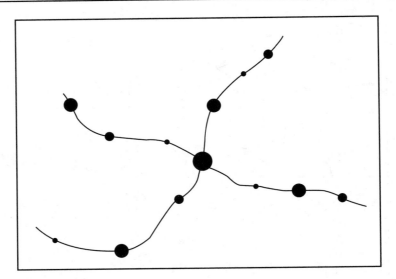

Figure 14.9 Sectoral plan.

Speculation as to the most appropriate form of future cities has tended to highlight the importance of potential energy shortages. In this context, van Til (1980) has suggested that future cities might emphasize regional subcenters within which most transportation is confined. These subcenters would develop where there is a particularly high density of commercial and industrial activity and in those locations that are central to the circulation system of an energy-poor society. They would be as autonomous as possible and would become the most preferred places of residence in a new, relatively immobile urban world.

Similar sentiments to those of van Til have been expressed by Birdsall (1980), who also foresees multiple commercial nodes and travel patterns that reflect higher energy costs. In addition, Birdsall expects a different type of urban political organization to emerge in the future, with the essential administrative functions of government encompassing the entire urbanized areas. Within this overall political structure, however, smaller communities within the city would develop and be responsible for the provision of local services, such as recreation, health care facilities, and primary education. The higher political jurisdiction would attempt to coordinate these local efforts and collect and redistribute taxes in order to minimize neighborhood fiscal disparities.

References

Abler, R., Adams, J., and Gould, P. 1971. *Spatial organization: The geographer's view of the world.* Englewood Cliffs, N.J.: Prentice Hall.

Adams, J. 1987. *Housing America in the 1980's.* New York: Russell Sage.

———. 1988. Growth of U.S. cities and recent trends in urban real estate values. In *Cities and their vital systems*, eds. J. Ausubel and R. Herman, 108–45. Washington, D.C.: National Academy Press.

———. 1991. Housing submarkets in an American metropolis. In *Our changing cities*, ed. J. Hart, 108–26. Baltimore: Johns Hopkins University Press.

Agnew, J., and Duncan, J. 1981. The transfer of ideas into Anglo-American geography. *Progress in Human Geography* 5:42–57.

Agresti, A., and Agresti, B. 1979. *Statistical methods for the social sciences.* San Francisco: Dellen.

Aitken, S., and Prosser, R. 1990. Residents' spatial knowledge of neighborhood continuity and form. *Geographical Analysis* 22:301–25.

Alba, R. 1987. Interpreting the parameters of log-linear models. *Sociological Methods and Research* 16:45–77.

Alexander, S. 1983. A model of population change with new and return migration. *Environment and Planning A* 15:1231–57.

Allison, P. 1977. Testing for interaction in multiple regression. *American Journal of Sociology* 83:144–53.

———. 1984. *Event history analysis: Regression for longitudinal event data.* Beverly Hills, Calif.: Sage.

Alonso, W. 1971. A theory of the urban land market. In *Internal structure of the city: Readings on space and environment*, ed. L.Bourne, 154–9. New York: Oxford University Press.

Alperovich, G. 1983. Lagged response in intra-urban migration of home owners. *Regional Studies* 17:297–304.

Alwin, D., and Hauser, R. 1981. The decomposition of effects in path analysis. In *Linear models in social research*, ed. P. Marsden, 123–40. Beverly Hills, Calif: Sage.

Amedeo, D., and Golledge, R. 1975. *An introduction to scientific reasoning in geography.* New York: Wiley.

Anderson, J. 1985. The changing structure of a city: Temporal changes in cubic spline urban density patterns. *Journal of Regional Science* 25:413–25.

Anderson, N. 1981. *Foundations of information integration theory.* New York: Academic Press.

———. 1982. *Methods of information integration theory.* NewYork: Academic Press.

Appelbaum, R., and Gilderbloom, J. 1990. The redistributional impact of modern rent control. *Environment and Planning A* 22:601–14.

Appleyard, D. 1969. City designers and the pluralistic city. In *Planning urban growth and regional development,* ed. L. Rodwin, 422–52. Cambridge, Mass.: MIT Press.

———. 1970. Styles and methods of structuring a city. *Environment and Behavior* 2:100–17.

Archer, M. 1982. Morphogenesis versus structuration: On combining structure and action. *British Journal of Sociology* 33:455–83.

Artle, R. 1965. *The structure of the Stockholm economy.* Ithaca, N.Y.: Cornell University Press.

Baerwald, T. 1981. The site selection process of suburban residential builders. *Urban Geography* 2:339–57.

———. 1989. Changing sales patterns in major American metropoles, 1963–1982. *Urban Geography* 10:355–74.

Baird, J., Wagner, M., and Noma, E. 1982. Impossible cognitive spaces. *Geographical Analysis* 14:204–16.

Ball, M. 1985. The urban rent question. *Environment and Planning A* 17:503–25.

Ballard, K., and Clark, G. 1981. The short-run dynamics of interstate migration: A space-time economic adjustment model of in-migration to fast growing states. *Regional Studies* 15:213–28.

Barff, R. 1990. The migration response to the economic turnaround in New England. *Environment and Planning A* 22:1497–516.

Barlow, I. 1991. *Metropolitan government.* New York: Routledge.

Barlow, J. 1990. Housing market constraints on labour mobility: Some comments on Owen and Green. *Geoforum* 21:85–96.

Barrow, C. 1993. *Critical theories of the state: Marxist, neo-Marxist, post-Marxist.* Madison, Wis.: University of Wisconsin Press.

Bassett, K., and Short, J. 1980. *Housing and residential structure:Alternative approache.* Boston: Routledge.

———. 1989. Development and diversity in urban geography. In *Horizons in human geography,* eds. D. Gregory and R. Walford, 175–93. Totowa, N.J.: Barnes and Noble.

Bates, J. 1988. Stated preference techniques and the analysis of consumer choice. In *Store choice, store location and market analysis,* ed. N. Wrigley, 187–202. London: Routledge.

Batty, M. 1978. Urban models in the planning process. In *Geography and the urban environment,* Volume 1, eds. D. Herbert and R. Johnston, 63–134. New York: Wiley.

———. 1989. Urban modelling and planning: Reflections, retrodictions and prescriptions. In *Remodelling Geography,* ed. B. Macmillan, 147–69. Oxford: Blackwell.

Batty, M., and Kim, K. 1992. Form follows function: Reformulating urban population density functions. *Urban Studies* 29:1043–70.

Bentler, P. 1980. Multivariate analysis with latent variables: Causal modeling. *Annual Review of Psychology* 31:419–56.

————. 1982. Linear systems with multiple levels and types of latent variables. In *Systems under indirect observation*, pt. 1, ed. K. Jöreskog and H. Wold, 101–30. Amsterdam: North-Holland.

Bentler, P., and Weeks, D. 1980. Linear structural equations with latent variables. *Psychometrika* 45:289–308.

Berry, B. 1963. *Commercial structure and commercial blight: Retail patterns and processes in the city of Chicago*. Research Paper 85, Department of Geography, University of Chicago, Chicago.

————. 1973. *The human consequences of urbanization: Divergent paths in the urban experience of the twentieth century*. New York: St. Martin's Press.

————. 1988. Counterurbanization. *International Regional Science Review* 11:221–6.

————. 1991. Long waves in American urban evolution. In *Our changing cities*, ed. J. Hart, 31–50. Baltimore: Johns Hopkins University Press.

Berry, B., and Bednarz, R. 1975. A hedonic model of prices and assessments for single-family homes: Does the assessor follow the market or the market follow the assessor? *Land Economics* 51:21–40.

Berry, B., and Garrison, W. 1958. The functional bases of the central place hierarchy. *Economic Geography* 34:145–54.

Berry, B., and Kim, H. 1993. Challenges to the monocentric model. *Geographical Analysis* 25:1–4.

Berry, B., and Parr, J. 1988. *Market centers and retail location: Theory and applications*. Englewood Cliffs, N.J.: Prentice Hall.

Berry, B. Conkling, E., and Ray, D. 1976. *The geography of economic systems*. Englewood Cliffs, N.J.: Prentice Hall.

————. 1987. *Economic geography*. Englewood Cliffs, N.J.: Prentice Hall.

Bertuglia, C., Leonardi, G., and Wilson, A., eds. 1990. *Urban dynamics: Designing an integrated model*. New York: Routledge.

Bielby, W., and Hauser, R. 1977. Structural equation models. *Annual Review of Sociology* 3:137–61.

Birdsall, S. 1980. Alternative prospects for America's urban future. In *The American metropolitan system: Present and future*, eds. S. Brunn and J. Wheeler, 201–11. New York: Wiley.

Black, J., and Blunden, W. 1984. *The land-use/transport system*. 2nd ed. New York: Pergamon.

Blackley, P., and Greytak, D. 1986. Comparative advantage and industrial location: An intrametropolitan evaluation. *Urban Studies* 23:221–30.

Blalock, H., Jr. 1964. *Causal inferences in nonexperimental research*. Chapel Hill: University of North Carolina Press.

————. 1979. *Social statistics*, rev. 2nd ed. New York: McGraw-Hill.

Boal, F. 1976. Ethnic residential segregation. In *Social areas in cities*, Vol. 1: Spatial processes and forms, eds. D. Herbert and R. Johnston, 41–79. New York: Wiley.

Bondi, L. 1990. Feminism, postmodernism, and geography: Space for women? *Antipode* 22:156–67.

————. 1991. Gender divisions and gentrification: A critique. *Transactions of the Institute of British Geographers*, n.s., 16:190–8.

Bondi, L., and Domosh, M. 1992. Other figures in other places: On feminism, postmodernism and geography. *Environment and Planning D: Society and Space* 10:199–213.

Boomsma, A. 1982. The robustness of LISREL against small sample sizes in factor analysis models. In *Systems under indirect observation*, pt. 1, eds. K. Jöreskog and H. Wold, 149–73. Amsterdam: North-Holland.

Boots, B., and Getis, A. 1988. *Point pattern analysis*. Beverly Hills, Calif.: Sage.

Borchert, J. 1983. Instability in American metropolitan growth. *Geographical Review* 73:127–49.

Borgers, A., and Timmermans, H. 1988. A context-sensitive model of spatial choice behavior. In *Behavioral modelling in geography and planning*, eds. R. Golledge and H. Timmermans, 159–78. London: Croom Helm.

Bourne, L. 1981. *The geography of housing*. New York: Wiley.

––––––. 1989. Are new urban forms emerging? Empirical tests for Canadian urban areas. *Canadian Geographer* 33:312–28.

––––––. 1993. The demise of gentrification? A commentary and prospective view. *Urban Geography* 14:95–107.

Bovy, P., and Stern, E. 1990. *Route choice: Wayfinding in transport networks*. Dordrecht: Kluwer.

Bradbury, J. 1985. Regional and industrial restructuring processes in the new international division of labor. *Progress in Human Geography* 9:38–63.

Branch, M. 1985. *Comprehensive city planning: Introduction and explanation*. Chicago: American Planning Association.

Brigham, E. 1965. The determinants of residential land values. *Land Economics* 41:325–34.

Browett, J. 1984. On the necessity and inevitability of uneven development under capitalism. *International Journal of Urban and Regional Research* 8:155–77.

Brown, K. 1981. Race, class and culture: Towards a theorization of the "choice/constraint" concept. In *Social interaction and ethnic segregation*, eds. P. Jackson and S. Smith, 185–203. Institute of British Geographers Special Publication no. 12. London: Academic Press.

Brown, S. 1992. The wheel of retail gravitation? *Environment and Planning A* 24:1409–29.

Browne, M. 1982. Covariance structures. In *Topics in applied multivariate analysis*, ed. D. Hawkins, 72–141. Cambridge: Cambridge University Press.

Brummell, A. 1981. A method of measuring residential stress. *Geographical Analysis* 13:248–61.

Buck, W. 1980. The industrial park. In *Planning industrial development*, ed. D. Walker, 291–321. New York: Wiley.

Bunge, M. 1959. *Causality*. Cambridge, Mass.: Harvard University Press.

Bunting, T., and Guelke, L. 1979. Behavioral and perception geography: A critical appraisal. *Annals of the Association of American Geographers* 69:448–62.

Burgess, E. 1925. The growth of the city. In *The city*, eds. R. Park, E. Burgess, and R. Mackenzie, 47–62. Chicago: University of Chicago Press.

Burnell, J., and Galster, G. 1992. Quality-of-life measurements and urban size: An empirical note. *Urban Studies* 29:727–35.

Burns, L. 1982. Metropolitan growth in transition. *Journal of Urban Economics* 11:112–29.

Burroughs, W., and Sadalla, E. 1979. Asymmetries in distance cognition. *Geographical Analysis* 11:414–21.

Cadwallader, M. 1973. A methodological examination of cognitive distance. In *Environmental design research*, Vol. 2, ed. W. Preiser, 193–9. Stroudsburg, Pa.: Dowden, Hutchinson and Ross.

––––––. 1975. A behavioral model of consumer spatial decision making. *Economic Geography* 51:339–49.

––––––. 1976. Cognitive distance in intraurban space. In *Environmental knowing: Theories, research, and methods*, eds. G. Moore and R. Golledge, 316–24. Stroudsburg, Pa.: Dowden, Hutchinson, and Ross.

———. 1977. Frame dependency in cognitive maps: An analysis using directional statistics. *Geographical Analysis* 9:284–92.

———. 1978a. The model-building process in introductory college geography: An illustrative example. *Journal of Geography* 77:100–2.

———. 1978b. Urban information and preference surfaces: Their patterns, structures, and interrelationships. *Geografiska Annaler* 60B:97–106.

———. 1979a. Problems in cognitive distance: Implications for cognitive mapping. *Environment and Behavior* 11:559–76.

———. 1979b. Neighborhood evaluation in residential mobility. *Environment and Planning A* 11:393–401.

———. 1979c. The process of neighborhood choice. Paper presented at the International Conference on Environmental Psychology, University of Surrey, Guildford, Eng.

———. 1981a. A unified model of urban housing patterns, social patterns, and residential mobility. *Urban Geography* 2:115–30.

———. 1981b. Towards a cognitive gravity model: The case of consumer spatial behavior. *Regional Studies* 15:275–84.

———. 1982. Urban residential mobility: A simultaneous equations approach. *Transactions of the Institute of British Geographers*, n.s., 7:458–73.

———. 1985. Structural equation models of migration: An example from the Upper Midwest USA. *Environment and Planning A* 17:101–13.

———. 1986a. Structural equation models in human geography. *Progress in Human Geography* 10:24–47.

———. 1986b. Migration and intra-urban mobility. In *Population geography: Progress and prospect*, ed. M. Pacione, 257–83. London: Croom Helm.

———. 1987. Linear structural relationships with latent variables: The LISREL model. *Professional Geographer* 39:317–26.

———. 1988. Urban geography and social theory. *Urban Geography* 9:227–51.

———. 1989a. A conceptual framework for analysing migration behavior in the developed world. *Progress in Human Geography* 13:494–511.

———. 1989b. A synthesis of macro and micro approaches to explaining migration: Evidence from inter-state migration in the United States. *Geografiska Annaler* 71B:85–94.

———. 1991. Metropolitan growth and decline in the United States: An empirical analysis. *Growth and Change* 22:1–16.

———. 1992a. Log-linear models of residential choice. *Area* 24:289–94.

———. 1992b. *Migration and residential mobility: Macro and micro approaches*. Madison, Wisc.: University of Wisconsin Press.

———. 1993. Inter-metropolitan housing value differentials: The United States, 1960–1980. *Geoforum* 24:307–13.

Can, A. 1990. The measurement of neighborhood dynamics in urban house prices. *Economic Geography* 66:254–72.

———. 1992. Specification and estimation of hedonic housing price models. *Regional Science and Urban Economics* 22:453–74.

Canter, D. 1977. *The psychology of place*. New York: St. Martin's Press.

Canter, D. , and Tagg, S. 1975. Distance estimation in cities. *Environment and Behavior* 7:59–80.

Carmines, E. 1986. The analysis of covariance structure models. In *New tools for social scientists: Advances and applications in research methods*, eds. W. Berry and M. Lewis-Beck, 23–55. Beverly Hills, Calif.: Sage.

Carter, H. 1981. *The study of urban geography*, 3rd ed. London: Edward Arnold.

Castells, M. 1977. *The urban question: A Marxist approach*. London: Edward Arnold.

———. 1983. *The city and the grassroots: A cross-cultural theory of urban social movements*. London: Edward Arnold.

Cebula, R. 1980a. Geographic mobility and the cost of living: An explanatory note. *Urban Studies* 17:353–5.

———. 1980b. Voting with one's feet: A critique of the evidence. *Regional Science and Urban Economics* 10:91–107.

———. 1983. *Geographic living-cost differentials*. Lexington,Mass.: Lexington Books.

Chalmers, J., and Greenwood, M. 1985. The regional labor market adjustment process: Determinants of changes in rates of labor force participation, unemployment, and migration. *Annals of Regional Science* 19:1–17.

Champion, A., and Townsend, A. 1990. *Contemporary Britain: A geographical perspective*. London: Edward Arnold.

Champion, A., Green, A., and Owen, D. 1988. House prices and local labour market performance: An analysis of Building Society data for 1985. *Area* 20:253–63.

Chapman, K., and Walker, D. 1991. *Industrial location: Principles and Policies*, 2nd ed. Oxford: Blackwell.

Charney, A. 1983. Intraurban manufacturing location decisions and local tax differentials. *Journal of Urban Economics* 14:184–205.

———. 1993. Migration and the public sector: A survey. *Regional Studies* 27:313–26.

Chatfield, C. 1984. *The analysis of time series: An introduction*, 3rd ed. London: Chapman and Hall.

Chisolm, M. 1979. *Rural settlement and land use: An essay in location*, 3rd ed. London: Hutchinson.

Chouinard, V., Fincher, R., and Webber, M. 1984. Empirical research in scientific human geography. *Progress in Human Geography* 8:347–80.

Church, A., and Hall, J. 1989. Local initiatives for economic regeneration. In *Social problems and the city: New perspectives*, eds., D. Herbert and D. Smith, 345–69. Oxford: Oxford University Press.

Clark, C. 1951. Urban population densities. *Journal of the Royal Statistical Society*, Series A, 114:490–96.

Clark, G. 1982a. Dynamics of interstate labor migration. *Annals of the Association of American Geographers* 72:297–313.

———. 1982b. Volatility in the geographical structure of U.S. interstate migration. *Environment and Planning A* 14:145–67.

———. 1983. *Interregional migration, national policy, and social justice*. Totowa, N.J.: Rowman and Allanheld.

———. 1992. 'Real' regulation: The administrative state. *Environment and Planning A* 24:615–27.

Clark, G., and Dear, M. 1981. The state in capitalism and the capitalist state. In *Urbanization and urban planning in capitalist society*, eds. M. Dear and A. Scott, 45–61. New York: Methuen.

———. 1984. *State apparatus: Structures and language of legitimacy*. Boston: Allen and Unwin.

Clark, G., and Gertler, M. 1983. Migration and capital. *Annals of the Association of American Geographers* 73:18–34.

Clark, W. 1981. On modelling search behavior. In *Dynamic spatial models*, eds. D. Griffith and R. Mackinnon, 102–31. Alphen aan den Rijn, Neth.: Sijthoff en Noordhoff.

————. 1982a. Recent research on migration and mobility: A review and interpretation. *Progress in Planning* 18:1–56.

————. 1982b. A revealed preference analysis of intraurban migration choices. In *Proximity and preference: Problems in the multidimensional analysis of large data sets*, eds. R. Golledge and J. Rayner, 144–68. Minneapolis: University of Minnesota Press.

————, ed. 1982c. *Modelling housing market search.* London: Croom Helm.

————. 1986a. Residential segregation in American cities: A review and interpretation. *Population Research and Policy Review* 5:95–127.

————. 1986b. *Human migration.* Beverly Hills, Calif.: Sage.

————. 1992a. Residential preferences and residential choices in a multiethnic context. *Demography* 29:451–66.

————. 1992b. Comparing cross-sectional and longitudinal analyses of residential mobility and migration. *Environment and Planning A* 24:1291–1302.

Clark, W., and Burt, J. 1980. The impact of workplace on residential relocation. *Annals of the Association of American Geographers* 70:59–67.

Clark, W., and Cadwallader, M. 1973a. Residential preferences: An alternate view of intraurban space. *Environment and Planning* 5:693–703.

————. 1973b. Locational stress and residential mobility. *Environment and Behavior* 5:29–41.

Clark, W., Deurloo, M., and Dieleman, F. 1984. Housing consumption and residential mobility. *Annals of the Association of American Geographers* 74:29–43.

————. 1986. Residential mobility in Dutch housing markets. *Environment and Planning A* 18:763–88.

————. 1988. Modeling strategies for categorical data: Examples from housing and tenure choice. *Geographical Analysis* 20:198–219.

Clark, W., and Onaka, J. 1983. Life cycle and housing adjustment as explanations of residential mobility. *Urban Studies* 20:47–57.

Clark, W., and Rushton, G. 1970. Models of intra-urban consumer behavior and their implications for central place theory. *Economic Geography* 46:486–97.

Clark, W., and Smith, T. 1982. Housing market search behavior and expected utility theory: 2. The process of search. *Environment and Planning A* 14:717–37.

Clarke, M., and Wilson, A. 1985. The dynamics of urban spatial structure: The progress of a research programme. *Transactions of the Institute of British Geographers*, n.s., 10:427–51.

Clayton, C. 1977a. Interstate population migration process and structure in the United States, 1935 to 1970. *Professional Geographer* 29:177–81.

————. 1977b. The structure of interstate and interregional migration, 1965–1970. *Annals of Regional Science* 11:109–22.

Converse, P. 1949. New laws of retail gravitation. *Journal of Marketing* 14:379–84.

Cooke, P. 1990. Modern urban theory in question. *Transactions of the Institute of British Geographers*, n.s., 15:331–43.

Coshall, J. 1985. Urban consumers' cognition of distance. *Geografiska Annaler* 67B:107–19.

Costner, H. 1985. Theory, deduction, and rules of correspondence. In *Causal models in the social sciences*, 2nd ed., ed. H. Blalock, Jr., 229–50. New York: Aldine.

Couclelis, H., and Golledge, R. 1983. Analytic research, positivism, and behavioral geography. *Annals of the Association of American Geographers* 73:331–9.

Coulson, N. 1991. Really useful tests of the monocentric model. *Land Economics* 67:299–307.

Cox, K. 1981. Bourgeois thought and the behavioral geography debate. In *Behavioral problems in geography revisited*, eds. K. Cox and R. Golledge, 256–79. New York: Methuen.

Cox, K., and Jonas, A. 1993. Urban development, collective consumption and the politics of metropolitan fragmentation. *Political Geography* 12:8–37.

Cox, K., and Nartowicz, F. 1980. Jurisdictional fragmentation in the American metropolis: Alternative perspectives. *International Journal of Urban and Regional Research* 4:196–209.

Cromartie, J., and Stack, C. 1989. Reinterpretation of black return and non-return to the South, 1975–1980. *Geographical Review* 79:297–310.

Cushing, B. 1987a. Location-specific amenities, topography, and population migration. *Annals of Regional Science* 21:74–85.

———. 1987b. A note on specification of climate variables in models of population migration. *Journal of Regional Science* 27:641–8.

Daniels, P., and Warnes, A. 1980. *Movement in cities: Spatial perspectives on urban transport and travel.* New York: Methuen.

Darden, J. 1989. Blacks and other racial minorities: The significance of colour in inequality. *Urban Geography* 10:562–77.

DaVanzo, J. 1978. Does unemployment affect migration? Evidence from micro data. *Review of Economic Statistics* 60:504–14.

DaVanzo, J., and Morrison, P. 1981. Return and other sequences of migration in the United States. *Demography* 18:85–101.

Davies, R. 1991. The analysis of housing and migration careers. In *Migration models: Macro and micro approaches*, eds. J. Stillwell and P. Congdon, 207–27. London: Belhaven Press.

Davies, R., and Pickles, A. 1985. A panel study of life-cycle effects in residential mobility. *Geographical Analysis* 17:199–216.

Davies, W. 1984. *Factorial ecology.* Aldershot, Eng.: Gower.

Davies, W., and Herbert, D. 1993. *Communities within cities: An urban social geography.* London: Belhaven Press.

Davies, W., and Murdie, R. 1991. Consistency and differential impact in urban social dimensionality: Intra-urban variations in the 24 metropolitan areas of Canada. *Urban Geography* 12:55–79.

Day, R. 1976. Urban distance cognition: Review and contribution. *Australian Geographer* 13:193–200.

Day, R., and Walmsley, D. 1981. Residential preferences in Sydney's inner suburbs: A study in diversity. *Applied Geography* 1:185–97.

Dear, M. 1988. The postmodern challenge: Reconstructing human geography. *Transactions of the Institute of British Geographers*, n.s., 13:262–74.

Dear, M., and Gleeson, B. 1991. Community attitudes toward the homeless. *Urban Geography* 12:155–76.

Dear, M., and Moos, A. 1986. Structuration theory in urban analysis: 2. Empirical application. *Environment and Planning A* 18:351–73.

Dear, M., and Wittman, I. 1980. Conflict over the location of mental health facilities. In *Geography and the urban environment*, Vol. 3, eds. D. Herbert and R. Johnston, 345–62. New York: Wiley.

Dear, M., and Wolch, J. 1987. *Landscapes of despair: From deinstitutionalisation to homelessness.* Princeton, N.J.: Princeton University Press.

Dear, M., Taylor, S., and Hall, G. 1980. External effects of mental health facilities. *Annals of the Association of American Geographers* 70:342–52.

Desbarats, J. 1983a. Spatial choice and constraints on behavior. *Annals of the Association of American Geographers* 73:340–57.

———. 1983b. Constrained choice and migration. *Geografiska Annaler* 65B:11–22.

Deurloo, M., Dieleman, F., and Clark, W. 1988. Generalized log-linear models of housing choice. *Environment and Planning A* 20:55–69.

De Vise, P. 1976. The suburbanization of jobs and minority employment. *Economic Geography* 52:348–63.

Dewhurst, J., Hewings, G., and Jensen, R., eds. 1991. *Regional input–output modelling: New developments and interpretations.* Brookfield, Vt.: Avebury Press.

Dierx, A. 1988. A life-cycle model of repeat migration. *Regional Science and Urban Economics* 18:383–97.

Doling, J., and Williams, P. 1983. Building societies and local lending behavior. *Environment and Planning A* 15:663–73.

Doorn, P., and Van Rietbergen, A. 1990. Lifetime mobility: Interrelationships of labor mobility, residential mobility and household cycle. *Canadian Geographer* 34:33–48.

Downs, R. 1981. Cognitive mapping: A thematic analysis. In *Behavioral problems in geography revisited,* eds. K. Cox and R. Golledge, 95–122. New York: Methuen.

Duncan, J. 1987. Review of urban imagery: Cognitive mapping. *Urban Geography* 8:264–72.

———. 1988. Commentary on Martin Cadwallader's urban geography and social theory. *Urban Geography* 9:265–8.

Duncan, J., and Ley, D. 1982. Stuctural Marxism and human geography: A critical assessment. *Annals of the Association of American Geographers* 72:30–59.

Duncan, O. 1975. *Introduction to structural equation models.* New York: Academic Press.

Dunn, R., Reader, S., and Wrigley, N. 1987. A non-parametric approach to the incorporation of heterogeneity into repeated polytomous choice models of urban shopping behavior. *Transportation Research A* 21:327–45.

Dwyer, J. 1983. *Statistical models for the social and behavioral sciences.* New York: Oxford University Press.

Eagle, T. 1988a. Context effects in consumer spatial behavior. In *Behavioral modelling in geography and planning,* eds. R. Golledge and H. Timmermans, 299–324. London: Croom Helm.

———. 1988b. Modelling context effects: Application of a weight shifting model in a spatial context. *Tijdschrift voor Economische en Sociale Geografie* 79:135–47.

Edel, M. 1981. Capitalism, accumulation and the explanation of urban phenomena. In *Urbanization and urban planning in capitalist society,* eds. M. Dear and A. Scott, 19–44. New York: Methuen.

Edmonston, B., Goldberg, M., and Mercer, J. 1985. Urban form in Canada and the United States: An examination of urban density gradients. *Urban Studies* 22:209–17.

Eldridge, J., and Jones, J. 1991. Warped space: A geography of distance decay. *Professional Geographer* 43:500–11.

Ellis, M., and Mackay, D. 1990. Modelling individual residential preferences using vector and ideal point logit models. *Tijdschrift voor Economische en Sociale Geographie* 81:123–32.

Ellis, M., Barff, R., and Markusen, A. 1993. Defense spending and interregional labor migration. *Economic Geography* 69:182–203.

Engelhardt, G., and Poterba, J. 1991. House prices and demographic change: Canadian evidence. *Regional Science and Urban Economics* 21:539–46.

England, K. 1991. Gender relations and the spatial structure of the city. *Geoforum* 22:135–47.

———. 1993. Suburban pink collar ghettos: The spatial entrapment of women? *Annals of the Association of American Geographers* 83:225–42.

Erickson, R. 1983. The evolution of the urban space economy. *Urban Geography* 4:95–121.

———. 1986. Multinucleation in metropolitan economies. *Annals of the Association of American Geographers* 76:331–46.

———. 1992. Enterprize zones: Lessons from the state government experience. In *Sources of metropolitan growth*, eds. E. Mills and J. McDonald, 161–82. New Brunswick, N.J.: Center for Urban Policy Research, Rutgers University.

Evans, A. 1983. The determination of the price of land. *Urban Studies* 20:119–29.

Everitt, B. 1984. *An introduction to latent variable models*. New York: Chapman and Hall.

Everitt, J. 1976. Community and propinquity in a city. *Annals of the Association of American Geographers* 66:104–16.

Everitt, J., and Cadwallader, M. 1981. Husband–wife role variation as a factor in home area definition. *Geografiska Annaler* 63:23–34.

Ewing, G. 1981. On the sensitivity of conclusions about the bases of cognitive distance. *Professional Geographer* 33:311–4.

Eyles, J. 1979. Area-based policies for the inner city: Context, problems, and prospects. In *Social problems and the city: Geographical perspectives*, eds. D. Herbert and D. Smith, 226–43. New York: Oxford University Press.

Fienberg, S. 1980. *The analysis of cross-classified categorical data*, 2nd ed. Cambridge, Mass.: M.I.T. Press.

Fincher, R. 1987. Social theory and the future of urban geography. *Professional Geographer* 39:9–12.

———. 1990. Women in the city: Feminist analyses of urban geography. *Australian Geographical Studies* 28:29–37.

Fingleton, B. 1981. Log-linear modelling of geographical contingency tables. *Environment and Planning A* 13:1539–51.

Fischer, M., and Aufhauser, E. 1988. Housing choice in a regulated market: A nested multinomial logit analysis. *Geographical Analysis* 20:47–69.

Fisher, G. 1988. Problems in the use and interpretation of product variables. In *Common problems / proper solutions: Avoiding error in quantitative research*, ed. J. Long, 84–107. Beverly Hills, Calif.: Sage.

Flowerdew, R. 1989. Some critical views of modelling in geography. In *Remodelling Geography*, ed. B. Macmillan, 245–52. Oxford: Blackwell.

Flowerdew, R., and Amrhein, C. 1989. Poisson regression models of Canadian census division migration flows. *Papers of the Regional Science Association* 67:89–102.

Flowerdew, R., and Salt, J. 1979. Migration between labour market areas in Great Britain, 1970–1971. *Regional Studies* 13:211–31.

Follain, J., and Jimenez, E. 1985. Estimating the demand for housing characteristics: A survey and critique. *Regional Science and Urban Economics* 15:77–107.

Forrest, D. 1991. An analysis of house price differentials between English regions. *Regional Studies* 25:231–8.

Fortura, P., and Kushner, J. 1986. Canadian inter-city house price differentials. *Journal of the American Real Estate and Urban Economics Association* 14:525–36.

Fotheringham, S. 1981. Spatial structure and distance-decay parameters. *Annals of the Association of American Geographers* 71:425–36.

———. 1983. A new set of spatial interaction models: The theory of competing destinations. *Environment and Planning A* 15:15–36.

———. 1988. Market share analysis techniques: A review and illustration of current U.S. practice. In *Store choice, store location and market analysis*, ed. N. Wrigley, 120–59. London: Routledge.

———. 1991. Migration and spatial structure: The development of the competing destinations model. In *Migration models: Macro and micro approaches*, eds. J. Stillwell and P. Congdon, 57–72. London: Belhaven Press.

Fotheringham, S., and O'Kelly, M. 1989. *Spatial interaction models: Formulations and applications*. Boston: Kluwer.

Frey, D. 1989. A structural approach to the economic base multiplier. *Land Economics* 65:352–8.

Frey, W., and Speare, A., Jr. 1988. *Regional and metropolitan growth and decline in the United States*. New York: Russell Sage Foundation.

Fuguitt, G. 1985. The nonmetropolitan population turnaround. *Annual Review of Sociology* 11:159–280.

Gaile, G., and Hanink, D. 1985. Relative stability in American metropolitan growth. *Geographical Analysis* 17:341–8.

Gale, N., Golledge, R., Halperin, W., and Couclelis, H. 1990. Exploring spatial familiarity. *Professional Geographer* 42:299–313.

Galster, G. 1988. Residential segregation in American cities: A contrary review. *Population Research and Policy Review* 7:93–112.

Gertler, M. 1984. Regional capital theory. *Progress in Human Geography* 8:50–81.

———. 1992. Flexibility revisited: Districts, nation-states, and the forces of production. *Transactions of the Institute of British Geographers*, n.s., 17:259–78.

Gescheider, G. 1988. Psychophysical scaling. *Annual Review of Psychology* 39:169–200.

Giddens, A. 1981. *A contemporary critique of historical materialism*. London: Macmillan.

———. 1984. *The constitution of society: Outline of the theory of structuration*. Cambridge, Eng.: Polity Press.

———. 1985. Time, space and regionalization. In *Social relations and spatial structures*, eds. D. Gregory and J. Urry, 265–95. London: Macmillan.

Gober, P., McHugh, K., and Reid, N. 1991. Phoenix in flux: Household instability, residential mobility, and neighborhood change. *Annals of the Association of American Geographers* 81:80–8.

Goldberg, M., and Horwood, P. 1980. *Zoning: Its costs and relevance for the 1980s*. Vancouver, British Columbia: The Fraser Institute.

Goldberg, M., and Mercer, J. 1986. *The myth of the North American city*. Vancouver: University of British Columbia Press.

Golledge, R. 1978. Learning about urban environments. In *Making sense of time*, eds. T. Carlstein, D. Parkes, and N. Thrift, 76–98. London: Edward Arnold.

———. 1980. A behavioral view of mobility and migration research. *Professional Geographer* 32:14–21.

———. 1981a. Misconceptions, misinterpretations, and misrepresentations of behavioral approaches in human geography. *Environment and Planning A* 13:1325–44.

———. 1981b. The geographical relevance of some learning theories. In *Behavioral problems in geography revisited*, eds. K. Cox and R. Golledge, 43–66. New York: Methuen.

———. 1990. The conceptual and empirical basis of a general theory of spatial knowledge. In *Spatial choices and processes*, eds. M. Fischer, P. Nijkamp, and Y. Papageorgiou, 147–68. Amsterdam: North-Holland.

Golledge, R., and Hubert, L. 1982. Some comments on non-Euclidean mental maps. *Environment and Planning A* 14:107–18.

Golledge, R., and Rushton, G. 1984. A review of analytic behavioral research in geography. In *Geography and the urban environment: Progress in research and applications*, vol. 6, eds. D. Herbert and R. Johnston, 1–43. New York: Wiley.

Golledge, R., and Stimson, R. 1987. *Analytical behavioral geography*. London: Croom Helm.

Golledge, R., Briggs, R., and Demko, D. 1969. The configuration of distances in intra-urban space. *Proceedings of the Association of American Geographers* 1:60–5.

Goodman, J., Jr. 1982. Linking local mobility rates to migration rates: Repeat movers and place effects. In *Modelling housing market search*, ed. W. Clark, 209–23. London: Croom Helm.

Goodman, L. 1979. A brief guide to the causal analysis of data from surveys. *American Journal of Sociology* 84: 1078–95.

Gordon, P., Richardson, H., and Wong, H. 1986. The distribution of population and employment in a polycentric city: The case of Los Angeles. *Environment and Planning A* 18:161–73.

Goss, E., and Chang, H. 1983. Changes in elasticities of interstate migration: Implications of alternative functional forms. *Journal of Regional Science* 23:223–32.

Gould, P. 1975. Acquiring spatial information. *Economic Geography* 51:87–99.

———, and White, R. 1986. *Mental maps*, 2nd ed. Boston: Allen and Unwin.

Graves, P. 1980. Migration and climate. *Journal of Regional Science* 20:227–37.

Greenwood, M. 1981. *Migration and economic growth in the United States: National, regional, and metropolitan perspectives*. New York: Academic Press.

Greenwood, M., and Hunt, G. 1989. Jobs versus amenities in the analysis of metropolitan migration. *Journal of Urban Economics* 25:1–16.

Gregory, D. 1981. Human agency and human geography. *Transactions of the Institute of British Geographers*, n.s., 6:1–18.

Gregson, N. 1986. On duality and dualism: The case of stucturation and time geography. *Progress in Human Geography* 10:184–205.

———. 1987. Stucturation theory: Some thoughts on the possibility for empirical research. *Environment and Planning D: Society and Space* 5:73–91.

Griffith, D., and Amrhein, C. 1991. *Statistical analysis for geographers*. Englewood Cliffs, N.J.: Prentice Hall.

Guest, A. 1972. Patterns of family location. *Demography* 9:159–72.

Haggett, P. 1983. *Geography: A modern synthesis*, rev. 3rd ed. New York: Harper & Row.

———. 1990. *The geographer's art*. Oxford: Blackwell.

Haggett, P., and Bassett, K. 1970. The use of trend-surface parameters in inter-urban comparisons. *Environment and Planning* 2:225–37.

Haggett, P., and Chorley, R. 1967. Models, paradigms, and the new geography. In *Models in geography*, eds. R. Chorley and P. Haggett, 19–41. London: Methuen.

Haggett, P., Cliff, A., and Frey, A. 1977. *Locational analysis in human geography*, 2nd ed. London: Edward Arnold.

Hall, P. 1979. Planning: A geographer's view. In *Resources and planning*, eds. B. Goodall and A. Kirby, 3–15. New York: Pergamon Press.

———. 1992. *Urban and regional planning*, 3rd ed. New York: Routledge.

Halperin, W. 1988. Current topics in behavioral modelling of consumer choice. In *Behavioral modelling in geography and planning*, eds. R. Golledge and H. Timmermans, 1–26. London: Croom Helm.

Hamnett, C. 1979. Area-based explanations: A critical appraisal. In *Social problems and the city: Geographical perspectives*, eds. D.Herbert and D. Smith, 244–60. New York: Oxford University Press.

———. 1984. Gentrification and residential location theory: A review and assessment. In *Geography and the urban environment: Progress in research and applications*, Vol. 6, eds. D. Herbert and R. Johnston, 283–319. New York: Wiley.

———. 1991. The blind men and the elephant: The explanation of gentrification. *Transactions of the Institute of British Geographers*, n.s., 16:173–89.

Harkman, A. 1989. Migration behavior among the unemployed and the role of unemployment benefits. *Papers of the Regional Science Association* 66:143–50.

Hartshorn, T. 1992. *Interpreting the city*, 2nd ed. New York: Wiley.

Harvey, D. 1973. *Social justice and the city*. London: Edward Arnold.

———. 1975. Class structure in a capitalist society and the theory of residential differentiation. In *Processes in physical and human geography*, eds. R. Peel, M. Chisholm, and P. Haggett, 354–69. London: Heinemann.

———. 1976. The Marxian theory of the state. *Antipode* 8:80–9.

———. 1978. The urban process under capitalism: A framework for analysis. *International Journal of Urban and Regional Research* 2:101–31.

———. 1981. Conceptual and measurement problems in the cognitive–behavioral approach to location theory. In *Behavioral problems in geography revisited*, eds. K. Cox and R. Golledge, 18–42. New York: Methuen.

———. 1982. *The limits to capital*. Chicago: University of Chicago Press.

———. 1985. *The urbanization of capital: Studies in the history and theory of capitalist urbanization*. Baltimore: Johns Hopkins University Press.

———. 1989a. *Condition of postmodernity: An enquiry into the origins of cultural change*. Oxford: Blackwell.

———. 1989b. From models to Marx: Notes on the project to "remodel" contemporary geography. In *Remodelling geography*, ed. B. Macmillan, 211–6. Oxford: Blackwell.

———. 1991. The urban face of capitalism. In *Our changing cities*, ed. J. Hart, 51–66. Baltimore: Johns Hopkins University Press.

Harvey, D., and Scott, A. 1989. The practice of human geography: Theory and empirical specificity in the transition from Fordism to flexible accumulation. In *Remodelling geography*, ed. B. Macmillan, 217–29. Oxford: Blackwell.

Harvey, R., and Clark, W. 1971. The nature and economics of urban sprawl. In *Internal structure of the city: Readings on space and environment*, ed. L. Bourne, 475–82. New York: Oxford University Press.

Hay, A. 1979. Positivism in human geography: Response to critics. In *Geography and the urban environment*, vol. 2, eds. D. Herbert and R. Johnston, 1–26. New York: Wiley.

———. 1985. Scientific method in geography. In *The future of geography*, ed. R. Johnston, 129–42. New York: Methuen.

Haynes, K., and Fotheringham, A. 1984. *Gravity and spatial interaction models*. Beverly Hills, Calif.: Sage.

Healey, M., and Ilberry, B. 1990. *Location and change: Perspectives on economic geography*. New York: Oxford University Press.

Heertje, A., Rushing, F., and Skidmore, F. 1983. *Economics*. Chicago: Dryden Press.

Heikkila, E., Gordon, P., Kim, J., Peiser, R., Richardson, H., and Dale-Johnson, D. 1989. What happened to the CBD-distance gradient?: Land values in a policentric city. *Environment and Planning A* 21:221–32.

Herbert, D. 1993. Neighborhood incivilities and the study of crime in place. *Area* 25:45–54.

Herbert, D. and Thomas, C. 1982. *Urban geography: A first approach*. New York: Wiley.

—–——. 1990. *Cities in space: City as place*. Savage, Md.: Barnes and Noble.

Herington, J. 1989. *Planning processes: An introduction for geographers*. New York: Cambridge University Press.

Herting, J. 1985. Multiple indicator models using LISREL. In *Causal models in the social sciences*, 2nd ed., ed. H. Blalock, Jr., 321–93. New York: Aldine.

Herting, J., and Costner, H. 1985. Respecification in multiple indicator models. In *Causal models in the social sciences*, 2nd ed., ed. H. Blalock, Jr., 321–93. New York: Aldine.

Herzog, H., and Schlottmann, A. 1984. Labor force mobility in the United States: Migration, unemployment, and remigration. *International Regional Science Review* 9:43–58.

Hewings, G. 1985. *Regional input–output analysis*. Beverly Hills, Calif.: Sage.

Hill, F. 1973. Spatio-temporal trends in urban population density: A trend surface analysis. In *The form of cities in central Canada: Selected papers*, eds. L. Bourne, R. MacKinnon, and J. Simmons, 103–19. Department of Geography Research Publication no. 11, University of Toronto.

Ho, L. 1992. Rent control: Its rationale and effects. *Urban Studies* 29:1183–90.

Hoch, C. 1991. The spatial organization of the urban homeless: A case study of Chicago. *Urban Geography* 12:137–54.

Holland, D., and Cooke, S. 1992. Sources of structural change in the Washington economy: An input–output perspective. *Annals of Regional Science* 26:155–70.

Holmes, J. 1987. The urban system. In *Australia—A geography*, Vol.2, Space and Society, ed. D. Jeans, 49–74. Sydney: Sydney University Press.

Holzer, H. 1991. The spatial mismatch hypothesis: What has the evidence shown? *Urban Studies* 28:105–22.

Honey, R. 1976. Metropolitan governance. In *Urban policymaking and metropolitan dynamics: A comparative geographical analysis*, ed. J. Adams, 425–62. Cambridge, Mass.: Ballinger.

Horowitz, J., and Louviere, J. 1990. The external validity of choice models based on laboratory choice experiments. In *Spatial choices and processes*, eds. M. Fischer, P. Nijkamp, and Y. Papageorgiou, 247–63. Amsterdam: North-Holland.

Hourihan, K. 1984. Residential satisfaction, neighborhood attributes, and personal characteristics: An exploratory path analysis in Cork, Ireland. *Environment and Planning A* 16:425–36.

Hoyt, H. 1939. *The structure and growth of residential neighborhoods in American cities*. Washington, D.C.: Federal Housing Administration.

Huff, D. 1964. Defining and estimating a trading area. *Journal of Marketing* 28:34–8.

Huff, J. 1982. Spatial aspects of residential search. In *Modelling housing market search*, ed. W. Clark, 106–29. London: Croom Helm.

—–——. 1984. Distance-decay models of residential search. In *Spatial statistics and models*, eds. G. Gaile and C. Willmott, 345–66. Dordrecht, Neth.: D. Reidel.

———. 1986. Geographic regularities in residential search behavior. *Annals of the Association of American Geographers* 76:208–27.

Humphreys, J. 1990. Place learning and spatial cognition: A longitudinal study of urban newcomers. *Tijdschrift voor Economische en Sociale Geografie* 81:364–80.

Huxley, M., and Winchester, H. 1991. Residential differentiation and social reproduction: The interrelations of class, gender, and space. *Environment and Planning D: Society and Space* 9:233–40.

Intriligator, M. 1978. *Econometric models, techniques, and applications.* Englewood Cliffs, N.J.: Prentice Hall.

Isserman, A. 1985. Economic-demographic modeling with endogenously determined birth and migration rates: Theory and prospects. *Environment and Planning A* 17:25–45.

Jackson, E., and Johnson, D. 1991. Geographic implications of mega-malls, with special reference to West Edmonton Mall. *Canadian Geographer* 35:226–32.

Jackson, P., and Smith, S., eds. 1981. Social interaction and ethnic segregation. *Institute of British Geographers,* Special Publication no. 12. London: Academic Press.

Jessop, B. 1990. *State theory: Putting the capitalist state in its place.* Cambridge, Eng.: Polity Press.

Johnson, M. 1991. An empirical update on the product-cycle explanation and branch-plant location in the nonmetropolitan U.S. South. *Environment and Planning A* 23:397–410.

Johnston, R. 1971. *Urban residential patterns: An introductory review.* New York: Praeger.

———. 1973. Spatial patterns in suburban evaluations. *Environment and Planning* 5:385–95.

———. 1980. Political geography without politics. *Progress in Human Geography* 4:439–46.

———. 1982a. *Geography and the state: An essay in political geography.* New York: St. Martin's Press.

———. 1982b. *The American urban system: A geographical perspective.* New York: St. Martin's Press.

———. 1983. *Philosophy and human geography: An introduction to contemporary approaches.* London: Edward Arnold.

———. 1984. Marxist political economy, the state and political geography. *Progress in Human Geography* 8:473–92.

———. 1985. To the ends of the earth. In *The future of geography,* ed. R. Johnston, 326–38. New York: Methuen.

———. 1989. Philosophy, ideology and geography. In *Horizons in human geography,* eds. D. Gregory and R. Walford, 48–65. Totowa, N.J.: Barnes and Noble.

———. 1991. *A question of place: Exploring the practice of human geography.* Oxford: Blackwell.

Jones, C., and Maclennan, D. 1987. Building societies and credit rationing: An empirical examination of redlining. *Urban Studies* 24:205–16.

Jones, K. 1991. Specifying and estimating multi-level models for geographical research. *Transactions of the Institute of British Geographers,* n.s., 16:148–59.

Jones, K., and Simmons, J. 1990. *The retail environment.* New York: Routledge.

Jöreskog, K. 1982. The LISREL approach to causal model-building in the social sciences. In *Systems under indirect observation,* pt.1, eds. K. Jöreskog and H. Wold, 81–99. Amsterdam: North-Holland.

Jöreskog, K., and Sorbom, D. 1981. *LISREL: Analysis of structural relationships by the method of maximum likelihood.* Version 5. Chicago: National Educational Resources.

Jud, G., and Frew, J. 1986. Real estate brokers, housing prices, and the demand for housing. *Urban Studies* 23:21–31.

Kachigan, S. 1986. *Statistical analysis: An interdisciplinary introduction to univariate and multivariate methods.* New York: Radius Press.

Katz, D. 1982. *Econometric theory and applications*. Englewood Cliffs, N.J.: Prentice Hall.

Kenny, D. 1979. *Correlation and causality*. New York: Wiley.

King, L. 1984. *Central place theory*. Beverly Hills, Calif.: Sage.

King, R., ed. 1986. *Return migration and regional economic problems*. London: Croom Helm.

Kinzy, S. 1992. An analysis of the supply of housing characteristics by builders within the Rosen framework. *Journal of Urban Economics* 32:1–16.

Kirby, A. 1982. *The politics of location: An introduction*. London: Methuen.

Klecka, W. 1980. *Discriminant analysis*. Beverly Hills, Calif.: Sage.

Knapp, T., and Graves, P. 1989. On the role of amenities in models of migration and regional development. *Journal of Regional Science* 29:71–87.

Knox, P. 1982. *Urban social geography: An introduction*. New York: Longman.

———. 1987. *Urban social geography: An introduction*, 2nd ed. New York: Longman.

———. 1989. The vulnerable, the disadvantaged, and the victimized: Who they are and where they live. In *Social problems and the city: New perspectives*, eds. D. Herbert and D. Smith, 32–47. Oxford: Oxford University Press.

———. 1991. The restless urban landscape: Economic and sociocultural change and the transformation of metropolitan Washington D.C. *Annals of the Association of American Geographers* 81:181–209.

———, ed. 1993. *The restless urban landscape*. Englewood Cliffs, N.J.: Prentice Hall.

Knox, P., and MacLaran, A. 1978. Values and perceptions in descriptive approaches to urban social geography. In *Geography and the urban environment*, vol. 1, eds. D. Herbert and R. Johnston, 197–246. Chichester, Eng.: Wiley.

Kowalski, J., and Paraskevopoulos, C. 1990. The impact of location on urban industrial land prices. *Journal of Urban Economics* 27:16–24.

Krumm, R. 1983. Regional labor markets and the household migration decision. *Journal of Regional Science* 23:361–75.

LaGory, M., and Pipkin, J. 1981. *Urban social space*. Belmont, Calif.: Wadsworth.

Lake, R., ed. 1987. *Resolving locational conflict*. New Brunswick, N.J.: Center for Urban Policy Research, Rutgers University.

Landale, N., and Guest, A. 1985. Constraints, satisfaction and residential mobility: Speare's model reconsidered. *Demography* 22:199–222.

Latham, R., and Yeates, M. 1970. Population density growth in metropolitan Toronto. *Geographical Analysis* 2:177–85.

Law, R., and Wolch, J. 1991. Homelessness and economic restructuring. *Urban Geography* 12:105–36.

Le Bourdais, C., and Beaudry, M. 1988. The changing residential structure of Montreal 1971–81. *Canadian Geographer* 32:98–113.

Lee, T. 1970. Perceived distance as a function of direction in the city. *Environment and Behavior* 2:40–51.

Lee, Y., and Schmidt, C. 1988. Evolution of urban spatial cognition: Patterns of change in Guangzhou, China. *Environment and Planning A* 20:339–51.

Leonard, S. 1982. Urban managerialism: A period of transition? *Progress in Human Geography* 6:190–215.

Ley, D. 1974. *The Black inner city as frontier outpost: Images and behavior of a Philadelphia neighborhood*. Washington, D.C.: Association of American Geographers.

———. 1981. Behavioral geography and the philosophies of meaning. In *Behavioral problems in geography revisited*, eds. K. Cox and R. Golledge, 209–30. New York: Methuen.

————. 1983. *A social geography of the city*. New York: Harper & Row.

————. 1986. Alternative explanations for inner-city gentrification: A Canadian assessment. *Annals of the Association of American Geographers* 76:521–35.

————. 1987. Styles of the times: Liberal and neo-conservative landscapes in inner Vancouver, 1968–1986. *Journal of Historical Geography* 13:40–56.

————. 1993. Postmodernism, or the cultural logic of advanced intellectual capital. *Tijdschrift voor Economische en Sociale Geografie* 84:171–4.

Liaw, K. 1990. Joint effects of personal factors and ecological variables on the interprovincial migration pattern of young adults in Canada: A nested logit analysis. *Geographical Analysis* 22:189–208.

Linneman, P., and Graves, P. 1983. Migration and job change: A multinomial logit approach. *Journal of Urban Economics* 14:263–79.

Lloyd, R. 1989a. Cognitive maps: Encoding and decoding information. *Annals of the Association of American Geographers* 79:101–24.

————. 1989b. The estimation of distance and direction from cognitive maps. *American Cartographer* 16:109–22.

Lloyd, R., and Heivly, C. 1987. Systematic distortions in urban cognitive maps. *Annals of the Association of American Geographers* 77:191–207.

Lloyd, R., and Jennings, D. 1978. Shopping behavior and income: Comparisons in an urban environment. *Economic Geography* 54:157–67.

Lloyd, W. 1991. Changing suburban retail patterns in metropolitan Los Angeles. *Professional Geographer* 43:335–44.

Loehman, E., and Emerson, R. 1985. A simultaneous equation model of local government expenditure decisions. *Land Economics* 61:419–32.

Long, J. 1981. Estimation and hypothesis testing in linear models containing measurement error: A review of Jöreskog's model for analysis of covariance stuctures. In *Linear models in social research*, ed. P. Marsden, 209–56. Beverly Hills, Calif.: Sage.

————. 1983. *Covariance structure models: An introduction to LISREL*. Beverly Hills, Calif: Sage.

Long, L. 1988. *Migration and residential mobility in the United States*. New York: Russell Sage Foundation.

Louviere, J. 1982. Applications of functional measurement to problems in spatial decision making. In *Proximity and preference: Problems in the multidimensional analysis of large data sets*, eds. R. Golledge and J. Rayner, 191–213. Minneapolis: University of Minnesota Press.

Louviere, J., and Timmermans, H. 1990a. A review of recent advances in decompositional preference and choice models. *Tijdschrift voor Economische en Sociale Geografie* 81:214–24.

————. 1990b. Hierarchical information integration applied to residential choice behavior. *Geographical Analysis* 22:127–44.

Lundberg, U. 1973. Emotional and geographical phenomena in psychophysical research. In *Image and environment: Cognitive mapping and spatial behavior*, eds. R. Downs and D. Stea, 322–37. Chicago: Aldine.

Lynch, K. 1960. *The image of the city*. Cambridge, Mass.: M.I.T. Press.

Macgill, S. 1989. Modelling in human geography: Evaluating the quality and feeling the width. In *Remodelling geography*, ed. B. Macmillan, 108–14. Oxford: Blackwell.

MacLaran, A. 1981. Area-based positive discrimination and the distribution of well-being. *Transactions of the Institute of British Geographers*, n.s., 6:53–67.

Maclennan, D., and Williams, N. 1980. Revealed-preference theory and spatial choices: Some limitations. *Environment and Planning A* 12:909–19.

Macmillan, B. 1989. Quantitative theory construction in human geography. In *Remodelling geography*, ed. B. Macmillan, 89–107. Oxford: Blackwell.

Maier, G., and Weiss, P. 1991. The discrete choice approach to migration modelling. In *Migration models: Macro and micro approaches*, eds. J. Stillwell and P. Congdon, 17–33. London: Belhaven Press.

Marden, P. 1992. The deconstructionist tendencies of postmodern geographies: A compelling logic? *Progress in Human Geography* 16:41–57.

Mark, J. 1977. Determinants of urban house prices: A methodological comment. *Urban studies* 14:359–63.

Mark, J., and Goldberg, M. 1988. Multiple regression analysis and mass assessment: A review of the issues. *Appraisal Journal* 56:89–109.

Marsden, P. 1981. Conditional effects in regression models. In *Linear models in social research*, ed. P. Marsden, 97–116. Beverly Hills, Calif.: Sage.

Marshall, J. 1985. Geography as a scientific enterprise. In *The future of geography*, ed. R. Johnston, 113–28. New York: Methuen.

———. 1989. *The structure of urban systems.* Toronto: University of Toronto Press.

Marston, S., Towers, G., Cadwallader, M., and Kirby, A. 1989. The urban problematic. In *Geography in America*, eds. G. Gaile and C. Willmott, 651–72. Columbus, Ohio: Merrill.

Massey, D. 1985. New directions in space. In *Social relations and spatial structures*, eds. D. Gregory and J. Urry, 9–19. London: Macmillan.

Massey, D.S., and Denton, N. 1987. Trends in the residential segregation of blacks, Hispanics, and Asians: 1970–1980. *American Sociological Review* 52:802–25.

McDonald, J. 1989. Econometric studies of urban population density: A survey. *Journal of Urban Economics* 26:361–85.

McDonald, J., and McMillen, D. 1990. Employment subcenters and land values in a polycentric urban area: The case of Chicago. *Environment and Planning A* 22:1561–74.

McDowell, L. 1992. Doing gender: Feminism, feminists and research methods in human geography. *Transactions of the Institute of British Geographers*, n.s., 17:399–416.

McGibany, J. 1991. The effect of property tax rate differentials on single-family housing starts in Wisconsin, 1978–1989. *Journal of Regional Science* 31:347–59.

McHugh, K. 1984. Explaining migration intentions and destination selection. *Professional Geographer* 36:315–25.

———. 1987. Black migration reversal in the United States. *Geographical Review* 77:171–82.

———. 1988. Determinants of black interstate migration, 1965–1970 and 1970–1980. *Annals of Regional Science* 22:36–48.

McMillen, D., and McDonald, J. 1991. Urban land value functions with endogenous zoning. *Journal of Urban Economics* 29:14–27.

Mercer, J. 1975. Metropolitan housing quality and an application of causal modeling. *Geographical Analysis* 7:295–302.

Meyer, R. 1980. A descriptive model of constrained residential search. *Geographical Analysis* 12:21–32.

Meyer, R., and Cooper, L. 1988. A longitudinal choice analysis of consumer response to a product innovation. In *Behavioral modelling in geography and planning*, eds. R. Golledge and H. Timmermans, 424–50. London: Croom Helm.

Miliband, R. 1969. *The state in capitalist society*. London: Quartet.

———. 1977. *Marxism and politics*. London: Oxford University Press.

Mills, E. 1972. *Studies in the structure of the urban economy*. Baltimore: Johns Hopkins University Press.

———. 1992. The measurement and determinants of suburbanization. *Journal of Urban Economics* 32:377–87.

Milne, W. 1993. Macroeconomic influences on migration. *Regional Studies* 27:365–73.

Moore, E. 1986. Mobility intention and subsequent relocation. *Urban Geography* 7:497–514.

Moore, E., and Clark, W. 1986. Stable structure and local variation: A comparison of household flows in four metropolitan areas. *Urban Studies* 23:185–96.

Moore, L. 1988. Stated preference analysis and new store location. In *Store choice, store location and market analysis*, ed. N. Wrigley, 203–20. London: Routledge.

———. 1989. Modelling store choice: A segmented approach using stated preference analysis. *Transactions of the Institute of British Geographers*, n.s., 14:461–77.

Moos, A., and Dear, M. 1986. Structuration theory in urban analysis: 1. Theoretical exegesis. *Environment and Planning A* 18:231–52.

Morrill, R. 1987. The structure of shopping in a metropolis. *Urban Geography* 8:97–128.

———. 1988. Intra-metropolitan demographic structure: A Seattle example. *Annals of Regional Science* 22:1–16.

Mulligan, G., and Kim, H. 1991. Sectoral-level employment multipliers in small urban settlements: A comparison of five models. *Urban Geography* 12:240–59.

Murdie, R. 1986. Residential mortgage lending in metropolitan Toronto: A case study of the resale market. *Canadian Geographer* 30:98–110.

Muth, R. 1969. *Cities and housing: The spatial pattern of urban residential land use*. Chicago: University of Chicago Press.

———. 1985. Models of land-use, housing, and rent: An evaluation. *Journal of Regional Science* 25:593–606.

Myers, G., and Papageorgiou, Y. 1991. Homo economicus in perspective. *Canadian Geographer* 35:380–99.

Nakosteen, R., and Zimmer, M. 1980. Migration and income: The question of self-selection. *Southern Economic Journal* 46:840–51.

Namboodiri, N., Carter, L., and Blalock, H. Jr. 1975. *Applied multivariate analysis and experimental designs*. New York: McGraw-Hill.

Nellis, J., and Longbottom, J. 1981. An empirical analysis of the determination of house prices in the United Kingdom. *Urban Studies* 17:9–21.

Newling, B. 1969. The spatial variation of urban population densities. *Geographical Review* 59:242–52.

Nicholson, B., Brinkley, I., and Evans, A. 1981. The role of the inner city in the development of manufacturing industry. *Urban Studies* 18:57–71.

Northam, R. 1979. *Urban geography*, 2nd ed. New York: Wiley.

Noyelle, T., and Stanback, T. 1984. *The economic transformation of American cities*. Totowa, N.J.: Rowman and Allanheld.

Oates, W. 1969. The effects of property taxes and local public spending on property values: An empirical study of tax capitalization and the Tiebout hypothesis. *Journal of Political Economy* 77:957–71.

O'Brien, L., and Harris, F. 1991. *Retailing: Shopping, society, space*. London: David Fulton.

Odland, J. 1988. Sources of change in the process of population redistribution in the United States, 1955–1980. *Environment and Planning A* 20:789–809.

Odland, J., and Bailey, A. 1990. Regional out-migration rates and migration histories: A longitudinal analysis. *Geographical Analysis* 22:158–70.

Oppenheim, N. 1980. *Applied models in urban and regional analysis.* Englewood Cliffs, N.J.: Prentice Hall.

Orleans, P. 1973. Differential cognition of urban residents: Effects of social scale on mapping. In *Image and environment: Cognitive mapping and spatial behavior,* eds. R. Downs and D. Stea, 115–30. Chicago: Aldine.

Owen, D., and Green, A. 1989. Spatial aspects of labor mobility in the 1980's. *Geoforum* 20:107–26.

Ozanne, L., and Thibodeau, T. 1983. Explaining metropolitan housing price differences. *Journal of Urban Economics* 13:51–66.

Pacione, M. 1982. The use of objective and subjective measures of life quality in human geography. *Progress in Human Geography* 6:495–514.

Pahl, R. 1969. Urban social theory and research. *Environment and Planning* 1:143–53.

———. 1979. Socio-political factors in resource allocation. In *Social problems and the city,* eds. D. Herbert and D. Smith, 33–46. London: Oxford University Press.

Palm, R. 1981. *The geography of American cities.* New York: Oxford University Press.

———. 1985. Ethnic segmentation of real estate agent practice in the urban housing market. *Annals of the Association of American Geographers* 75:58–68.

Papageorgiou, Y. 1990. *The isolated city state: An economic geography of urban spatial structure.* New York: Routledge.

Parr, J., O'Neill, G., and Nairn, A. 1988. Metropolitan density functions. *Regional Science and Urban Economics* 18:463–78.

Pas, E. 1986. The urban transportation planning process. In *The geography of urban transportation,* ed. S. Hanson, 49–70. New York: Guilford Press.

Peach, C., Robinson, V, and Smith, S. 1981. *Ethnic segregation in cities.* London: Croom Helm.

Peet, R., and Thrift, N. 1989. Political economy and human geography. In *New models in geography: The political economy perspective,* eds. R. Peet and N. Thrift, 3–29. London: Unwin Hyman.

Phelps, N. 1992. External economies, agglomeration and flexible accumulation. *Transactions of the Institute of British Geographers,* n.s., 17:35–46.

Phillips, E., and LeGates, R. 1981. *City lights: An introduction to urban studies.* New York: Oxford University Press.

Phipps, A. 1979. Scaling problems in the cognition of urban distances. *Transactions of the Institute of British Geographers,* n.s., 4:94–102.

———. 1983. Utility function switching during residential search. *Geografiska Annaler* 65B:23–38.

———. 1989. Residential stress and consumption disequilibrium in the Saskatoon housing market. *Papers of the Regional Science Association* 67:71–87.

Phipps, A., and Carter, J. 1984. An individual-level analysis of the stress-resistance model of household mobility. *Geographical Analysis* 16:176–89.

———. 1985. Individual differences in the residential preferences of inner-city homeowners. *Tijdschrift voor Economische en Sociale Geografie* 76:32–42.

Phipps, A., and Laverty, W. 1983. Optimal stopping and residential search behavior. *Geographical Analysis* 15:187–204.

Pipkin, J. 1981. Cognitive behavioral geography and repetitive travel. In *Behavioral problems in geography revisited*, eds. K. Cox and R. Golledge, 145–81. New York: Methuen.

———. 1986. Disaggregate travel models. In *The geography of urban transportation*, ed. S. Hanson, 179–206. New York: Guilford Press.

Plaut, T. 1981. An econometric model for forecasting regional population growth. *International Regional Science Review* 6:53–70.

Pocock, D., and Hudson, R. 1978. *Images of the urban environment*. London: Macmillan.

Pogodzinski, J., and Sass, T. 1990. The economic theory of zoning: A critical review. *Land Economics* 66:294–314.

Polzin, P. 1990. The verification process and regional science. *Annals of Regional Science* 24:61–7.

Poot, J. 1987. Estimating duration-of-residence distributions: Age, sex and occupational differentials in New Zealand. *New Zealand Geographer* 43:23–32.

Porell, F. 1982. Intermetropolitan migration and quality of life. *Journal of Regional Science* 22:137–58.

Pratt, G., and Hanson, S. 1988. Gender, class, and space. *Environment and Planning D: Society and Space* 6:15–35.

———. 1991. On theoretical subtlety, gender, class, and space: A reply to Huxley and Winchester. *Environment and Planning D: Society and Space* 9:241–6.

Pred, A. 1964. The intrametropolitan location of American manufacturing. *Annals of the Association of American Geographers* 54:165–80.

———. 1977. *City-systems in advanced economies: Past growth, present processes, and future development options*. New York: Wiley.

Preston, V. 1982. A multidimensional scaling analysis of individual differences in residential area evaluation. *Geografiska Annaler* 64B:17–26.

———. 1984. A path model of residential stress and inertia among older people. *Urban Geography* 5:146–64.

———. 1986. A case study of context effects and residential area evaluation in Hamilton, Canada. *Environment and Planning A* 18:41–52.

Ravenstein, E. 1885. The laws of migration. *Journal of the Royal Statistical Society* 48:167–227.

———. 1889. The laws of migration. *Journal of the Royal Statistical Society* 52:214–301.

Rees, P. 1970. Concepts of social space: Toward an urban social geography. In *Geographic perspectives on urban systems*, eds. B. Berry and F. Horton, 306–94. Englewood Cliffs, N.J.: Prentice Hall.

———. 1983. Multiregional mathematical demography: Themes and issues. *Environment and Planning A* 15:1571–83.

Reilly, W. 1931. *The law of retail gravitation*. New York: W. Reilly.

Reitsma, R., and Vergoossen, D. 1988. A causal typology of migration: The role of commuting. *Regional Studies* 22:331–40.

Renas, S., and Kumar, R. 1981. The cost of living, labor market opportunities, and the migration decision: Some additional evidence. *Annals of Regional Science* 15:74–9.

———. 1983. Climatic conditions and migration: An econometric inquiry. *Annals of Regional Science* 17:69–78.

Rex, J., and Moore, R. 1967. *Race, community and conflict*. London: Oxford University Press.

Richardson, H. 1985. Input–output and economic base multipliers: Looking backward and forward. *Journal of Regional Science* 25:607–61.

————. 1988. Monocentric vs. policentric models: The future of urban economics in regional science. *Annals of Regional Science* 22:1–12.

Richardson, H., Gordon, P., Jun, M., Heikkila, E., Peiser, R., and Dale-Johnson, D. 1990. Residential property values, the CBD, and multiple nodes: Further analysis. *Environment and Planning A* 22:829–33.

Roback, J. 1982. Wages, rents, and the quality of life. *Journal of Political Economy* 90:1257–78.

Rogers, A. 1990. Requiem for the net migrant. *Geographical Analysis* 22:283–300.

Rogers, A., and Belanger, A. 1990. The importance of place of birth in migration and population redistribution analysis. *Environment and Planning A* 22:193–210.

Rogers, A., and Willekens, F., eds. 1986. *Migration and settlement: A comparative study*. Dordrecht, Neth.: D. Reidel.

Rogerson, P. 1984. New directions in the modelling of interregional migration. *Economic Geography* 60:111–21.

————. 1987. Changes in U.S. national mobility levels. *Professional Geographer* 39:344–51.

————. 1990. Migration analysis using data with time intervals of differing widths. *Papers of the Regional Science Association* 68:97–106.

Rose, D. 1984. Rethinking gentrification: Beyond the uneven development of Marxist urban theory. *Environment and Planning D: Society and Space* 2:47–74.

Roseman, C., and McHugh, K. 1982. Metropolitan areas as redistributors of population. *Urban Geography* 3:22–33.

Rossi, P. 1980. *Why families move*, 2nd ed. Beverly Hills, Calif.: Sage.

Rushton, G. 1981. The scaling of locational preferences. In *Behavioral problems in geography revisited*, eds. K. Cox and R. Golledge, 67–92. New York: Methuen.

Sack, R. 1980. *Conceptions of space in social thought: A geographic perspective*. London: Macmillan Press.

Sandefur, G., and Scott, W. 1981. A dynamic analysis of migration: An assessment of the effects of age, family and career variables. *Demography* 18:355–68.

Sandefur, G., and Tuma, N. 1987. How data type affects conclusions about individual mobility. *Social Science Research* 16:301–28.

Sandefur, G., and Tuma, N., and Kephart, G. 1991. Race, local labour markets and migration in the United States, 1975–1983. In *Migration models: Macro and micro approaches*, eds. J. Stillwell and P. Congdon, 187–206. London: Belhaven Press.

Sargent, C., Jr. 1976. Land speculation and urban morphology. In *Urban policy making and metropolitan dynamics: A comparative geographcial analysis*, ed. J. Adams, 21–57. Cambridge, Mass.: Ballinger.

Saunders, P. 1979. *Urban politics: A sociological interpretation*. London: Hutchinson.

————. 1981. *Social theory and the urban question*. New York: Holmes and Meier.

Saunders, P., and Williams, P. 1986. The new conservatism: Some thoughts on recent and future developments in urban studies. *Environment and Planning D: Society and Space* 4:393–9.

Sayer, A. 1984. *Method in social science: A realist approach*. London: Hutchinson.

————. 1985. Realism and geography. In *The future of geography*, ed. R. Johnston, 159–73. New York: Methuen.

Schatzki, T. 1991. Spatial ontology and explanation. *Annals of the Association of American Geographers* 81:650–70.

Schofield, J. 1987. *Cost–benefit analysis in urban and regional planning*. Boston: Allen and Unwin.

Schwind, P. 1975. A general field theory of migration: United States, 1955–1960. *Economic Geography* 51:1–16.

Scott, A. 1980. *The urban land nexus and the state*. London: Pion.

––––––. 1982. Locational patterns and dynamics of industrial activity in the modern metropolis. *Urban Studies* 19:111–41.

––––––. 1983. Location and linkage systems: A survey and reassessment. *Annals of Regional Science* 17:1–39.

––––––. 1988. *New industrial spaces*. London: Pion.

Scott, A., and Paul, A. 1991. Industrial development in southern California, 1970–1987. In *Our changing cities*, ed. J. Hart, 189–217. Baltimore: Johns Hopkins University Press.

Sell, R. 1983. Analyzing migration decisions: The first step—whose decisions? *Demography* 20:299–311.

Selye, H. 1956. *The stress of life*. New York: McGraw-Hill.

Seyfried, W. 1963. The centrality of urban land values. *Land Economics* 39:275–84.

Shaw, G., and Wheeler, D. 1985. *Statistical techniques in geographical analysis*. New York: Wiley.

Shaw, R. 1985. *Intermetropolitan migration in Canada: Changing determinants over three decades*. Toronto: NC Press.

Sheppard, E. 1984. The distance-decay gravity model debate. In *Spatial statistics and models*, eds. G. Gaile and C. Willmott, 367–88. Dordrecht, Neth.: D. Reidel.

––––––. 1986. Modeling and predicting aggregate flows. In *The geography of urban transportation*, ed. S. Hanson, 91–118. New York: Guilford Press.

––––––. 1988. The search for flexible theory: Comments on Cadwallader. *Urban Geography* 9:255–64.

Shevky, E., and Bell, W. 1955. *Social area analysis*. Stanford, Calif.: Stanford University Press.

Shevky, E., and Williams, M. 1949. *The social areas of Los Angeles*. Los Angeles: University of California Press.

Simmons, J. 1986. The urban system: Concepts and hypotheses. In *Urban systems in transition*, eds. J. Borchert, L. Bourne, and R. Sinclair, 23–31. Utrecht: Netherlands Geographical Studies.

––––––. 1991. The regional mall in Canada. *Canadian Geographer* 35:232–40.

Simon, H. 1957. *Models of man*. New York: Wiley.

Slater, P. 1984. A partial hierarchical regionalization of 3140 U.S. counties on the basis of 1965–1970 intercounty migration. *Environment and Planning A* 16:545–50.

––––––. 1989. A field theory of spatial interaction. *Environment and Planning A* 21:121–6.

Smith, C. 1983. A case study of structuration: The pure-bred beef business. *Journal for the Theory of Social Behavior* 13:3–17.

Smith, C. J. 1988. *Public problems: The management of urban distress*. New York: Guilford Press.

Smith, G. 1988. The spatial shopping behaviour of the urban elderly: A review of the literature. *Geoforum* 19:189–200.

––––––. 1992. The cognition of shopping centers by the central area and suburban elderly: An analysis of consumer information fields and evaluative criteria. *Urban Geography* 13:142–63.

Smith, G., and Ford, R. 1985. Urban mental maps and housing estate preferences of council tenants. *Geoforum* 16:25–36.

Smith, N. 1982. Gentrification and uneven development. *Economic Geography* 58:139–55.

––––––. 1984. *Uneven development: Nature, capital and the production of space*. Oxford: Blackwell.

––––––. 1989. Uneven development and location theory: Towards a synthesis. In *New models in geography: The political economy perspective*, ed. R. Peet and N. Thrift, 142–63. London: Unwin Hyman.

Smith, N., and Williams, P., eds. 1986. *Gentrification of the city*. Boston: Allen and Unwin.

Smith, S. 1989. The challenge of urban crime. In *Social problems and the city: New perspectives*, eds. D. Herbert and D. Smith, 271–88. Oxford: Oxford University Press.

Smith, T., and Clark, W. 1980. Housing market search: Information constraints and efficiency. In *Residential mobility and public policy*, eds. W. Clark and E. Moore, 100–25. Beverly Hills, Calif.: Sage.

Smith, T., and Clark, W., and Onaka, J. 1982. Information provision: An analysis of newspaper real estate advertisements. In *Modelling housing market search*, ed. W. Clark, 160–86. London: Croom Helm.

Soja, E. 1989. *Postmodern geographies: The reassertion of space in critical social theory*. New York: Verso.

Speare, A., Jr. 1974. Residential satisfaction as an intervening variable in residential mobility. *Demography* 11:173–88.

Spencer, C., and Weetman, M. 1981. The microgenesis of cognitive maps: A longitudinal study of new residents of an urban area. *Transactions of the Institute of British Geographers*, n.s., 6:375–84.

Stabler, J., and St. Louis, L. 1990. Embodied inputs and the classification of basic and nonbasic activity: Implications for economic base and regional growth analysis. *Environment and Planning A* 22:1667–75.

Stapleton, C. 1980. Reformulation of the family life-cycle concept: Implications for residential mobility. *Environment and Planning A* 12:1103–18.

Stapleton-Concord, C. 1982. Sex differentials in recent U.S. migration rates. Urban Geography 3:142–65.

———. 1984. A mover/stayer approach to residential mobility of the aged. *Tijdschrift voor Economische en Sociale Geografie* 75:249–62.

Steed, G. 1973. Intrametropolitan manufacturing: Spatial distribution and locational dynamics in Greater Vancouver. *Canadian Geographer* 17:235–58.

Stillwell, J. 1991. Spatial interaction models and the propensity to migrate over distance. In *Migration models: Macro and micro approaches*, eds. J. Stillwell and P. Congdon, 34–56. London:Belhaven Press.

Stolzenberg, R. 1979. The measurement and decomposition of causal effects in nonlinear and nonadditive models. In *Sociological methodology, 1980*, ed. K. Schuessler, 459–88. San Francisco: Jossey-Bass.

Storper, M., and Walker, R. 1983. The theory of labor and the theory of location. *International Journal of Urban and Regional Research* 7:1–43.

———. 1984. The spatial division of labor: Labor and the location of industries. In *Sunbelt/snowbelt: Urban development and regional restructuring*, eds. L. Sawers and W. Tabb, 19–47. New York: Oxford University Press.

———. 1989. *The capitalist imperative: Territory, technology, and industrial growth*. New York: Blackwell.

Szyrmer, J. 1992. Input–output coefficients and multipliers from a total-flow perspective. *Environment and Planning A* 24:921–37.

Tabuchi, T. 1988. Interregional income differentials and migration: Their interrelationships. *Regional Studies* 22:1–10.

Talarchek, G. 1982. Sequential aspects of residential search and selection. *Urban Geography* 3:34–57.

Taylor, M. 1986. The product-cycle model: A critique. *Environment and Planning A* 18:751–61.

Taylor, P. 1977. *Quantitative methods in geography: An introduction to spatial analysis*. Boston: Houghton Mifflin.

Thill, J. 1992. Choice set formation for destination choice modelling. *Progress in Human Geography* 16:361–82.

Thill, J., and Sui, D. 1993. Mental maps and fuzziness in space preferences. *Professional Geographer* 45:264–76.

Thomas, R. 1990. Quantitative methods: Input–output analysis as macroeconomic geography. *Progress in Human Geography* 14:404–19.

———, and Huggett, R. 1980. *Modelling in geography: A mathematical approach.* New York: Barnes & Noble.

Thompson, W. 1965. *A preface to urban economics.* Baltimore: Johns Hopkins University Press.

Thorndyke, P. 1981. Distance estimation from cognitive maps. *Cognitive Psychology* 13:526–50.

Thrall, G. 1988. Statistical and theoretical issues in verifying the population density function. *Urban Geography* 9:518–37.

Thrift, N. 1983. On the determination of social action in space and time. *Environment and Planning D: Society and Space* 1:23–57.

Timmermans, H. 1982. Consumer choice of shopping center: An information integration approach. *Regional Studies* 16:171–82.

———. 1984. Decompositional multiattribute preference models in spatial choice analysis: A review of some recent developments. *Progress in Human Geography* 8:189–221.

———. 1988. Multipurpose trips and individual choice behavior: An analysis using experimental design data. In *Behavioral modelling in geography and planning*, eds. R. Golledge and H. Timmermans, 356–67. London: Croom Helm.

Timmermans, H., and van der Waerden, P. 1992. Modelling sequential choice processes: The case of two-stop trip chaining. *Environment and Planning A* 24:1483–90.

Timmermans, H., Borgers, A., and van der Waerden, P. 1992. Choice experiments versus revealed choice models: A before–after study of consumer spatial shopping behavior. *Professional Geographer* 44:406–16.

Timmermans, H., van der Heijden, R., and Westerveld, V. 1983. Conjoint measurement of individual preference functions: Some tests of validity. *Area* 15:245–50.

———. 1984. Decisionmaking between multiattribute choice alternatives: A model of spatial shopping-behavior using conjoint measurements. *Environment and Planning A* 16:377–87.

Timms, D. 1971. *The urban mosaic: Towards a theory of residential differentiation.* Cambridge, Eng.: Cambridge University Press.

Todd, D. 1982. Subjective correlates of small-town population change. *Tijdschrift voor Economische en Sociale Geografie* 73:109–21.

Tolman, E. 1948. Cognitive maps in rats and men. *Psychological Review* 55:189–208.

Townsend, A. 1980. The role of returned migrants in England's poorest region. *Geoforum* 11:353–69.

Trowbridge, C. 1913. On fundamental methods of orienting and imaginary maps. *Science* 38:888–97.

Tversky, B. 1992. Distortions in cognitive maps. *Geoforum* 23:131–8.

Unwin, D. 1981. *Introductory spatial analysis.* New York: Methuen.

Urry, J. 1987. Some social and spatial aspects of services. *Environment and Planning D: Society and Space* 5:5–26.

Van Arsdol, M., Jr., Camilleri, S., and Schmid, C. 1958. The generality of urban social area indexes. *American Sociological Review* 23:277–84.

Vandell, K., and Carter, C. 1993. Retail store location and market analysis: A review of the research. *Journal of Real Estate Literature* 1:13–45.

van der Heijden, R., and Timmermans, H. 1988. The spatial transferability of a decompositional multi-attribute preference model. *Environment and Planning A* 20:1013–25.

Vanderkamp, J. 1989. Regional adjustment and migration flows in Canada, 1971 to 1981. *Papers of the Regional Science Association* 67:103–19.

van Til, J. 1980. A new type of city for an energy-short world. *The Futurist* 14:64–70.

Waddell, P., and Shukla, V. 1993. Manufacturing location in a polycentric urban area: A study in the composition and attractiveness of employment subcenters. *Urban Geography* 14:277–96.

Waddell, P., Berry, B., and Hoch, I. 1993. Residential property values in a multinodal urban area: New evidence on the implicit price of location. *Journal of Real Estate Finance and Economics* 7:117–41.

Walker, R. 1981. A theory of suburbanization: Capitalism and the construction of urban space in the United States. In *Urbanization and urban planning in capitalist society*, eds. M. Dear and A. Scott, 383–429. New York: Methuen.

Walker, R., and Storper, M. 1981. Capital and industrial location. *Progress in Human Geography* 5:473–509.

Walmsley, D. 1982. Personality and regional preference structures: A study of introversion–extroversion. *Professional Geographer* 34:279–88.

Wang, K., Grissam, T., Webb, J., and Spellman, L. 1991. The impact of rental properties on the value of single-family residences. *Journal of Urban Economics* 30:152–66.

Ward, D. 1989. *Poverty, ethnicity, and the American city, 1840–1925: Changing conceptions of the slum and the ghetto.* Cambridge, Eng.: Cambridge University Press.

Watts, H. 1987. *Industrial geography.* Harlow, Eng.: Longman.

Webber, M. 1982. Location of manufacturing activity in cities. *Urban Geography* 3:203–23.

———. 1983. Life-cycle stages, mobility, and metropolitan change: 1. Theoretical issues. *Environment and Planning A* 15:293–306.

———. 1984. *Industrial location.* Beverly Hills, Calif.: Sage.

Weicher, J., and Thibodeau, T. 1988. Filtering and housing markets: An empirical analysis. *Journal of Urban Economics* 23:21–40.

White, S. 1980. A philosophical dichotomy in migration research. *Professional Geographer* 32:6–13.

Wieand, K. 1987. An extension of the monocentric urban spatial equilibrium model to a multicenter setting: The case of the two-center city. *Journal of Urban Economics* 21:259–71.

Williams, C. 1992. The contribution of regional shopping centres to local economic development: Threat or opportunity? *Area* 24:283–8.

Williams, P. 1978. Urban managerialism: A concept of relevance? *Area* 10:236–40.

———. 1982. Restructuring urban managerialism: Towards a political economy of urban allocation. *Environment and Planning A* 14:95–105.

———. 1984. Economic processes and urban change. An analysis of contemporary patterns of residential restructuring. *Australian Geographical Studies* 22:39–57.

Wilson, A. 1989a. Mathematical models and geographical theory. In *Horizons in human geography*, eds. D. Gregory and R. Walford, 29–47. Totowa, N.J.: Barnes and Noble.

———. 1989b. Classics, modelling and critical theory: Human geography as structured pluralism. In *Remodelling geography*, ed. B. Macmillan, 61–9. Oxford: Blackwell.

Wilson, D. 1989. Toward a revised urban managerialism: Local managers and community development block grants. *Political Geography Quarterly* 8:21–41.

Wilson, F. 1987. Metropolitan and nonmetropolitan migration streams: 1935–1980. *Demography* 24:211–28.

Winship, C., and Mare, R. 1983. Structural equations and path analysis for discrete data. *American Journal of Sociology* 89:54–110.

Woods, R. 1982. *Theoretical population geography*. Harlow, Eng.: Longman.

Wright, E. 1983. Giddens' critique of Marxism. *New Left Review* 139:11–35.

Wright, S. 1985. Path coefficients and path regressions: Alternative or complementary concepts? In *Causal models in the social sciences*, 2nd ed., ed. H. Blalock, Jr., 39–53. New York: Aldine.

Wrigley, N. 1985. *Categorical data analysis for geographers and environmental scientists*. New York: Longman.

Wrigley, N., and Longley, P. 1984. Discrete choice modelling in urban analysis. In *Geography and the urban environment*, vol. 6, eds. D. Herbert and R. Johnston, 45–94. New York: Wiley.

Wrigley, N., Longley, P., and Dunn, R. 1988. Some recent developments in the specification, estimation and testing of discrete choice models. In *Behavioral modelling in geography and planning*, eds. R. Golledge and H. Timmermans, 96–123. London: Croom Helm.

Yeates, M. 1965. Some factors affecting the spatial distribution of Chicago land values, 1910–1960. *Economic Geography* 41:55–70.

———. 1990. *The North American city*. 4th ed. New York: Harper & Row.

Yeates, M., and Garner, B. 1971. *The North American city*. New York: Harper & Row.

———. 1980. *The North American city*, 3rd ed. New York: Harper & Row.

Yinger, J. 1992. City and suburb: Urban models with more than one employment center. *Journal of Urban Economics* 31:181–205.

Young, W. 1984. Modelling residential location choice. *Australian Geographer* 16:21–8.

Zeller, R., and Carmines, E. 1980. *Measurement in the social sciences: The link between theory and data*. Cambridge, Eng.: Cambridge University Press.

Index